EDITING HUMANITY

ALSO BY KEVIN DAVIES:

Breakthrough (with Michael White)

Cracking the Genome

The $1,000 Genome

DNA: The Story of the Genetic Revolution
(with James D. Watson and Andrew Berry)

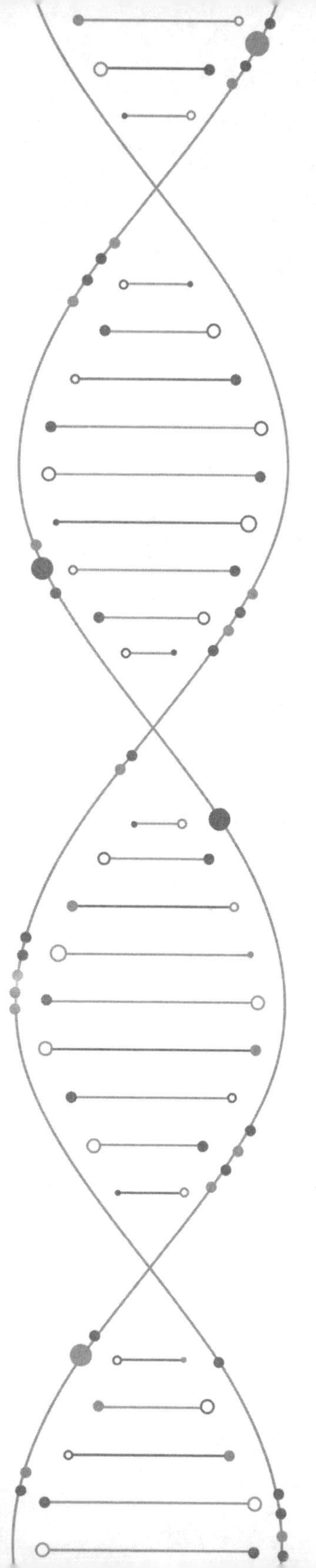

EDITING HUMANITY

THE CRISPR REVOLUTION AND THE NEW ERA OF GENOME EDITING

KEVIN DAVIES

PEGASUS BOOKS
NEW YORK LONDON

EDITING HUMANITY

Pegasus Books, Ltd.
148 W 37th Street, 13th Floor
New York, NY 10018

First Pegasus Books cloth edition October 2020

All images, unless otherwise credited, are courtesy of the author.

Interior design by Maria Fernandez

Library of Congress Cataloging-in-Publication Data is available.

ISBN: 978-1-64313-308-9

10 9 8 7 6 5 4 3 2

Printed in the United States of America
Distributed by Simon & Schuster
www.pegasusbooks.com

To my parents

and

In memory of Michael White (1959–2018),

author, musician, friend.

CONTENTS

PROLOGUE

Thanksgiving weekend, November 2018: I settled in for a fifteen-hour flight from New York to Hong Kong to attend a scientific conference. It was a long trek for a three-day gathering of bioethicists, and I wasn't expecting any major fireworks. Even Alta Charo, a member of the organizing committee, worried it might be "a really boring summit."[1]

As we headed towards the Arctic Circle, I folded my laptop, ordered a Pot Noodle, and flicked through the in-flight entertainment menu before settling on an action film called *Rampage*. Following an explosion aboard the international space station, canisters containing a mysterious chemical crash to earth. Naturally, one lands in the San Diego Zoo, where Dwayne "The Rock" Johnson happens to work as a primatologist, contaminating his beloved albino gorilla, George. Two other animals—Lizzie the crocodile and Ralph the wolf—mutate overnight before converging on Chicago. This was all utterly ridiculous if mildly entertaining, but the material wreaking mutational mischief was the reason I was on this journey. "Have you heard of CRISPR?" Naomie Harris asks our chiseled hero.

He hadn't. Nor indeed had I until a few years earlier. The summit in Hong Kong was the sequel to a gathering in 2015 in Washington, DC, to discuss the ethics of human genome editing. The debate had been prompted by the emergence of a revolutionary technology for fixing or modulating genes with a catchy acronym: CRISPR. Earlier that year, scientists in China had altered the genes of a human embryo in a dish for the first time. It was a preliminary experiment on non-viable embryos (no women were pregnant), but it sparked fears that someone, somewhere might attempt to rewrite the genetic code of the human species. One of the undisputed pioneers of the gene-editing technology, Jennifer Doudna, called for international debate on how this technology should be used, whether it should be restrained, or even banned altogether. The chair of the DC conference, Nobel laureate David Baltimore, referenced *Brave New World* to warn that we were facing "the prospect of new and powerful means to control the nature of the human population." Three years later, I presumed many of the same characters would be sounding similarly abstract warnings.

I landed in Hong Kong in the early afternoon of Monday, November 26, and turned on my phone. As I scrolled through Twitter, it took a few seconds in my groggy state to register what I was reading. A science reporter named Antonio Regalado had published a sensational scoop that a Chinese scientist was overseeing the pregnancy of a gene-edited fetus. It strongly suggested that babies genetically altered using CRISPR might already have been born.[2] Within hours, those rumors were confirmed and then some by the Associated Press. The AP story revealed the birth of genetically edited twins.[3] #CRISPRbabies was trending.

There was more: in a series of YouTube videos, a thirty-four-year-old Chinese scientist named He Jiankui described his historic feat. "Two beautiful little Chinese girls, named Lulu and Nana, came crying into this world a few weeks ago," he said in halting English. "The girls are home now, with their mom Grace, and their dad, Mark." The name He Jiankui didn't mean anything to me but he was due to speak at the conference. Would he still show up? Would the organizers allow him to appear?

Two days later, He Jiankui did indeed try to explain what he had done, and more importantly, why. It was one of the most intensely watched scientific presentations in history, in front of hundreds of journalists and press photographers and close to two million viewers of the live webcast. And I had a front row seat. As He Jiankui entered the packed auditorium and walked across the stage, the only sound was the staccato clatter of about two hundred camera shutters. On social media, an American scientist was screaming this was a travesty and that the organizers were crowning a science celebrity. On the contrary, I felt we were watching a dead man walking. Indeed, by the time He Jiankui had left the stage and returned home to Shenzhen, his dreams of fame and national glory were in tatters. Instead, he faced house arrest, ignominy, and, one year later, prison.[4]

The #CRISPRbabies story marks an extraordinary turning point in human history. There's a reason that the Ken Burns documentary *The Gene*, based on Siddhartha Mukherjee's superb book of the same name, opens with a colleague of He Jiankui injecting CRISPR circumspectly into a human embryo. "The baby crisperer," as the *Economist* dubbed him,[5] had wrested control of heredity from nature, at least for one of the 20,000 genes that make up the human genome.

It was just fifteen years earlier that an international consortium of scientists completed the Human Genome Project (HGP), piecing together more or less every page of the book of life, a script of 3.2 billion letters comprised of a four-letter alphabet. A group of researchers at the University of Leicester printed the complete sequence as an encyclopedia of more than one hundred volumes, each chromosome bound in a different color. Those volumes of information are miraculously bundled into twenty-three pairs of chromosomes that reside in trillions of cells in your body. With the human genome sequence at hand, scientists could set about identifying the genes that go awry in literally thousands of rare hereditary diseases, as well as the DNA variants or misspellings that shape our predisposition to common diseases like diabetes, heart disease, and mental illness. Even before this genetic revolution, researchers dreamed of using specific DNA sequences as gene therapy, injecting healthy genes into patients' cells to compensate for the faulty genes. But the idea of performing DNA surgery—fixing broken genes by cutting and pasting DNA directly into a patient's genome—was a fantasy.

All that changed when the CRISPR craze erupted in the summer of 2012. Two scientists—French microbiologist Emmanuelle Charpentier and Doudna, an American biochemist—published a groundbreaking discovery. They stood on the shoulders of many researchers around the world who had been toiling out of the public eye to understand the biological purpose of CRISPR. These *Clustered Regularly Interspaced Short Palindromic Repeats* (the official acronym) were known to be the critical component of a natural bacterial immune system, a microbial missile defense shield to neutralize attack by certain viruses. Researchers in the Doudna and Charpentier labs reconfigured the molecular machinery to produce an ingenious method for precisely targeting and cutting genes and other DNA targets. Six months later, several groups, led by Feng Zhang at the Broad Institute and George Church at Harvard Medical School, showed that the DNA of mammalian cells could be edited using CRISPR. The prospect of being able to precisely edit almost any sequence of DNA, be it human, bacterial or of any other organism, was extraordinary. CRISPR's ease of use was unlike anything seen before—a technical if not conceptual breakthrough that would transform science and medicine, and perhaps the very fabric of humanity.

Thanks to these researchers and many other scientists around the world, we can now exercise control of heredity with unprecedented ease and precision. We can erase or rewrite disease genes, one person or embryo at a time. We can change the genomes of livestock, plants, and parasites to improve the lives of millions of people, especially in developing countries ravaged by climate change. We can save species from extinction—and maybe even re-create some that have already left this mortal coil. And although we don't know nearly enough about the complexity of the gene networks that underpin our predisposition to diabetes, heart disease, and mental illness, let alone shape our behavior, personality, and intelligence, we can imagine a day when we might be able to augment or manipulate some of those characteristics, too.

The beauty of CRISPR is that it is easier, faster, and much cheaper than earlier genome editing platforms. Researchers have edited a Noah's Ark of plant and animal life: fruits and vegetables, insects and parasites, crops and livestock, cats and dogs, fruit flies and zebrafish, mice and men. Do-it-yourself biohackers began experimenting on themselves and their pets. A tsunami of research papers appeared in the most prestigious science journals as scientists CRISPR'd anything they could get their hands on. "The CRISPR Craze," as *Science* dubbed it, swept over the popular press.[6] The *Economist*'s cover featured an innocent crawling baby with a menu of potentially editable traits, including perfect pitch, 20/20 vision, and no baldness.[7] The *Spectator* riffed on that with "Eugenics is back," featuring a cartoon baby sitting on a petri dish ("not ginger" was the hair preference).[8]

MIT Technology Review labeled CRISPR "the biggest biotech discovery of the century."[9] And a *WIRED* cover story by Amy Maxmen stated: "No hunger. No pollution. No disease. And the end of life as we know it."[10]

When I was training as a geneticist in the 1980s, I was part of a team desperately searching the human genome to find the faulty genes that cause Duchenne muscular dystrophy (DMD) and cystic fibrosis (CF). We were literally genetic detectives, hunting for clues to the whereabouts of genes and mutations that compromise and curtail human life.* We met patients in our lab at St Mary's Hospital in London, including teenagers with CF who would be lucky to see their 20th birthday. The identification of the genes mutated in patients with CF, DMD, and other disorders gave us hope that cures were just around the corner. "From gene to drug" and "From bench to bedside" were the memes of the day, heralding a revolution in molecular medicine. Every few years the future of medicine would get a new name—personalized medicine, precision, genomic, or individualized medicine, as if changing the name would change fortunes.

In 1990—the year of the launch of the HGP—I hung up my lab coat for the last time. My personal eureka moment came when I chanced upon a classified advertisement for a job with *Nature* magazine. Well, I thought, that's one way to get my name in the pages of the world's most famous science journal. I was offered the job by the editor, Sir John Maddox. Two years later, Maddox handed me the chance to helm the first spin-off journal bearing the *Nature* nameplate—*Nature Genetics.*[11] At a conference in 1993 to mark our first birthday, I mingled with a handpicked all-star cast, including Francis Collins, Craig Venter, and Mary-Claire King, who over dinner inspired me to write my first book.

Written with my friend the late Michael White, *Breakthrough* told of the brilliant Berkeley geneticist who, in 1990, mapped the so-called breast

* In the mid-1980s, Francis Collins posed for a newspaper photographer in a lab coat holding a needle in front of a large haystack—the literal metaphor for DNA detectives scouring the genome for a single spelling mistake among three billion letters.

cancer gene, *BRCA1*. A few weeks after Myriad Genetics scooped King in the race to isolate the gene in 1994, the organizers of a major genetics conference in Montreal convened a plenary session to celebrate Myriad's highly publicized success. King stole the show, presenting the results from multiple families in which her team had documented specific *BRCA1* mutations capable of wreaking what the British journalist John Diamond called the "cytological anarchy of cancer and death."

King insisted she didn't care who won the race to identify *BRCA1*: win or lose, her team would be in the lab studying mutations in families. It was important, she said, to distinguish reality from the media frenzy. "Fantasy has been *New York Times* profiles, *60 Minutes*, guys on motorcycles in *Time* magazine"—a jab at Francis Collins, her former collaborator. Reality, she said,

> is having the gene, not knowing what it does, and the realization that in the twenty years since we have been working on this project, more than a million women have died of breast cancer. We very much hope that something we do in the next twenty years will preclude another million women dying of the disease.[12]

King received a standing ovation. Two decades later, a lawsuit that stemmed from a dispute over *BRCA1* genetic testing resulted in the U.S. Supreme Court outlawing gene patents in a unanimous decision.[13]

My next book, *Cracking the Genome*, covered the biological equivalent of the moon landing—the HGP—which became a fierce partisan feud between the international consortium led by Collins and a privately funded hostile takeover led by Craig Venter.[14] The command center at his company, Celera Genomics, looked more like the bridge of the starship *Enterprise*, with two massive video screens streaming DNA sequences rather than photon torpedoes. With the draft sequence in hand, we had the parts list of the human body and could systematically identify the mutations that underlie not only dominant and recessive (Mendelian) disorders but also begin to crack the genetic basis of more common diseases such as asthma and depression.

The ink had barely dried on the genome project when I heard Venter call for a new sequencing technology to speed-read DNA that could deliver the "$1,000 genome." As the first draft had cost $2 billion, this really did sound like science fiction. But the seeds were sown one Sunday afternoon in February 2005, when Clive Brown emailed his colleagues at a British biotech company called Solexa with the subject line: "WE'VE DONE IT!!!!" Using a new technology invented by a pair of Cambridge University chemistry professors,* Brown's team had sequenced the genome of the smallest virus known, ΦX174. The next year, another company, Illumina, acquired Solexa, setting it on course to reach the mythical $1,000 threshold a decade later.[15] By then, we saw the first cases of genome sequencing saving lives, ending the diagnostic odysseys of patients like Nicholas Volker suffering mystery genetic diseases. The plummeting cost of sequencing was accompanied by major advances in speed. For example, Stephen Kingsmore at Rady Children's Hospital in San Diego recently set a Guinness World Record by sequencing and processing the complete genome of a newborn baby in just twenty hours.[16]

Each of these stories chronicled a massive leap forward in genetics propelled by advances in sequencing DNA. We are on course for the $100 genome, with new sequencing platforms offering incredible new possibilities for speed-reading DNA.** [17] One in particular, nanopore sequencing, is housed in a portable device smaller than a smartphone and has found its way onto the International Space Station.

Along with *reading* DNA, we are also seeing tremendous progress in synthesizing or *writing* DNA. Church and other scientists have digitally encoded books and films in DNA sequences[18] and engineered a yeast cell by fusing the organism's natural complement of sixteen chromosomes into a single mosaic chromosome.[19] Synthetic biology has an exciting future designing DNA circuits and customizing organisms for a host of applications from bioengineered fragrances and petrochemicals to the

---

* Both professors, Shankar Balasubramanian and David Klenerman, have since been knighted.

** In nanopore sequencing, the DNA is unzipped to allow a single strand to be threaded through a nanopore—a bacterial protein shaped like a ring donut. By measuring the electrical current as the DNA speeds through the pore like a subway train, Oxford Nanopore can translate those electrical squiggles into the underlying DNA sequence.

next generation of antibiotics and antimalarial drugs. Scientists have even expanded the original four-letter genetic alphabet by synthesizing novel chemical building blocks that can substitute for the naturally occurring ones in the double helix. This paves the way for designing synthetic proteins containing novel building blocks.[20]

Advances in reading and writing are very important. But if I could only read and write this book without an ability to edit, search, and replace, the result would never see the light of day. So too with genome engineering, or *editing*: It allows scientists and even nonscientists to rewrite the genetic code as easily as I can change *money* to *honey* or *gnome* to *genome* on my computer.

In 2014, I was invited by the Nobel laureate Jim Watson to help update a popular science book that he'd written with Andrew Berry a decade earlier, simply called *DNA*.[21] As I reflected on the major advances in genetics, there was no escaping CRISPR. In November 2014, at the annual Breakthrough Prize ceremony, televised live from a NASA hangar in California, I watched Cameron Diaz handing Doudna and Charpentier the most lucrative science prize in the world, worth $3 million apiece.* Barely thirty months after their landmark study, the scientific establishment and the titans of Silicon Valley had crowned the women as scientific royalty.

Neither Doudna nor Charpentier were physicians, but CRISPR holds the prospect to taking gene therapy to a new level. Both women launched biotech companies to deliver CRISPR-based therapies designed to fix mutations that cause sickle-cell disease, blindness, DMD, and many other disorders. "For the past decade, I've been making GMO humans," says Fyodor Urnov, a colleague of Doudna's who helped develop genome editing at Sangamo Therapeutics in California. In 2019—seventy years after Linus Pauling proposed sickle-cell anemia as the first "molecular disease"—Victoria Gray, an African American mother from Mississippi, became the first American patient to

* The annual Breakthrough Prize of $3 million per recipient is worth about ten times a one-third share of the Nobel Prize, which has a total purse of $900,000.

receive a gene-editing therapy for sickle-cell disease.[22] A year later, she is healthy, blessedly free of complications, her blood cells rejuvenated. Many more genome editing trials are getting underway, mostly using CRISPR. We are truly on the verge of a new era in medicine.

But genome editing has gone much further. The actions of He Jiankui crossed a red line, a scientific Rubicon that virtually all scientists deemed sacrosanct. Heritable genome (or germline) editing is no longer the domain of dystopian science fiction movies and panicked stories of designer babies. The genie is well and truly out of the bottle and cannot be put back. Will germline editing find a niche in treating genetic disorders? Will couples want to use CRISPR to genetically enhance their children? Why should we stop at correcting genetic diseases? Can we not entertain the idea of applying CRISPR technology for enhancement? How about tuning a gene to reduce the amount of sleep we need, or provide protection against the onset of dementia, or shield astronauts against radiation poisoning? Or is heritable genome editing, as Urnov argues, "a solution in search of a problem"?

CRISPR "is a remarkable technology with many great uses," said Broad Institute director Eric Lander. "But if you are going to do anything as fateful as rewriting the germline, you'd better be able to tell me there is a strong reason to do it. And you'd better be able to say that society made a choice to do this—that unless there's broad agreement, it is not going to happen."[23]

Editing Humanity is the story of one of the most remarkable scientific revolutions we have ever seen—the CRISPR revolution. My original intent in this book, supported by a science writing fellowship from the Guggenheim Foundation, was to focus on CRISPR—the science and the scientists. In 2017, I conceived the launch of a new journal called *The CRISPR Journal*, and began meeting the scientists leading this exciting research. Our journey begins with the stories of a band of unheralded microbiologists and biochemists—the true "heroes of CRISPR"—trying to fathom the function of obscure lines of genetic code in bacterial DNA. CRISPR demonstrates emphatically the immense value of funding basic academic and investigator-driven research. Big science consortia like the HGP can do great things, but lest we forget, so too can humble scientists with modest financial support. Few could have predicted that studies of how bacteria vanquish their viral nemeses would spawn a multi-billion-dollar industry that could cure disease and alleviate world hunger.

In Part II, I discuss the rise and fall of genetic therapy, which is undergoing a renaissance after years of despair. One of the great hopes of genome editing is the possibility of treating patients with a wide range of debilitating diseases including muscular dystrophy, hemophilia, blindness, and sickle-cell disease. The term "Holy Grail" is overused in science, but if fixing a single letter in the genetic code of a fellow human being isn't the coveted chalice of salvation, I don't know what is.

In the second half of the book, I turn to the CRISPR babies and the extraordinary story behind this reckless experiment. I lift the veil on He Jiankui's secretive ambitions. I question whether He Jiankui was the rogue scientist that he has been painted and assess the fallout of his actions. In the closing chapters, I present some of the exciting new directions that CRISPR might take us, from gene drives to eradicate malaria to de-extinction to resurrect species we have lost, and separate truth from fiction in the debate over designer babies.

This is biology's century. As Urnov points out, there is a huge gap between a genius idea and its practical realization. In 1505, Leonardo da Vinci designed a model of an ornithopter—a flying machine. It was almost four centuries later, in December 1903, when Orville Wright defied gravity for twelve seconds and the length of a baseball diamond (120 feet). It took several decades more for that landmark flight to usher in commercial air travel, let alone carry the first man into earth orbit and beyond.

Siddhartha Mukherjee says it well in *The Gene*: "The challenge with all these technologies is that DNA is not just a genetic code, it is in some sense also a moral code. It doesn't just ask questions about what we will become. Now that we have these tools, we have the capacity to ask the question, what can we become?"

This book is about the origins, development, uses and misuses of CRISPR, the technology adapted from some of the most ancient organisms on earth, which bring us to the precipice of editing humanity.

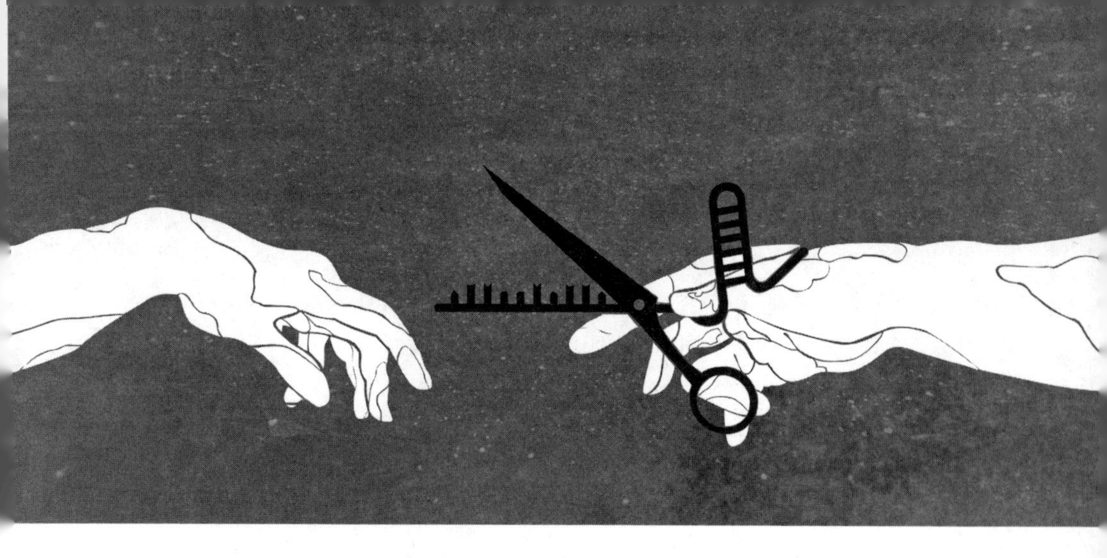

PART I

O, wonder! How many goodly creatures are there here!
How beauteous mankind is!
O brave new world,
That has such people in't!

—William Shakespeare, *The Tempest*

"You can't just boss bacteria around like that," said the younger Mrs. Hempstock. "They don't like it."

"Stuff and silliness," said the old lady. "You leave wigglers alone and they'll be carrying on like anything. Show them who's boss and they can't do enough for you. You've tasted my cheese."

—Neil Gaiman, *The Ocean at the End of the Lane*

"Progress in science depends on new techniques, new discoveries and new ideas—probably in that order."

—Sydney Brenner

CHAPTER 1
THE CRISPR CRAZE

"This is CRISPR?"

Bill Whitaker, a correspondent for the *60 Minutes* television show, sounds puzzled as he points to a small white plastic tube containing a few drops of a colorless liquid that might as well be water. Indeed, it mostly is water.

Holding the vial under the bright television lights is Feng Zhang, a thirtysomething scientist who has helped ignite a biological revolution.

"It has CRISPR in it," Zhang says helpfully.[1]

The interview is being filmed in Zhang's laboratory at the Broad Institute in Cambridge, Massachusetts, one of the most elite biomedical research centers in the country. The Broad (rhymes with *road*) is named after philanthropist Eli Broad and his wife Edythe. It jockeys for space with other ivory towers established by billionaire businessmen including the Koch brothers (cancer), publisher Pat McGovern (brain research), and medical device inventor Jack Whitehead (cell biology). When I first moved to the United States in the late 1980s, working at the Whitehead Institute on the fringe of the MIT campus, the only other sign of civilization was a Legal Seafood restaurant. Today, Kendall Square is the center of the biotech universe, as much a Boston landmark as Fenway Park. A nearby plaque proclaims this to be the "Most innovative square mile on the planet." Few would argue.

Zhang looks, if anything, younger than his thirty-four years, with black cropped hair and a smooth cherubic face. In a lab photo with about twenty students and research fellows, you could easily mistake him for a grad student. But since 2013, Zhang has become accustomed to media requests and the intrusion of film crews. This is literally a case of déjà vu: a few months earlier he was answering the same questions in front of the same CBS cameras. But the interviewer on that occasion, Charlie Rose, had since been sacked for good reasons,* so *60 Minutes* producer Nichole Marks opted to reshoot the entire segment.

Whitaker stares at the tiny tube disbelievingly. "So this is what's revolutionizing science and biomedicine? . . . That's wild!"

Even if the *60 Minutes* cameras could zoom into the contents of that tiny vial between Zhang's finger and thumb—like a scene from *Fantastic Voyage*—it would be hard to see what the fuss was about. But CRISPR is a very big deal—a tempest in a test tube. The term CRISPR has a precise scientific definition (more on that later) but in the space of just a few years, this obscure acronym has become a household word, both noun and occasionally verb that epitomizes the revolution in genome editing—the ability to pinpoint and alter a given DNA sequence in any organism.

But Whitaker isn't letting this go. "So the CRISPR is not the liquid, the CRISPR is in the . . . ?"

"It's dissolved in the liquid," Zhang explains patiently. "There are probably billions of molecules of CRISPR in here."

"*Billions . . .* ?"

Marks had been pondering a story on CRISPR for a year or two, visiting the Broad to meet Zhang and the institute's founding director Eric Lander, and attending the major genome editing ethics conference in Washington, DC, in 2015. CRISPR was being trumpeted in the media as the next miracle biotechnology. But for all its medical potential, CRISPR had not yet entered the clinic, let alone cured anyone of a disease.

The tipping point for Marks came in the summer of 2017, when researchers in Oregon led by Shoukhrat Mitalipov became the first American group to successfully edit a gene in a human embryo using CRISPR. Mitalipov insisted he had no plans to use gene editing to produce actual human beings. But it was hard to ignore the possibility that, in the

---

* #MeToo

biological equivalent of the Doomsday Clock, we had moved a big step closer to the alarming prospect of designer babies.

The scientific possibilities of CRISPR seemingly know no bounds. By harnessing the components of a prehistoric bacterial immune system, scientists have developed a remarkable molecular cursor that can scan the 3 billion letters that make up the human genome for a specific sequence, cut it, then repair or change it. The Human Genome Project (HGP) was all about reading humankind's genetic makeup for the first time. We identified more than 20,000 genes that make up the parts list of the human body. We catalogued mutations in fully one third of those genes that are known to give rise to a plethora of genetic diseases. We spent about $2 billion over thirteen years to spell out this rambling string of As, Cs, Ts, and Gs.* The fruits of that labor are blooming, spurring the development of precision treatments for cancer and many diseases.

But now, with CRISPR, scientists have a powerful, easy, affordable tool that puts researchers in a position to surgically rewrite the code when there's a glitch. To "play God." They can design and engineer the DNA sequence of organisms big and small, from viruses and bacteria to plants (crops, flowers, trees), worms, fish, rodents, dogs, monkeys—and humans. There were other genome editing technologies developed "B.C."—before CRISPR—that have entered the clinic to treat HIV and rare genetic disorders. And there are already enhancements to CRISPR, even more precise versions called base editing and prime editing that take us closer to the Holy Grail of safe, pinpoint control of the DNA sequence.

There have been some momentous medical revolutions over the past few centuries: sanitation and clean water, anesthesia, vaccines, antibiotics, small-molecule drugs, biologics, in vitro fertilization (IVF) and prenatal diagnosis. In the basic sciences, new tools and technologies continually drive science. Tools to control neurons, map the architecture of the cell nucleus, and conduct a liquid biopsy of DNA fragments circulating in the bloodstream. But CRISPR has changed science in a profound way: the technique caught

* The genome project's dirty little secret is that the human sequence is still incomplete. Portions of the genome have proven just too difficult to read using current DNA sequencing technologies.

fire, its simplicity, flexibility, and affordability catching the imagination of researchers around the world in a dazzling democratization of technology.

CRISPR wasn't the result of a dedicated applied engineering effort. Instead it is the culmination of decades of investment in basic biomedical research, supporting dozens of dedicated scientists working in unfashionable fields, conducting research for the thrill of discovery to better understand the natural world around us. As Nobel laureate Bill Kaelin noted in a piece in the *Washington Post* championing basic research in cancer rather than razzle-dazzle moonshots: "The CRISPR gene-editing technology that will revolutionize medicine and agriculture emerged from studies of bacteria and their resistance to viruses."[2] It is hard to envision a less trendy area of research—or it was until the CRISPR breakthrough.

What can CRISPR do? Treat cancer and thousands of genetic diseases. Simple, cheap, mobile diagnostic tools to detect outbreaks of deadly infectious diseases including the COVID-19 pandemic. Designing heartier, more nutritious strains of crops to feed the world. Creating new breeds of disease-resistant livestock and animals for organ transplantation. Conjuring the notion of "de-extinction," a way to resurrect extinct species such as the woolly mammoth, while providing a new tool for conservationists to save endangered species. Shaping evolution to control or even eliminate the scourge of infectious diseases. And changing the human gene pool, for better or worse, by editing the DNA of human embryos in a scene straight out of a science fiction movie.

Indeed, it didn't take long for scriptwriters and novelists to become enthralled by CRISPR. In the 2016 finale of the *X-Files* reboot, Mulder and Scully search for an antidote to a CRISPR bioweapon that knocks out a crucial gene in the immune system, thereby jeopardizing the human race. Jennifer Lopez was reportedly working on a pilot television show with the working title *C.R.I.S.P.R.*[3] Billed as a police thriller set in the near future, the series would see "mentor and protégé battle for control over the human genome in a game of cat and mouse in which the future of our species may rest." Tragically, J. Lo has yet to realize her vision of our Crispered future. Writer Neal Baer featured a CRISPR bioweapon pandemic plot line in the third season of *Designated Survivor,* starring Kiefer Sutherland.[4]

Back in the real world, Lander calls CRISPR "the most surprising discovery, and maybe most consequential discovery, in this century so far." He's a little biased, it must be said: Zhang is one of the star faculty at the Broad Institute, Zhang led one of the first demonstrations of CRISPR gene editing in human cells, and has co-founded five companies in five years. Prestige, patents, and prizes are all at stake. Beyond that, a sense of scientific immortality perhaps—the chance to be remembered as the inventor of one of the great discoveries in science and medicine, to be catapulted into the pantheon of science—Pasteur, Einstein, Fleming, Crick, Franklin, Hawking.

But most of the international recognition and early awards for the discovery of CRISPR belong to a pair of female scientists who collaborated to produce what one scientist called an "immortal" paper that appeared in June 2012. Like a short-lived supergroup, Emmanuelle Charpentier and Jennifer Doudna teamed up to program a bacterial enzyme to target and cut any DNA sequence according to the investigators' whim, laying the groundwork for a game changing, genome editing tool with boundless applications.

According to the doyen of DNA, Jim Watson, what Doudna and Charpentier did was "the biggest advance in science since the discovery of the double helix." But it's important to use it so that it's equitable. "If it's only used to solve the problems and desires of the top 10 percent, that will be horrible," Watson warned. "We have evolved more and more in the past few decades into an inequitable society, and this would make it much worse."[5]

In a profile of the dynamic duo for the *Time 100*, Mary-Claire King called their work "a tour de force of elegant deduction and experiment" that affords scientists "the power to remove or add genetic material at will." King christened CRISPR "a true breakthrough, the implications of which we are just beginning to imagine."[6]

Over the past few years, I've watched the impact of CRISPR spread like wildfire around the world, commanding attention not only from scientists and the media but also from royalty, politicians, and even the Pope.

Every two years, the Norwegian capital hosts the Kavli Prize, awarded by the nonprofit foundation set up by the late Fred Kavli, a Norwegian inventor who made a small fortune in Southern California. In collaboration with the Norwegian Academy of Sciences, the Kavli Foundation bestows three $1-million awards for spectacular science that is big (astrophysics), small (nanoscience), and complex (neuroscience). The awards may not quite match the luster of the Nobel prize or the purse of Silicon Valley's Breakthrough Prize, but they are among the most coveted in science.

In September 2018, the Kavli Foundation awarded the nanoscience prize to a trio of CRISPR pioneers—Charpentier and Doudna were joined by Lithuanian molecular biologist Virginijus Šikšnys. The first two came as no surprise. The inclusion of the Lithuanian was belated recognition of his own pioneering work, despite being scooped by the Charpentier-Doudna team in the summer of 2012. My efforts to peek into the jury deliberations were swatted away with amusement by a Kavli program officer, who said I'd have to wait fifty years for the jury notes to be unsealed.

The streets of Oslo are adorned with banners marking the Kavli celebrations. En route to the University of Oslo campus, I ask my Uber driver if he's heard of the Kavli Prize. He starts to shake his head, but then he remembers: "Oh wait, I read about one of them—he's from my country! He's our man! Tell him to book a ride with Raymondias." Against the odds, I'm being chauffeured by the one Oslo Uber driver who has heard of Virgis Šikšnys.

That evening, the laureates mingle with guests at a reception held in the Norwegian Academy of Science and Letters. A young attaché to the American Embassy asks me for an introduction to Doudna, a late arrival with her husband, fellow Berkeley professor Jamie Cate, and their teenage son Andrew. "How do you pronounce her name? Is it Dood-na?" (It's not.) Doudna smiles graciously but looks like she needs some sleep. In the buffet line I'm cornered by a Norwegian philosophy professor, who says I should expect plenty of speeches at the formal banquet the following evening. He explains with a little Scandinavian joke. "A Dane, a Swede and a Norwegian find themselves on death row. Each is given one last request. The Dane says, 'I want a feast, a roast pork dinner with all the trimmings.' The Norwegian says, 'I want to give a long speech.' As for the Swede, he begs to be shot before the Norwegian's speech." (Perhaps you had to be there.)

The next day, hundreds of guests file into the Norwegian City Theater for the official prize ceremony. The audience chatter halts abruptly as King Harald V enters the stage. Part Oscars, part Eurovision Song Contest, we're treated to a gloriously eclectic selection of musical entertainment. The show opens with Mathias Rugsveen, a fifteen-year-old "sorcerer of the accordion," who crushes a selection from *The Barber of Seville*. Before the neuroscience prize, Norway's answer to Adele belts out a version of "Crazy." At one point, cohost Alan Alda loses his place on the teleprompter but recovers effortlessly: "Hold on, my ad lib is here somewhere!" Finally, it is time for King Harald to present the nanoscience prize to Charpentier, Doudna, and Šikšnys. Only at the ceremony's conclusion, as Doudna beckons family and friends to join her on stage, do the laureates visibly relax and hug each other.

Later that evening, the laureates make a grand entrance down a long marble staircase in the magnificent Oslo City Hall to the forty-nine bells of the carillon. King Harald joins the laureates at the VIP table. We feast on a menu of salmon and halibut sashimi followed by filet of deer. I'm reliably informed by the president of the Norwegian Student Union that deer has a completely different taste than reindeer, which he hunted as a boy.

As promised, there are speeches—no fewer than six. The keynote is Marcia McNutt, an ocean scientist and president of the U.S. National Academy of Sciences. She approves that Kavli recognized more female scientists than men. "For every little girl who dreams of rising to the pinnacle of scientific achievement, your future awaits."* The director general for research for the European Commission, Jean-Eric Paquet, praises the winners' curiosity, tenacity, and willingness to take risks. "Fortune favors the bold," he says, citing another famous risk-taker, polar explorer Ernest Shackleton. Before his first expedition to Antarctica, Shackleton ran a newspaper ad: "Men wanted for hazardous journey. Low wages. Bitter cold. Long hours of complete darkness. Safe return doubtful."

* There will be no such speech in 2020: all seven Kavli laureates were men.

As it was for Shackleton, instant success was rare for the Kavli laureates. "Tonight, we see only the result, the exceptional achievement," Paquet remarks. "What we don't see tonight are the years of hard work, the setbacks, the failures, and the times when they each had to pick themselves up and carry on. In the end, these men and women achieved so much not because they avoided the risk but because they allowed their curiosity to guide them in spite of it—just as Shackleton did."[7]

The party continues into the wee hours with cognac and a live jazz band, but the Doudnas—having already enjoyed audiences on the prize circuit with the emperor of Japan and the king of Spain—have an early wake-up call. Doudna's son has to get back to California for school. For the Kavli laureates, there are more events as they head north to Trondheim to give talks at the country's leading scientific university. Speculation is mounting that, as the scene shifts a few hundred miles to Stockholm, CRISPR will soon be crowned with a Nobel Prize. It's only a matter of time, but it is not to be this year.

Two weeks after her excursion to Norway, Doudna was back home on the big island of Hawaii. She receives a rapturous welcome as she walks on stage wearing a traditional Hawaiian garland, or lei, featuring the red ohia lehua of the island. It's a much more down-to-earth affair, followed by a celebratory dinner at Ken's House of Pancakes.[8] It was just another whistle-stop engagement for the scientist who, perhaps more than anyone, embodies CRISPR—the woman who literally has DNA in her name.

In the summer of 2017, U.S. senator Lamar Alexander was enjoying a fishing vacation in Canada, off the grid and only listening to the radio for the weather forecast. One day however, he happened to catch a story about CRISPR. We should have a hearing on this, he thought to himself. Luckily, as the chair of the Senate Health, Education, Labor and Pensions (HELP) committee, that was his prerogative.[9]

After running between U.S. Senate buildings trying to obtain my press credentials on a crisp November day, I made it to the hearing room just in time to hear Senator Alexander hailing CRISPR as "just one of the amazing discoveries that has come from basic research funded in part by the federal government."[10] But he was alarmed by "designer baby" headlines and a report from James Clapper, the former U.S. head of national security, that genome editing had been classified as a potential weapon of mass destruction.

Testifying before the HELP committee were three CRISPR experts—a CEO, a physician, and a bioethicist. "There are a few times in our lives when science astonishes us—this is one of those moments," said Katrine Bosley, the CEO at the time of Editas Medicine, the first CRISPR biotech company to go public. There are some 6,000 known genetic diseases, Bosley testified. "What if we could repair those broken genes?" she asked. "We owe a responsibility to patients and their families." But Bosley cautioned it would be a long road ahead. "Cures" was a big word, and Bosley, speaking on behalf of the biotech industry, didn't want to overpromise.

Most of the Senators stayed only long enough to ask their individual questions before ducking out for more pressing business. Susan Collins, senator from Maine, came out swinging: "It would be possible for genes to be edited that could affect intelligence or athletic ability. We live in a global world; it seems that the scientific advancements have outpaced the policy in this area. How do we ensure this exciting breakthrough in gene editing is used for good by scientists in countries like China or Russia, as well as in our own country?"

I rolled my eyes and winked at science writer Emily Mullin. We'd barely started the questions and already the predictable "designer baby" trope had come up. But Jeffrey Kahn, a bioethicist at Johns Hopkins University, agreed that science advancement usually outpaces policy. "We do have robust structures for oversight that these technologies are used for purposes we intend. International dialogue is happening," he said reassuringly. "A much smarter approach is to restrict control to allow careful responsible science to go forward, within our borders, not to push them out."

African American senator Tim Scott, from South Carolina, wanted to know about the prospects of curing sickle-cell disease (SCD). The third witness, Stanford University physician-scientist Matthew Porteus, said he was preparing to enroll some "very brave" volunteers in a groundbreaking clinical trial. His team would harvest the patient's stem cells, use CRISPR to edit the gene that carries the sickle-cell mutation—correcting a single letter in the three billion letters of the human genome. They would next return about a billion edited stem cells via IV to the patient. These fixed cells would find their way back to the bone marrow and reconstitute the patient's blood. With any luck, the patient will no longer have their disease. "That's pretty cool," said Scott. And then he left.

Virginia senator Tim Kaine asked if CRISPR had any relevance for Alzheimer's disease, but Bosley wasn't so optimistic. New therapies were further out, she said, because the genetics was more complicated than say SCD. Kaine also asked about regulatory oversight. Kahn pointed to the UK, which has strict control processes that allows the country to license emerging biotechnologies. "People don't want to go to jail for ten years" for violations, Kahn said. "We lose in multiple ways when we drive science underground." Nor did the U.S. want to cede its leadership and competitiveness in gene editing to anyone else.

No one could have predicted that, one year later, Kahn's hypothetical warnings would come kicking and screaming to life.

Six months later, in April 2018, the CRISPR road show landed in a most unlikely location—the Vatican. The "Unite to Cure" conference is the brainchild of Robin Smith, president of the Cura Foundation, in conjunction with the Pontifical Council for Culture. Smith's rolodex is ridiculous: among the guests were TV personalities Dr. Mehmet Oz and CNN's Sanjay Gupta, with cameo appearances from A-list celebrities including Katy Perry (the power of meditation) and Jack Nicklaus (stem cells). At times, the meeting felt uncomfortably like a late-night infomercial. Billionaire Ed Bosarge bought a clinic in the Bahamas and was receiving experimental injections that would cross the blood-brain barrier to combat memory loss and aging. "My goal is to be healthy and fit and playing tennis at 120," he said with deadly seriousness.

Then there was Peter Gabriel, the first rock vocalist I ever heard (on Genesis's *Selling England by the Pound*). Gabriel treated us to an unplugged concert, although he was a bit rusty as evidenced by a false start on "Solsbury Hill." He emotionally dedicated the set to his wife Meabh, who had recovered from an aggressive form of non-Hodgkin's lymphoma following CAR-T therapy. Later he said: "Rich people will live forever and poor people will die in their billions. The trickle-down model isn't going to change that."

With so many celebrities in attendance, a panel discussion between the CEOs of the three public CRISPR biotech companies was almost an afterthought. All three companies were launched by scientists at the heart of the CRISPR drama: Zhang and Doudna co-founded Editas Medicine, but Doudna quit over a patent dispute and later joined the founders of Intellia Therapeutics. Meanwhile, Charpentier launched her own company, CRISPR Therapeutics. Katrine Bosley said CRISPR was "the biggest thing to happen in biology in a generation," moving beyond science fiction to treat genetic diseases. The CEO of CRISPR Therapeutics, Sam Kulkarni, said CRISPR had captured the imagination of millions of people because it was so easy to use. "It has completely democratized [gene-editing] technology." Intellia's CEO, John Leonard, said CRISPR clinical trials were imminent, and he predicted CRISPR would be the "standard of care for sickle-cell in short order."

Bosley conceded there was a dark side to CRISPR. Technology is neither good nor bad, "it's what we do with it," she said. That was also the theme of a speech by Pope Francis, delivered in a private audience with a few hundred conference attendees in the stunning Audience Hall, the interior of which resembles a snake's visage. The Pontiff spoke of the need to protect the environment and exercise caution in the application of gene editing. He acknowledged "the great strides made by scientific research in discovering and making available new cures," especially in rare autoimmune and neurodegenerative diseases. But science, he said, has opened up new methods

> to intervene in ways so profound and precise as to make it possible to modify our DNA. Here we see the need for an increased awareness of our ethical responsibility toward humanity and

> the environment in which we live. While the Church applauds every effort in research and application directed to the care of our suffering brothers and sisters, she is also mindful of the basic principle that 'not everything technically possible or doable is thereby ethically acceptable.'[11]

It was a well-manicured statement, crafted in consultation with Bosley and other CRISPR company executives. And it was soon forgotten as Katy Perry and other VIPs lined up for a chance to kiss the Papal ring. Nobody—not even the Pope—knew that halfway around the world, a Chinese woman was in the first weeks of pregnancy. She was carrying twins whose DNA had been sculpted not by God but by the undivine hands of an ambitious genome engineer with an assist from his embryologist.

It was a most maculate conception.

Fears that a renegade scientist might dare to rewrite the genome's Holy Script have existed for decades, albeit more in the fictional realm. But a 2015 study by a Chinese group showed for the first time that scientists were prepared to fix a disease gene in a human embryo just hours after in vitro fertilization. In ethical terms, it was a giant leap over the mythical red line that supposed that human beings would never play God with their genetic destiny. Over the next three years, various august medical societies and committees published dozens of erudite reports on the ethical pros and cons of genome editing.[12] While scientists and ethicists debated, Australian geneticist Daniel MacArthur tweeted, "My grandchildren will be embryo-screened, germline-edited. Won't 'change what it means to be human.' It'll be like vaccination."

Several groups, mostly in China, reported experiments on human embryos, without any plans of implanting those edited embryos. But a young Chinese scientist who had spent five years training in the United States dared to take the next fateful step. He Jiankui assumed that his groundbreaking work would be celebrated at home and abroad, published in the world's premier journal, putting him on a pedestal with his scientific hero, Nobel laureate Robert Edwards, the co-inventor of IVF.

But there was no celebration, no acclaim, and no *Nature* paper. Instead, outrage and fierce, near-universal condemnation. The work was sloppy, irresponsible, rash, unethical, and possibly criminal, leaving serious questions over the health of two babies. He Jiankui's career crumbled overnight as he was placed under house arrest, sacked by his university, and eventually sentenced to three years in prison. Undeterred, a Russian geneticist announced his intent to take matters into his own hands and use CRISPR gene editing to help couples with inherited deafness. "We keep advancing where this line is and, in effect, there is no line," said Regalado.[13]

The director of the National Institutes of Health, Francis Collins, is steadfastly opposed to any attempt to tamper with the DNA of human embryos. "Evolution has been working toward optimizing the human genome for 3.85 billion years. Do we really think that some small group of human genome tinkerers could do better without all sorts of unintended consequences?" he said.[14] Many leading scientists have called for a temporary moratorium to give scientists and other stakeholders time to explore the rationale and circumstances under which germline editing might be approved.[15] But others find little to fear about the prospect of germline editing. Harvard's George Church, a veteran genome engineer, is keeping an open mind. "I just don't think that blue eyes and [an extra] 15 IQ points is really a public health threat," he told a British newspaper. "I don't think it's a threat to our morality."[16]

In his final book, the great physicist Stephen Hawking predicted that we were heading toward an era of what he termed self-designed evolution. "We will be able to change and improve our DNA," Hawking wrote. "We have now mapped DNA, which means we have read 'the book of life,' so we can start writing in corrections."[17] But for Hawking, the slippery slope didn't stop at curing devastating diseases such as his own affliction, a slowly progressive form of amyotrophic lateral sclerosis. Hawking believed scientists would use techniques such as CRISPR to modify or enhance traits like intelligence, memory, and longevity—violating the law if necessary. These "superhumans" would be available to wealthy elites, putting them in conflict with natural humans. Hawking continued:

> Once such superhumans appear, there are going to be significant political problems with the unimproved humans, who won't be able to compete. Presumably, they will die out, or become unimportant. Instead, there will be a race of self-designing beings who are improving themselves at an ever-increasing rate.

There was an immediate fear that the abhorrent actions of one scientist could derail the remarkable progress in using CRISPR and other editing techniques for gene therapy in children and adults. As I talked to friends and scientists in the audience, I heard one genome editing luminary portray the numerous concerns over the CRISPR babies as an "existential threat" to the future of therapeutic genome editing.

Thankfully, those fears have not yet come to fruition. Although still early days, genome editing is showing genuine promise in the clinic for patients with cancer, blood diseases, hereditary forms of blindness and many other disorders. The training wheels are coming off, says Fyodor Urnov. "The world gets to see what CRISPR can really do for the world in the most positive sense."[18]

CHAPTER 2
A CUT ABOVE

On June 26, 2000, President Bill Clinton walked into the East Room of the White House flanked by two famous scientists, Francis Collins and Craig Venter. Clinton announced a landmark in the Human Genome Project (HGP)—the first rough draft of the human genome sequence, the book of life. Beaming in via satellite from 10 Downing Street was British prime minister Tony Blair, waving the flag for the British team that chipped in about a third of the sequence.

For two years, two teams—small armies more like—of scientists had been dueling to reach this epic milestone. In one corner was Collins, the field marshal of the international government-funded alliance to decode human DNA. The prize was a treasure map of the human genome, displaying the order of 3 billion letters of the DNA alphabet (a four-letter code of chemicals abbreviated by A, C, T, and G) bundled into twenty-three pairs of chromosomes.

In the opposite corner was Venter, a maverick scientist/entrepreneur who brazenly launched a hostile takeover of the genome project. He divulged his plans to Collins in the United Airlines lounge at Washington Dulles airport, and then to the world via the front page of the *New York Times*. His new company, Celera Genomics, vowed to cut through years of government inefficiency and bureaucracy to compile the sequence faster and cheaper using a warehouse full of the latest automated sequencing machines—each named after a sci-fi character—plus a massive Compaq supercomputer to crunch the data. As a consolation prize, Venter said

Collins could sequence the mouse genome instead. Almost overnight, the tables had been turned: the "Darth Vader" of genomics had the weaponry and momentum; Collins and his allies were the plucky overmatched rebels with their backs against the wall.

As public bickering between the factions degenerated into outright hostility, it threatened to tarnish the reputations of the project leaders, not to mention the purpose of the mission. The White House helped orchestrate a temporary ceasefire to facilitate a historic celebration.[1] Clinton hailed the achievement as "the most important, most wondrous map ever produced by humankind . . . The language in which God created life." GENETIC CODE OF HUMAN LIFE IS CRACKED BY SCIENTISTS was the banner headline on the front page of the *New York Times*.[2]

But who was the owner of said cracked code? The NIH consortium had collected DNA from dozens of anonymous volunteers who answered a March 1997 newspaper ad placed in the *Buffalo News* by molecular geneticist Pieter de Jong (the Master Chef of building DNA libraries). Years later, genetic analysis revealed that the largest single contributor, code-named RP11, was likely to be African American.[3] Like everyone else, RP11 and the other DNA donors were mutants, each carrying hundreds or thousands of DNA variants predisposing to rare and common diseases, including type 1 diabetes and hypertension.[4] Celera had selected DNA from five volunteers of diverse ethnic backgrounds; Venter later admitted he was one of the chosen few.

Reading the book of life—even if at this stage there were many pages missing or torn or out of order—was a monumental achievement. This was the moonshot of biology, arguably the biggest event since Crick and Watson assembled the double helix in 1953. We had become the first species to translate the instruction manual, even if we couldn't describe how much of it works. Textbook chapters proclaiming that humans possess more than 100,000 genes were rendered obsolete as we were humbled to learn that our genome contains barely 20,000.

One of the biggest champions of the HGP was Sir John Maddox, the editor emeritus of *Nature*. In 1999, Maddox published an ambitious book few would dare undertake, entitled *What Remains to Be Discovered*. Maddox wrote:

> It is likely that the deeper knowledge of the working of the human genome now being won will suggest ways in which the design of *Homo sapiens* provided by 4.5 million years of natural selection could be decisively improved upon by genetic manipulation. After all, people are now manipulating the genetic structure of genes so as to make plants resistant to infections. Why not manipulate the human genome to the same end? It is a reasonable guess that *Homo sapiens* will not always disclaim such opportunities.[5]

As he wrote those words, a band of scientists worlds away from the television cameras and presidential plaudits were taking the first steps toward developing a new technology that could tinker with the code we had just spent some $2 billion over a decade to spell out. It was the dawn of genome editing.

Editing is an essential step in creating works of literature, or music, or art. The fortunes of many blockbuster films might have been very different if producers had gone with their original titles. *Alien* was going to be called "Star Beast," *Back to the Future* was almost released as "Spaceman from Pluto," and the working title of *Pretty Woman* was "3,000." Jane Austen's "First Impressions" became *Pride and Prejudice.* Margaret Mitchell's Scarlett O'Hara was originally named Pansy. "Editing, of text literary or genetic, (almost) always makes things better," writes Fyodor Urnov.[6]

While I was watching the rapid progress in high-throughput DNA sequencing in the 2000s, scientists were devising a molecular word processor to edit the book of life—to search, cut, and paste words and letters, identifying typos, deleting misspellings, and pasting in corrections. Within a decade of crowning ourselves the first species to decode our genetic script, we were already testing our ability to engineer changes in any organism on a whim. Taken to its logical conclusion, we can now redirect and accelerate our own evolution, and that of almost every organism on earth.

"This is the nature of discovery," says geneticist Shirley Tilghman, former president of Princeton University. Every major scientific discovery has the capacity to be deployed for good and ill. "It's going to take wise societies to direct those discoveries down the right path."[7] The rapid development of genome editing is a daunting, unprecedented, and in some ways frightening responsibility. One that has already been violated.

Before we go any further, let's consider what is so special about this revolutionary technology with the funny name that sounds like a cross between a candy bar and a refrigerator drawer. In striving to paint a picture of CRISPR, writers have reached for one metaphor after another: the hand of God, a bomb disposal squad, a pencil eraser, a surgeon's scalpel, a retinal scanner, and frequently, a "molecular scissors."[8] *STAT* produced a top ten list of CRISPR analogies, culminating in the Offiziersmesser, better known as the Swiss Army knife of molecular biology. Likewise, CRISPR is more than just a single sharp blade for cutting DNA, but an ever-expanding array of molecular gadgets for editing and manipulating DNA with ever greater finesse and flexibility.

CRISPR is one of those once-in-a-generation breakthroughs that changes the way science is conducted almost overnight. Ironically, the technology harnessed from a bacterial antiviral immune system went viral. But it was not the first technique for genome editing. Earlier methods for gene editing were conceived in the early 2000s, refined, and even entered the clinic before the advent of CRISPR. Urnov and his colleagues at Sangamo coined the term "genome editing" in 2005 while refining a technology called zinc finger nucleases (ZFNs), which is still in clinical use. In 2011, the year before CRISPR burst into the scientific mainstream, the journal *Nature Methods* anointed genome editing its "Method of the Year." ZFNs and another gene-editing platform called TALENs have their admirers, but were too fussy and expensive to break out the way CRISPR has.

CRISPR takes the premise of other forms of genome editing and (in the parlance of *Spinal Tap*) turns it up to 11. From Australia to Zaire, researchers worldwide are using CRISPR to edit genes in almost any organism on planet earth. The ease of uptake stems from the fact that CRISPR is, in essence, a technology honed by evolution over hundreds of millions of years. CRISPR doesn't require expensive lab instruments such as $1-million state-of-the-art DNA sequencing machines—most of the reagents can be ordered over the Internet and handled in the lab without any special safety precautions, just as Zhang demonstrated for *60 Minutes*. High-school students can learn the fundamentals of CRISPR in a biology classroom.[9] A nonprofit in Boston called Addgene serves as a clearing house for CRISPR reagents. By early 2020, Addgene had distributed more than 180,000 CRISPR constructs to more than 4,000 laboratories around the world, according to director Joanne Kamens.[10]

In the summer of 2012, the groups of Charpentier and Doudna demonstrated that they could take the bacterial CRISPR system and, with some nifty molecular tweaking, transform it into an exquisitely tunable genetic cursor that could be used to cut more or less any specific stretch of DNA. Rodolphe Barrangou, the chief editor of *The CRISPR Journal*, calls that study a tipping point that showed that "you could repurpose this cool, idiosyncratic, revolutionary immune system in bacteria and turn that into a tool that people can use readily in the lab to cut DNA."[11] Six months later, Zhang's group, in collaboration with the Rockefeller University's Luciano Marraffini, and independently George Church's group, demonstrated that the CRISPR-Cas9 tool could effectively edit mammalian DNA. "That changed the world," says Barrangou.

Indeed, around the world researchers seized this simple, programmable gene-editing tool, producing new discoveries that flew into the pages of the top science and medical journals. Stanford law professor Hank Greely offers a nice analogy. "The Model T was cheap and reliable, and before long everybody had a car and the world changed. CRISPR has made gene editing cheap, easy and accessible . . . I think it's going to change the world," he says. "Exactly how beats me."[12]

The incandescent rivalry between the two giants of soccer in Buenos Aires—River Plate and Boca Juniors—has been called eternal. But there is a rivalry that has shaped life on earth from the beginning and rages all around us to the present day. The most important arms race on the planet takes place between two implacable enemies, the nuclear superpowers of the microbial world—bacteria and the viruses (or bacteriophages) intent on their mutual destruction. This war has raged for life eternal, a billion years at least.

We didn't need to experience the COVID-19 pandemic to know that viruses are the invisible menace, harbingers of sickness and death. "The single biggest threat to man's continued dominance on this planet is the virus," Nobel laureate Joshua Lederberg famously said. Beyond social distancing and some natural immunity, the human species mounts a variety of countermeasures, including vaccines and a battery of tailored or repurposed drugs and therapies. The threat is never extinguished, because viruses are able to mutate, evolve, capture genetic material from their hosts, and continually reinvent themselves.

Bacteria know how we feel. They face a constant viral threat of their own from bacteriophages—viruses that exclusively infect bacteria. There are an unfathomable 10 nonillion (10^{31}) phages on planet earth—one trillion for every grain of sand.* "Don't ask me how people calculate this number, but I believe them," says Marraffini.[13] Laid end to end, those submicroscopic phages would stretch 200 million light-years.[14] Under the electron microscope, many look quite menacing, like a cross between the lunar lander and a spider, legs splayed to hook onto the cell surface; others have the innocent charm of a circle lollipop with a long tail. Once attached, the virus impregnates the bacterium with its own genetic material, a short strand of either DNA or its chemical cousin RNA, hijacking the host's protein-manufacturing machinery. Within twenty to thirty minutes, scores of freshly assembled viral progeny burst out of the now defunct host cell like a hundred Aliens erupting out of John Hurt's stomach. "The cells explode, they pop," like a balloon, says Marraffini.

Surrounded by would-be phage invaders, bacteria have evolved a variety of defense systems to surveil and destroy this threat. When I was studying biochemistry in the 1980s, we learned that bacteria boast an army of potent enzymes that recognize and attack specific motifs in any foreign DNA. (The same sequences in the bacterial DNA are protected from those same nucleases

* The late Roger Hendrix, a renowned microbiologist, came up with the estimate of 10^{31} phage (10,000,000,000,000,000,000,000,000,000,000) on the planet, making them the most prevalent biological entity.

with chemical tags, like a child-safety electrical outlet cover.) Scientists seized on these restriction enzymes as a means to cut, swap, and ligate DNA fragments, for example pasting human genes into bacteria, giving birth to the biotechnology industry. But as we shall see later, we now know that bacteria possess another immune system. CRISPR is a small subsection of the bacterial genome that stores snippets of captured viral code for future reference, each viral fragment (or spacer) neatly separated by an identical repetitive DNA sequence. Think of it as an FBI filing cabinet of Most Wanted offenders.

CRISPR is more than just a vault of viral villainy; within reach is the armory for a potent ground-to-air missile defense system. When the cell detects an invading virus, the first step is to activate the CRISPR array, producing an RNA copy of the archived viral sequences. This RNA string is then sliced up into individual sequences, each fragment derived from a different virus and serving as a police artist's sketch of a possible offender. The RNA can't do any damage by itself, so it is weaponized by binding to a DNA-cutting enzyme called Cas (CRISPR-associated sequence), forming a ribonucleoprotein complex that is armed with a GPS signal and ready to do battle.

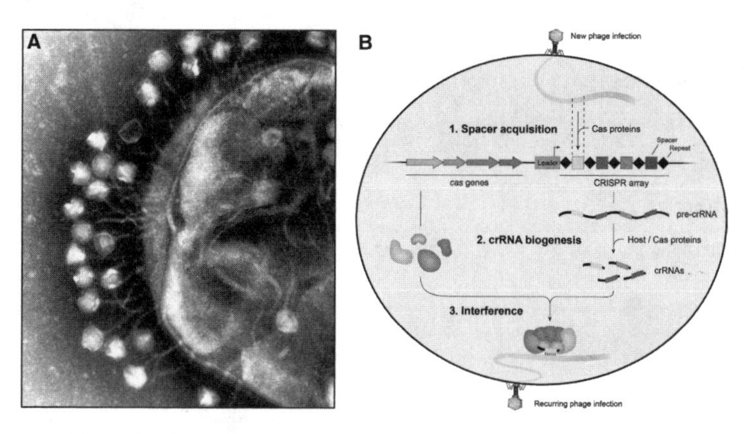

Phages and the CRISPR Pathway. (A) Caught in the Act: Phages land on the surface of *E. coli* to launch their attack. (B) CRISPR–Cas immunity. 1. Bacteria capture fragments of viral DNA and integrate these spacers into the expanding CRISPR array. 2. To combat a phage infection, the CRISPR array (pre-crRNA) is transcribed into RNA, then processed into mature crRNAs. 3. In the interference stage, the crRNA and Cas protein(s) form a complex that targets the corresponding phage sequences for degradation. Some CRISPR systems (Class 1) feature multiple Cas proteins as shown, whereas the simpler Class 2 systems require only a single nuclease such as Cas9. (Adapted from ref. 15.)

There are half-a-dozen different flavors or types of CRISPR system in the microbial universe, which are organized into two classes based on their architecture and other properties.[15] One of the simplest arrangements—Type II—features an enzyme called Cas9. This nuclease makes a clean break on both strands of the DNA double helix like a pair of nail clippers, but not indiscriminately. It grabs an RNA tag, holding it like a mugshot, searching the incoming DNA for a match. Once encountered, Cas9 will latch onto the viral DNA and cut it, neutralizing the threat. Cas9 is "truly wondrous," Urnov explains. "When Cas9 polices the intracellular neighborhood for invasions, it literally carries a copy of that most wanted poster with it. Asking everyone that comes in: "Excuse me, do you carry an exact match to this little most wanted poster that I'm carrying? Yes? Then I'll cut you."[16]

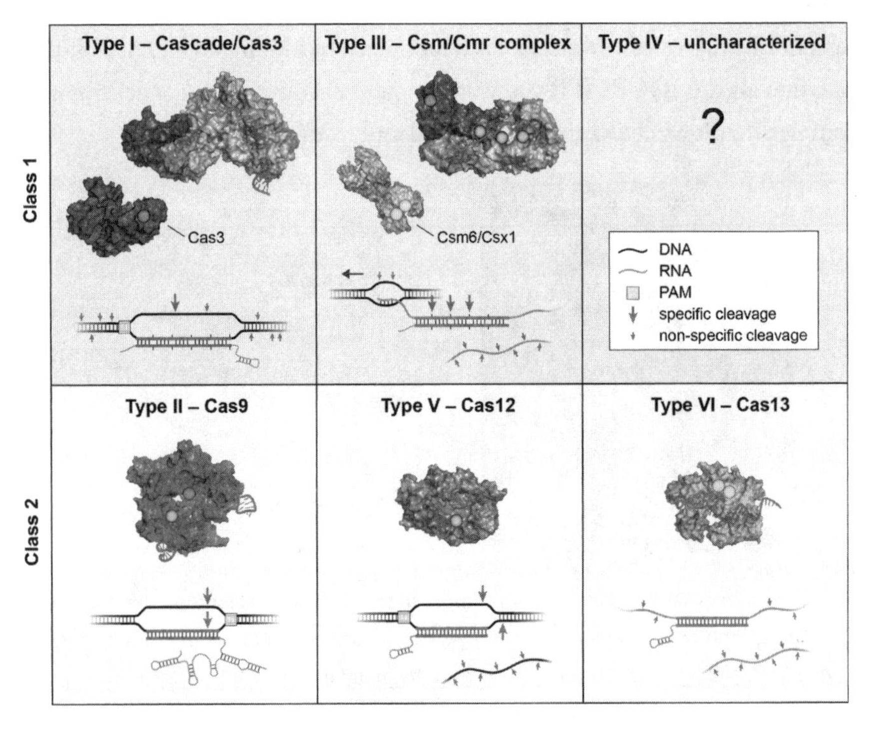

Classes of CRISPR. There are several flavors of CRISPR, which are categorized into two broad classes, 1 and 2. In Class 1, the DNA cleavage is performed by a complex of proteins, sometimes called Cascade. In Class 2, CRISPR systems feature a single Cas nuclease such as Cas9, Cas12 or Cas13. (For details, see ref. 15).

Marraffini reveals how the two bacterial defense systems complement each other. Restriction enzymes offer the first barrier of defense against the viral menace, shredding the viral DNA into pieces that can be incorporated into the CRISPR array. But if phages, ever evolving, dodge the first line of defense, CRISPR immunization kicks in. It is analogous to vaccination, Marraffini says. "When the phage DNA is dead, CRISPR can scavenge spacers to immunize the host." Only a very few infected bacteria actually acquire spacers—about 1 in 10 million—but that provides one cell with the power to vanquish the viral threat and rebuild the population.*

In Adam Bolt's 2019 documentary *Human Nature*, we meet David Sanchez, a charming boy who suffers from sickle-cell disease. As he learns about the potential of CRISPR to cure his disease, he asks perceptively: "How does this thing work and know how to target the right gene, not the gene that makes hair?"

The genius of the CRISPR revolution was to parcel Cas9 not with a virally derived RNA, as in nature, but with a synthetic guide RNA programmed by researchers that allows them to target more or less any DNA sequence in any gene in any organism. The result is we have hijacked a bacterial enzyme a billion years old and repurposed it into a 21st-century molecular scalpel for precision gene surgery. Whether we want to edit the genome of a hamster or a human, a mosquito or a mouse, a redcurrant or a redwood, the process essentially is the same. That's because all organisms in nature use the same inert DNA code, composed of the same four-letter alphabet.

In its natural state, Cas9 is rather disinterested in DNA, essentially colliding randomly and bouncing off. But once Cas9, which has a hand-shaped structure, clasps a guide RNA, a subtle reconfiguration of the protein's structure primes it to react with DNA as it goes in search of its matching target. According to Blake Wiedenheft, a professor at Montana State University, the Cas protein complexes "patrol the entire intracellular environment, find and bind this foreign [viral] DNA, and mark that

* In 2020, Rotem Sorek's group at the Weizmann Institute of Science in Israel reported a new bacterial back-up anti-phage defense system called retrons.

foreign DNA for destruction in a matter of minutes . . . that's a pretty remarkable task."[17]

The task of finding and binding the target sequence is a two-step process. First, Cas9 seeks out and interacts with a short motif in the DNA called the PAM* sequence—a beacon that provides the enzyme with a cue to briefly caress the DNA. "That ephemeral interaction results in a distortion of the DNA," explains Wiedenheft. By bending the DNA, Cas9 unzips the double helix to allow the guide RNA to slip into the resulting crevice (forming a so-called R-loop).[18] The guide conducts a quick sequence check against the target DNA. If a perfect match is found along all twenty or so bases, this marks the DNA sequence for destruction. Cas9 severs** both strands of the DNA as cleanly as a kitchen knife, creating a double-strand break (DSB) just a few bases away from the PAM sequence.[19]

This remarkable process was captured in a stunning video shot by University of Tokyo researchers Hiroshi Nishimasu and Osamu Nureki in 2017. Using a technique called high-speed atomic force microscopy, they were able to zoom in at the precise moment that Cas9 grasps the DNA. In the film, Cas9 looks like a gold-colored rock as it pauses over a strand of DNA for several seconds before guillotining the DNA in half.[20] The clip went viral after Nishimasu posted it on his Twitter account and it was shown on Japanese television.

But repurposing Cas9 to seek out a specific unique sequence in the human genome is literally a million times more complicated than cutting viral DNA. As the Cas9 complex enters the alien surroundings of a cell nucleus, it is confronted by a maze of DNA—twenty-three pairs of chromosomes, six billion letters of DNA—compared to a typical phage genome of just a few thousand bases. Once in the nucleus, each Cas9 molecule scours the densely packed coils of DNA to identify PAM sites, which occur on average once every full 360° rotation of the double helix. In principle, the enzyme has to interrogate 300–400 million bases to identify its precise target.

---

* PAM stands for Protospacer Adjacent Motif. Different Cas enzymes recognize different PAMs, ranging from three to six bases. The most commonly used Cas9, from *Streptococcus pyogenes*, recognizes a triplet sequence, NGG, where N can be any of the four bases.

** Cas9 actually has two active sites, providing two separate cutting actions, one for each strand of the double helix.

Johan Elf, a biophysicist at Uppsala University in Sweden, calculates that Cas9 normally takes about six hours to search through every PAM sequence in the bacterial genome, pausing at each prospective site for a mere twenty milliseconds to peer into the double helix to see if it has found the correct target.[21] But the packaging of DNA in a eukaryotic cell nucleus is far more complex than bacteria. During lectures to his students at the University of Edinburgh, Andrew Wood shows a diagram of a bacterial cell alongside a winding, looping mammalian DNA fiber. "Cas9 didn't evolve to work in the environment in which we now put it," he says. "It's mind-boggling that it is possible to interrogate hundreds of millions of nucleotides in a matter of hours."[22]

Once Cas9 has cut the DNA, the cell's DNA repair enzymes to reseal the break. Experts marvel that it works as well as it does.[23] Cas9 even surpasses the previously developed ZFN and TALEN* gene-editing platforms. "They both evolved to regulate eukaryotic DNA and yet Cas9 seems to outperform them," Wood says.

Let's pause to note that the PAM sequence has a critical role: by searching for a short PAM sequence rather than having to unzip and check essentially the entire genome, the task of Cas9 to latch onto its target sequence is greatly simplified. The PAM also answers the riddle of how Cas9 doesn't accidentally carve up the repeats in the CRISPR array. That's because when they are initially added to the bacterial CRISPR array, the PAM sequence is clipped off. Genome engineers refuse to be limited by the natural list of PAM sequences, so they are modifying the original Cas9 and Cas enzymes from other species to expand their PAM preferences.

With such an effective security system, one might reasonably ask: why aren't all viruses extinct? Viruses have sneakily evolved a multitude of escape mechanisms—a group of proteins that are able to disable the Cas nucleases, known as anti-CRISPR proteins. Bacteria and their viruses are like prey and predators locked in a perpetual battle that rages on after hundreds of millions of years.[24] CRISPR is found in 40 percent of bacterial genomes, and almost all archaeal genomes, but surprisingly not at all in the genomes of

* ZFN, zinc finger nuclease; TALEN, transcription activator-like effector nuclease (see chapter 8).

higher organisms. Although Cas9 is by far the most popular enzyme used in CRISPR applications—and subject to a bitter patent dispute I'll discuss later—this enzyme represents a blip in the diverse CRISPR systems seen in nature. A huge effort is underway to mine the biological diversity on earth to uncover new Cas family proteins with novel functions to expand the CRISPR toolbox.[25]

Once a researcher has identified the gene sequence they wish to target, they can go to any number of websites, key in the desired matching sequence, and order that custom short guide RNA sequence. If CRISPR is a molecular word processor, then the RNA acts as the "CTRL-F" function, targeting the gene sequence of interest. Cas9 acts as the "CTRL-X" keystroke. But genome editing isn't just about pointing the cursor to highlight and remove a typo. It's about deciding and managing what happens next—how to correct the typo.

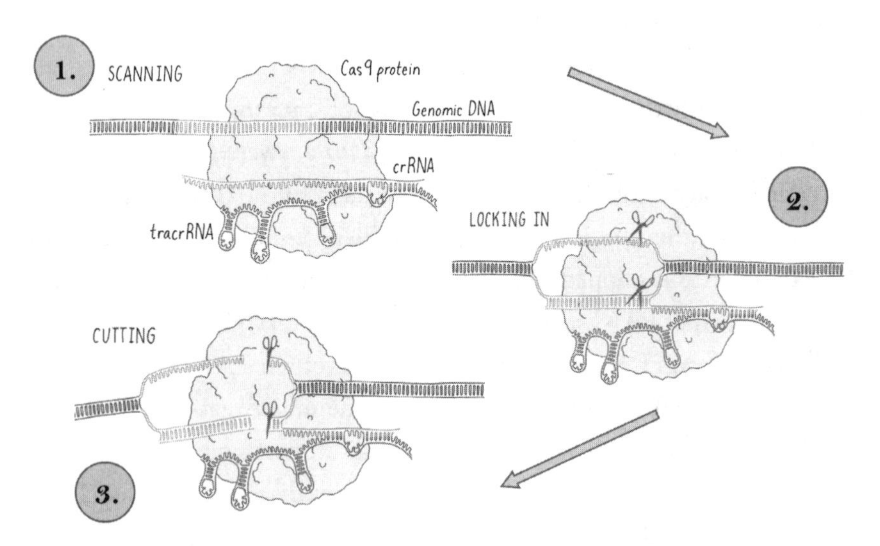

CRISPR Cutting of DNA. 1. Scanning: The Cas9 nuclease is bound to a guide RNA in a ribonucleoprotein complex. The guide consists of CRISPR RNA (crRNA) and the tracrRNA. The Cas9 complex scans the DNA in search of a PAM sequence, which is the cue to check for a sequence match. 2. Locking In: Cas9 binds to the DNA and unzips the double helix, allowing the crRNA to align to the single-stranded DNA. 3. Cutting: If there is a perfect DNA:RNA match, Cas9 undergoes a conformational change resulting in both DNA strands being cut in the same position. (Adapted from ref. 23.)

Cells possess multiple molecular pathways to repair breaks and other mutations in DNA; if they didn't, we wouldn't be alive. The two most common repair pathways are called non-homologous end joining (NHEJ) and homology-directed repair (HDR). NHEJ sloppily stitches the broken ends of DNA back together, but frequently results in small insertions or deletions at the repair site. This is ideal for investigators using CRISPR to deliberately disrupt the function of a gene by breaking it and introducing various random insertions and deletions. The other pathway, HDR, makes a faithful repair if a suitable template is available. In normal circumstances, the template is the corresponding gene on the sister chromosome. The beauty of the CRISPR genome editing is that the investigator can supply a suitable template containing the desired sequence to patch into the Cas9-induced break, thereby resulting in the desired edit.* [26]

In January 2020, about five hundred scientists flocked to Banff, a ski resort in the Canadian Rockies, for the first big CRISPR conference of the year. (It also turned out to be the last, as the COVID-19 pandemic shut down all conference travel.) The organizers invited Doudna to deliver the opening keynote address on a Sunday morning, a role to which she has grown accustomed. She began with a heartfelt apology for not being unable to stay and mingle for the next several days, but she had to get back to Berkeley to give a Monday morning lecture to six hundred undergraduates.

Doudna's lecture, delivered with the same humility and wonder at the march of science that she had at the start of her career more than three decades ago, was the biotech equivalent of a State of the Union address. Her opening summation—a tribute not only to her own work but also that of legions of researchers over the previous quarter century—was simple:

"Precision editing of any genome is within reach."[27]

---

* It will never catch on, but Patrick Harrison, a geneticist at Trinity College, Dublin, came up with a modified definition of CRISPR, a mnemonic that explains the editing/repair process: Cut—Resect—Invade—Synthesis—Proofread—Repair. On *Last Week Tonight*, comedian John Oliver had his own irreverent definition: Crunchy-Rectums-In-Sassy-Pink-Ray-Bans.

CHAPTER 3
WE CAN BE HEROES

Great moments in technology and science can emerge from the most unlikely sources. In 1966 Playtex, the company behind the iconic Cross Your Heart bra, entered a NASA competition to design the spacesuit for the first Apollo moon landing. The suits had to be able to withstand pressure and extreme temperature swings. They also had to be flexible, an attribute that Playtex handsomely demonstrated by filming one of its technicians playing American football for hours while sporting a spacesuit. Thus, it came to pass that four Playtex seamstresses sewed the twenty-one-layer A7L spacesuit that Neil Armstrong fashioned on the lunar runway.[1]

From the Sea of Tranquility on the moon to the salterns of Santa Pola off the Mediterranean. I'm visiting Alicante, a popular tourist resort on the Costa Blanca in southeast Spain. It is an unlikely candidate for one of the more extreme habitats of life on planet earth. But drive south about fifteen miles, you reach Las Salinas de Santa Pola. Salterns, or salt flats, are as the name suggests, a network of rectangular lagoons characterized by extreme salt concentrations resulting from intense sun and wind. At the perimeter, where the water meets land, the salt crystallizes out, forming a crusty white band like the rim of a perfect margarita.

This place has ecological, historical, and commercial significance. Flamingos and other wildlife abound. A watchtower dates back to the 16th century, where the lookouts of King Felipe II kept watch against the Moors. It looks like a nature reserve, but this is an industrial salt mine. Each lake is the size of a football field, concentrating the salt in stepwise fashion. Today, Bras del Port extracts on average 4,000 tons of salt daily from the Mediterranean Sea. A veritable mountain of salt sits ready for distribution: about 60 percent will be used for water treatment, the rest is for food.

For Francisco Mojica, a microbiologist at the University of Alicante, the study of the halophilic life that thrives in this peculiar habitat is his passion. Where he once toiled in obscurity, today he struggles to escape the media spotlight. In 2017, the leading Spanish newspaper, *El País*, speculated whether Mojica would make it from the salterns to the Nobel spa.

Mojica has kindly agreed to drive me in his unflashy Volkswagen Passat to las Salinas. It is a journey he makes quite frequently, usually in the company of photographers or film crews who direct him to survey the pristine pink waters or hold a flask of salt water up to the Spanish sun as if for the very first time, like admiring a glass of rioja. In a delicious irony, Mojica never actually collected samples for his own research because as a young graduate student, there were already samples in the lab, taken by his boss a decade earlier.

Mojica first came to the salterns after he had finished military service in 1989 and was looking for a research position. He was offered a PhD position in the microbiology department at the local university in a lab that studied a microbe called Haloferax. "I didn't have an interest in particular with these organisms. My boss decided the issue of my thesis work," he told me as we stroll around the salterns.[2]

Haloferax is not a bacterium (although confusingly it used to be called Halobacterium) but belongs to a distinct group of single-cell organisms called Archaea. To the naked eye there is little to distinguish the two clades. But that belies an evolutionary chasm of some three billion years. The appreciation that Archaea are not just a superficial off-branch of prokaryotes but an entirely separate "third domain" of life is due to the seminal work of evolutionary biologist Carl Woese. DNA sequencing revealed striking genetic differences between Archaea and bacteria, like comparing the operating systems of a Mac and a PC. Ed Yong put it nicely: "It was as if everyone was staring at a world map, and Woese had politely shown that a full third of it had been folded underneath."[3]

The water is pink and the salty crusts of the lagoons are drizzled with pinkish red bands that teem with microscopic life. The source of the reddish hue is the production of carotenoids, part of the microbial defense mechanism against salt and sunlight. "It's like a sunscreen," Mojica laughs. The color changes with salinity: red shifts to pink as the salt concentration rises from 10 to 30 percent. The same chemicals give the flamingos their trademark pink plumage as they feed on the tiny brine shrimp that, like the Haloferax, thrive in these salty waters.

The salt-loving Archaea of Alicante are by definition extremophiles—lifeforms that are adapted to live in unusually harsh habitats, whether it be underwater volcanic vents, parched deserts, or frozen tundra. For Haloferax, the level of salt in regular seawater just doesn't cut it: they require ten times as much salt to thrive. Trying to replicate those conditions in the lab is extremely difficult, thus they remain poorly understood compared to their bacterial distant cousins. The two main Haloferax species here are *H. mediterranei* and *H. volcanii* (the latter named not because they've mistaken a salt lake for a volcanic vent but after the Israeli scientist who discovered them, Benjamin Volcani). I can also smell the presence of anaerobic bacteria responsible for the strong sulfurous odor that wafts over the water.

Mojica's obsession with the salt-loving microbes of Santa Pola is the embodiment of basic research. "It was knowing by knowing, to expand knowledge," he says.[4] Buried in the circular genetic code of Haloferax, Mojica reasoned, must lie a clue to explain its love of salt. This was not so straightforward: the first complete microbial genome sequence wasn't reported until 1995, by Claire Fraser and Craig Venter's group, which also decoded the first complete Archaea genome two years later. Mojica's lab was not a flashy genome center with the latest DNA sequencing hardware. In the early 1990s, sequencing for many researchers was still a cumbersome manual process, which involved making a large gel sandwich between two glass plates, then separating radioactively labeled DNA fragments by size in an electric current. From the resulting ladder imaged on an X-ray film, Mojica could spell out the corresponding DNA sequence.

In one of his first sequencing attempts, in August 1992, Mojica saw something so surprising, he assumed he'd messed up the experiment.

He saw weird repetitive sequences, each about 30 bases long, which he duly noted in his first paper.[5] "We were absolutely lucky," he said, after sequencing less than 1 percent of the Haloferax genome. "It was the first paper where CRISPR was taken seriously!" he says.* Mojica also showed that the repeats were surprisingly transcribed into RNA, suggesting they had some sort of function.

"When you see something very peculiar, you have no alternative but to research it. I thought this was a nice thing to keep working on," he says as we continue our stroll along the salterns. He had a hunch the mystery repeats might be tied to salt adaptation, perhaps by changing their conformation and thus gene activity by sensing changes in the cell's osmotic pressure. "At the time, [DNA] supercoiling was the answer to everything in the regulation of gene expression!" It was a nice hypothesis, but wrong.**

Working after hours in the university library, Mojica eventually unearthed a Japanese report from 1987. Atsuo Nakata and Yoshizumi Ishino*** at Osaka University described a similar repeating motif in the genome of *E. coli*. While sequencing a gene of interest, the Japanese group had noticed an unusual nearby sequence with a distinctive repeat pattern, like crop circles carved into the DNA terrain. This region consisted of a series (the *cluster*) of *short repeated* stretches of *palindromic* sequence (reading the same forward and backward); each identical repeat, twenty-nine letters in length, was separated by a thirty-two-base stretch of unique sequence (the *interspaced* DNA). But because nothing similar had been seen before, and there were no clues as to their biological function, the researchers let it slide. The team dutifully wrote up and published their observations, which attracted scant attention at the time.[6]

---

* The term "CRISPR" wasn't coined until 2001. Although the true significance of Mojica's 1993 discovery was unknown at the time, twenty-five years later, the date would be immortalized on IMAX movie screens around the world as the opening credit in *Rampage*.

** The CRISPR repeats have nothing to do with adaptation to salinity. "It is still a mystery!" Mojica says. He has yet to receive funding to study the question further.

*** Ishino is now a biochemistry professor at Kyushu University. He studies DNA repair in thermophilic archaea, and is searching for new CRISPRs.

Two years later,[7] Mojica described a short sequence repeated hundreds of times in tandem, spanning more than 1,000 bases. Between each pair of repeats was a unique DNA sequence of unknown function. Mojica's boss suggested calling these repeats—which was also observed in another extremophile, a volcano-loving Archaea—TREPs, for tandem repeats. Surely there was a reason why prokaryotes devote up to 2 percent of their precious compact DNA to these strange repeats? Microorganisms "cannot allow themselves luxuries," Mojica thought, "they must have an important function."[8]

Other scientists also stumbled upon these repeats. German microbiologist Bernd Masepohl puzzled over a stretch of thirteen DNA repeats, found in a cyanobacterium, which he called LTRR, for "long tandemly repeated repetitive." But in focusing on the repeated DNA elements, Masepohl paid little attention to the unique sequences in between.[9] Another team also came close to solving the mystery of the DNA repeats. In 2002, Eugene Koonin, a Russian expat computational biologist at the National Center for Biotechnology Information at the NIH, and his colleague Kira Makarova, described a series of bacterial genes they suspected to be part of a DNA repair system.[10] What they didn't realize was that these genes were sitting adjacent to the CRISPR array and—as we shall soon see—play an essential role in the function of CRISPR and gene editing.

After a few years working in Oxford, Mojica returned to Alicante in 1997 to set up his own group. With little funding, Mojica tried to do some very cheap experiments, "even though I had no idea about bioinformatics." The nagging question was the origin of the spacer DNA, the sequencers interspersed between the repeats. "The easiest thing is to look at the databases and expect that something comes out, but we didn't get anything—until 2003." By now, the DNA databases were bursting with bacterial and archaea genomes, many of which carried versions of these repeats.

In 2000, Mojica renamed his obsession SRSRs (short regularly spaced repeats). That didn't last long. Later he exchanged emails with Ruud Jansen in the Netherlands, who was studying a family of genes adjacent to the mystery repeats. Jansen felt a new name was needed, so Mojica suggested "CRISPR." On November 21, 2001, Jansen emailed his enthusiastic approval:

> Dear Francis,
> What a great acronym is CRISPR. I feel that every letter that was removed in the alternatives made it less crispy so I prefer the snappy CRISPR over SRSR and SPIDR.[11]

CRISPR finally had a name, as did a group of unusual genes that seemed to piggyback with the CRISPR elements. Jansen reasonably, if unimaginatively, dubbed these "CRISPR-associated" genes, or Cas. For now, Mojica was still focused on the curious CRISPR spacers.

The breakthrough finally came one picture-postcard afternoon in August 2003. Mojica was vacationing with his wife close to home, near the salterns. Feeling the heat, Mojica made an excuse to pop back to his air-conditioned lab, where he could run a few more computer searches. This was routine, almost like playing a video game: Mojica would copy one of the mystery spacer sequences and paste it into a computer program called BLAST that would search for matches in the massive DNA database, GenBank. Mojica had run this program hundreds of times to no avail. His colleagues called his quest a waste of time. But on this day, to his astonishment, the computer flagged a match. This particular spacer from *E. coli* matched a stretch of viral DNA called P1 that crucially infects the same bacterium. Over the next few weeks, Mojica catalogued dozens of other examples of matches between CRISPR spacers and various viruses.

In October, Mojica submitted the biggest research paper of his life to the top journal—*Nature*. "I remember the title of the submission was: 'Prokaryotic repeats are involved in an immunity system.' To convince the editor and referees, we wrote that the existence of an acquired immune system in prokaryotes will have tremendous repercussions in biology and clinical sciences." And the result? "It wasn't even reviewed!"[12]

Perhaps something was lost in translation, but *Nature*'s editors didn't consider this to be a conceptual advance "of sufficient general interest" that merited publication in its prestigious pages. Mojica appealed, arguing this was the first description of a bacterial immune system with a memory function. *Nature* indicated it was willing to reconsider if he could describe the mechanism underlying this immunity. But Mojica's group couldn't generate any experimental proof for the hypothesis, just the smoking gun of the sequences. One reason, it turned out, is that CRISPR is repressed in the most popular lab workhorse, *E. coli.*[13] It was as if Mojica had incriminating physical evidence but no security camera footage.

Mojica licked his wounds and resubmitted to another journal . . . and another. Three more journals, including the *Proceedings of the National Academy of Sciences (PNAS)*, all passed on CRISPR. Each delay increased the chances he might get scooped. Finally, in October 2004, Mojica submitted his manuscript to a lesser known journal specializing in evolution. It took a full six months before he finally heard some encouragement from the editor. Three months later, the paper was accepted. "I remember those two years like a nightmare," he told me. "When you have something so big in your hands and you send it to very good journals—and all of them agreed it was not interesting enough to be published—you think, is it me who is crazy or something else?"[14]

Most scientific papers are team efforts, the fruit of months if not years of planning, reviewing, and repeating experiments, a continual exchange of ideas between student and mentor. If one member receives the spotlight, other members, rightly or wrongly, can feel aggrieved.

The second of four authors on Mojica's groundbreaking report was graduate student César Díez-Villaseñor. He watched the accolades showered on Mojica with a mixture of pride and envy. In the early days of CRISPR, "it seemed likely that spacers didn't have any function at all,"[15] he recalled, because each spacer had a unique sequence, minimizing the likelihood of any sort of common function. But Díez-Villaseñor was puzzled: "If the sequence of the spacers is not really relevant, why bother?" Perhaps their uniqueness meant they were somehow toxic to similar sequences. "The proposition was immediately dismissed, although it didn't leave my mind." He says wistfully, "it was hinting the right direction." Another possibility was that the spacers were being generated by some peculiarly sloppy form of DNA replication. But then adjacent spacers ought to be more similar to each other than distant ones, which was not the case. Díez-Villaseñor charted the CRISPR spacers from *E. coli* to show his boss, refuting the idea of mutation incorporation. He remembers it was the day before Mojica's eureka moment.

"I said with a bit of frustration that spacers had to be taken from previous sequences but, obviously, they had to be present inside the cell at that time." As he said that aloud, he realized it made perfect sense. "Of course—it must be an immune system!" There were known examples of RNA interference being used as an immune system. "Suddenly, everything that had looked so strange made perfect sense." Díez-Villaseñor asked Mojica if he was running sequence searches with the spacers. Mojica responded briskly: "Doing that is my job! Don't do anything."

To find a microbial immune system that could recognize invading phage was a very big discovery. The next day, doubts started to be dispelled. "Francis came to the lab elated and directly talked to me. He told me he had found the first homology of an *E. coli* spacer in phage P1." Later he told another professor that CRISPRs were like "memorabilia from past hosted genetic elements." Díez-Villaseñor says: "That was probably the most satisfying professional moment in my entire life."

Mere weeks after Mojica began his publication odyssey,[16,17] Gilles Vergnaud in Paris submitted his own CRISPR story, and experienced similar frustrations. With concerns growing about Saddam Hussein's use of biological weapons, Vergnaud, working for the French Ministry of Defense, was tasked with improving microbial detection methods. At the end of 2002, Vergnaud obtained access to DNAs from dozens of strains of

Yersinia pestis isolated during a plague outbreak in Vietnam in the mid-1960s. Using the best genetic methods available at the time, he found the strains were identical except for one region, which they called named minisatellite number 6 (MS06). When a graduate student, Gregory Salvignol, sequenced MS06 in more detail, it turned out to be a CRISPR. Moreover, the French team found that it acquired new spacers from viral DNA. They too proposed that these structures were part of a bacterial immune system. The first draft of the manuscript, written in July 2003, included the notion of a "defense mechanism"—a record of "past genetic aggressions."

But like Mojica, Vergnaud endured his own depressing runaround. He submitted his paper—the first to include "CRISPR" in the title—just behind Mojica, but in November 2003, *PNAS* dismissed it without review. It was the same story at the *Journal of Bacteriology* (twice), *Nucleic Acids Research*, and *Genome Research*. In July 2004, he submitted to *Microbiology*, which eventually published his paper in 2005.[18]

The early history of CRISPR would look very different if any one of those journals had said yes. As it was, Vergnaud applied for grants from the French National Research Agency over three consecutive years without success. A third report on CRISPR came out from Alexander Bolotin, a Russian microbiologist working for the French ministry of agriculture. Bolotin noted a correlation between the number of spacers and phage sensitivity, deducing that "spacer elements are the traces of past invasions by extrachromosomal elements."[19]

We now know that some 90 percent Archaea contain CRISPR elements, but only 40 percent bacteria. Mojica explains that bacteria have a larger repertoire of defense systems at their disposal. CRISPR serves as a genetic barrier of sorts to microbial evolution because it discourages horizontal gene transfer. "Do you prefer to have a barrier to genetic transfer or defend against viruses?" says Mojica.

Spain has only celebrated two Nobel Prize winners in science—Santiago Ramón y Cajal in 1906 and Severo Ochoa in 1959. That puts a lot of pressure on Mojica, even as he laughs off such idle speculation. "It's good to know that some people think I deserve it, I really appreciate that, but thinking about the possibility of getting the Nobel Prize is—how you say—crazy! There's no way one could expect to get the Nobel Prize."

Before heading back to Alicante, Mojica and I stop for a beer before lunch. He is about to embark on a three-week lecture tour of Australia, the sort of career success that most academics covet. I'm curious about how the microbiologist who has been nicknamed *el padrino*—the godfather—is handling his newfound celebrity. Mojica pauses, reaching for another olive. The restaurant is deserted, but in almost a whisper, he says, "I hate it . . . I hate it." This was not the reaction I was expecting. "I just want a quiet life," he says, shaking his head. "I want to do my research and go home to my wife."

Anyone taking bets on where the next pivotal step in the CRISPR story would occur could have found extremely long odds on a Danish yogurt company. But for scientists at Danisco, ensuring bacteria used in starter cultures can ward off the constant threat of phage infection is a commercial priority. The next CRISPR breakthrough came in two parts of the world where cheese making is revered—France and Wisconsin.

Philippe Horvath was born about fifty miles south of Strasbourg, close to the German border. As we walk to a restaurant in Vilnius, Lithuania, he stops to catch the score of a World Cup game displayed on a giant outdoor screen. Croatia are winning en route to their surprise appearance in the final of the 2018 tournament. "Did you know my name means 'from Croatia' in Hungarian?" Horvath asks.[20] (I mean, why would I know that?!)

During his PhD at the University of Strasbourg, Horvath studied the genome of *Lactobacillus plantarum*, which is traditionally used in food fermentation including sourdough, kimchi, pickles, and sauerkraut. I was skeptical this could sustain an entire PhD thesis, but Horvath shoots me a look. "This is not a trivial bacterium like *E. coli*," he says sternly. Sauerkraut is serious business in Alsace, where it forms the base of the famous choucroute garnie.

Horvath saw an ideal job advertisement for a molecular biologist in industry and sent off the only application letter he has ever written. He was hired in December 2000 by Rhodia Food (formerly Rhone-Poulenc, a famous French chemical company).* Horvath's expertise in bacterial genetics helped improve the quality of starter cultures—the seed bacteria used to ferment milk into yogurt and cheese, which Rhodia sold to food giants like General Mills, Danone, and Nestlé. Phages that prey on the bacteria used in fermentation are found naturally in milk. "When you have a tank containing 10,000 liters of milk and you add a starter culture that is sensitive to a phage that is present, it's a disaster!" says Horvath. "The milk remains milk."

A typical starter culture consists of a high density of three to eight strains of bacteria—about 1 trillion bacteria per gram. Common examples include *Lactobacillus acidophilus*, *Lactococcus lactis*, and *Streptococcus thermophilus*. Horvath explains a freeze-dried starter culture pouch or brick is added to some 2,000 liters of milk. The goal is to minimize the number of rounds of cell division the bacteria need to produce enough lactic acid to lower the pH of the milk. Acidification must occur quickly to protect the milk from spoilage bacteria such as *Salmonella* and *Listeria*. "The higher the number of bacteria, the fewer generations you'll need and the less risk you'll take in terms of phages," says Horvath. Humans have fermented milk in this fashion for millennia without knowing the molecular minutiae.

While starter culture customers prioritize acidification and phage resistance, they also value other qualities including texture and aroma. "You have to acidify quickly but you must also produce texture. You've experienced liquid yogurt?" Horvath asks me as we are about to tuck into dinner. "This is due to phages that have killed the texturing strain." Horvath's group developed starter cultures for a variety of fermentations including more than 1,000 different French cheeses. The starter culture for pizza cheese is far different to those used to produce Camembert. And because phages can manifest at any time, Horvath's group has to formulate different starter cultures for the same product with unrelated phage sensitivities.

Much of Horvath's work involves selecting daughter strains that are immune to the phages that attacked the parent. The process is

* In 1997, Rhone-Poulenc split its chemical and pharma businesses to form Rhodia and, two years later, Aventis. The food company remained with chemical division.

survival-of-the-fittest straightforward: add a phage to a sensitive strain in the lab, wait patiently and search for survivors—naturally occurring mutants called bacteriophage insensitive mutants (BIM)—perhaps because the phage is unable to attach to the bacterial cell surface.

In September 2002, while attending a symposium in the Netherlands on lactic acid bacteria, Horvath came across a poster presented by Alexander Bolotin. The poster mentioned a repetitive DNA motif called "SPIDR" (spaced interspersed direct repeats), which would later be renamed CRISPR. "We have identified a region with repeats that is very useful for strain identification," Bolotin stated. Horvath was so intrigued that he snuck a photograph.

Back in the lab, Horvath compared the sequence from one of his own group's *Streptococcus* strains (LMD-9) with the strain that Bolotin had studied and other strains. To Horvath's delight, there was a huge diversity of spacers across the SPIDR repeat regions. Every strain was different, resembling a DNA fingerprint. Horvath noticed something else: some of the sequences alternating with the SPIDR repeats matched the DNA of viruses, suggesting a link between spacers and phages. "By comparing spacer sequences with known [viral] sequences, we saw identities with phage sequences. Yes, in 2003!" At the time, only Mojica and a handful of other investigators were remotely interested in the CRISPR repeats. Horvath tried in vain to get his supervisors interested in a project he dubbed "CRISPy-SPIDRs," but researching obscure virus biology in the food division of a chemical company wasn't likely to win many converts. "We were told to stop working on that," Horvath said. He continued his CRISPR research on the side, running computer searches just like Mojica.

Attitudes changed after Rhodia was acquired by Danisco, catapulting the Danish food ingredient company to second in the starter culture market, trailing only another Danish company, Chrysanthum. In 2004, every other loaf of bread and a third of ice creams contained Danisco ingredients such as color, texturants, or emulsifiers. Suddenly flush with money, Horvath felt reborn. That December, he was finally able to buy a DNA sequencing instrument. "What did we do? CRISPR sequencing! The more we sequenced, the more obvious it became!"

Horvath was on the verge of a crucial discovery in the brief history of CRISPR. It came courtesy of a new colleague, a French expat based, appropriately, in Wisconsin—the cheese state.

"I used to be French,"[21] says Rodolphe Barrangou, professor of food science at North Carolina State University (NCSU). We're speaking in the dramatic setting of the university's 21st-century Hunt Library, like something out of *The Jetsons*. Barrangou's self-confidence borders on a swagger, an impression magnified by his penchant for wearing cowboy boots to alleviate strain on his back following a serious basketball injury. He drives a modest Honda Accord that he has maintained since graduate school, adorned with a personalized CRISPR vanity plate.

Barrangou was born in Paris, but after he moved to North Carolina for his PhD, he fell in love with the Tar Heel state. During his PhD, which was funded by Danisco, Barrangou's twin interests were developing next-generation probiotics and starter cultures to ferment milk ("Maybe in the spirit of Pasteur" he says, a little cheesily). His first publications were on the bacteria and attendant viruses involved in sauerkraut fermentation, including the SPIDR repeats.

In February 2005, Barrangou and his wife drove their Honda to Madison, Wisconsin, where Barrangou joined Danisco. The reports that year by Mojica and others marked the first signals of a direct connection between CRISPR elements and viruses. Barrangou began characterizing the genomes of starter cultures, checking in with Horvath. He used the CRISPR repeats to fingerprint the *S. thermophilus* starter cultures (a key ingredient in yogurt manufacturing), using "those peculiar loci to tell which strain is which and where it came from." The more they sequenced, comparing new strains with older cultures thawed from Horvath's freezer in Dangé-Saint-Roman, the more it became obvious that the CRISPR repeats could grow and evolve.

A trio of experiments clinched the association. First, Barrangou asked, what happens when a bacterial strain is exposed to a virus? "We saw the [bacterial] immune system activate itself and pick up new pieces of DNA from the viral genome and integrate them into the CRISPR locus in a particular order," he says. This strongly supported the notion of a link between the spacer content and phage resistance.

Barrangou handled the next experiment, as his lab in Wisconsin was the only lab at Danisco permitted to do this kind of genetic engineering. Horvath was comparing two bacterial strains: DGCC7710 and a daughter (mutant) strain called 7778, which had arisen following a phage challenge performed in 1990. "Suddenly, the black box was opened!" Horvath said. In the resistant daughter strain, Barrangou found two additional spacers in the CRISPR region. "We had the sequence of the phage used in 1990: spacers 1 and 2 were present in the phage. So what did we do? We engineered to prove it was sufficient to provide resistance." Removing both spacers in the daughter strain resulted in loss of resistance. "Add the two spacers in the parental strain, and without any challenge, it becomes resistant. Bingo!"

Understandably, Barrangou calls this his favorite experiment of all time. "When you swap the two immune systems between two strains, you swap their resistance to some of these sensitivities to viruses. That was essentially the proof that there is a direct link between the CRISPR genotype and the antiviral phenotype."

The third experiment, Barrangou admits, was a bit lucky. Adjacent to the CRISPR motifs are the CRISPR-associated, or Cas, genes, which encode the nucleases that actually cleave the viral DNA. Inactivating the two biggest Cas genes had a major impact on the CRISPR system: knocking out Cas9 abolished the immune potential, whereas inactivating another gene, Csn2, left the immune potential intact but scrapped the ability of the CRISPR array to acquire new spacers. "This is where sometimes you have to be serendipitous, right?"

The priority for Danisco wasn't so much to trumpet the results in a major science journal but to patent the CRISPR discovery. Filing the initial patent application on August 26, 2005, gave the company one year to provide additional examples to illustrate the patent. Horvath's group had to keep quiet and hope nobody scooped them. The inventors were listed as Horvath, Barrangou, their respective bosses, Christophe Fremaux and Dennis Romero, and Patrick Boyaval (the boss of Fremaux and Romero). With the clock ticking, Horvath and Fremaux reached out to Canadian virologist Sylvain Moineau, who was also an expert on *S. thermophilus*. Moineau was skeptical at first until he reproduced the Danisco team's results in his own lab. One of Moineau's postdocs, Hélène Deveau, came to his office one day: "You're not going to believe this," she said. DNA analysis showed the CRISPR array had increased in size as the bacteria she was studying gained resistance to phages. "It was just history from there," Moineau said.[22]

The patent was converted in August 2006, meaning the results were now publicly available. Danisco's head of innovation, Egan Beck Hansen, had a PhD in phage biology so he understood the significance of the discovery, and agreed with Horvath it was time to publish. Horvath typically favored specialized microbiology journals, but Barrangou boldly suggested they try *Science*. His colleagues scoffed at the idea but Barrangou persisted. Besides, if the paper was rejected, they would only lose a few weeks and they could always resubmit elsewhere.

Horvath was listed as the senior author but Barrangou drafted most of the manuscript. "He is good at painting the big picture, I am more in the details," Horvath told me. Barrangou and Horvath submitted their manuscript to *Science* in October 2006, but it was rejected. CRISPR was not the hot commodity it is today. One of the three reviewers objected that the results were not observed in the classic model bacteria such as *E. coli* or *Bacillus subtilus*. However, Caroline Ash, the editor handling the paper, told Horvath he could resubmit if they added some more data. Barrangou and Horvath set about disrupting more Cas genes to show they were involved in phage resistance, and generated more BIMs and spacer sequences. Ash accepted a revised version of the paper, which was published in March 2007—the 20th anniversary of the first report of the mysterious sequences that would become a household word.[23]

The Danisco team had shown experimentally that the biological function of CRISPR-Cas systems is to provide adaptive immunity in bacteria against viruses. It was the first appearance of CRISPR in the famous pages of *Science* magazine. More importantly, the predictions of Mojica, Vernaud, and Bolotin had been proven correct. Yogurt for the win!

Looking back, Horvath said the paper marked "a small revolution in microbiology, a new immune system but not the big revolution that occurred in 2012." The paper immediately resonated with biochemists from the Baltics to Berkeley. Horvath and Barrangou began presenting their results on the science circuit. Over the next few years, CRISPR became a bigger and bigger deal. "We knew it was cool," says Barrangou. "You could use it for genotyping and vaccination and cutting viral DNA." But there were complications. "In the dairy industry, as soon as you speak about DNA, they suspect GMO manipulations and genetic engineering," says Horvath. "It's a sensitive topic."

In May 2011, DuPont bought Danisco for $6.3 billion. Today, 100 percent of all commercial cultures at DuPont (and other companies) are enhanced using CRISPR screening. "Whether you have yogurt, a bite of cheese, whether you put that on your nachos or pizza or cheeseburger, in Beijing or Paris or London or New York or Buenos Aires, you are consuming a fermented dairy product that was manufactured using a CRISPR-enhanced starter culture," says Barrangou. Today, the legacy of Barrangou and Horvath's discovery can be found in products such as CHOOZIT SWIFT 600, one of Dupont's most successful starter cultures, especially designed for pizza cheese. DuPont sells hundreds of starter culture packs developed from studies using Horvath's library of some 7,000 phages.

One year after his team's landmark paper, Horvath was asked why the *Science* paper had become so frequently referenced. He predicted CRISPR would have a major impact on improving the quality of functional food ingredients and a detrimental impact on the population of bacterial viruses.[24] Asked whether their work had any broader ramifications, Horvath replied: "Our research does not have any social nor political implication."

The yogurt maker extraordinaire would soon eat his words.

CHAPTER 4
"THELMA AND LOUISE"

Fifteen years to the day after President Clinton's human genome celebration in the White House in June 2000, Jennifer Doudna posed in her laboratory at Berkeley for a *New York Times* photographer. The background is full of typical lab paraphernalia—cubicle freezers tucked under the bench, pipettes hanging on the wall, old yellow radioactive hazard tape. Doudna looks off to the side, wearing a pinstripe double-breasted jacket while holding a pristine white lab coat (creases still visible).

The photo reminded me of a classic *Time* cover featuring Craig Venter during the height of the genome wars, a lab coat over half of his dark business suit, creating a distinctly Jekyll-and-Hyde appearance. Like Venter, Doudna's brand had evolved in short order from a dedicated academic researcher working in an obscure branch of biochemistry to an international scientific celebrity credited with spearheading a transformative new field of genome science. Venter was the poster boy for fueling a biotech revolution in genome sequencing. Doudna is the face of the CRISPR revolution, developing the ubiquitous utility tool of molecular biology enabling scientists around the world to edit DNA, from classroom to clinic, and farm to pharma. Doudna's contemplative gaze might well have signified the ethical controversies hanging in the air, weighing heavily on her shoulders. The ensuing *Times* story was entitled "The CRISPR Quandary."[1]

I first met Doudna in 1998, her star already ascending years before she or anyone else had heard of CRISPR. Just thirty-five years old at the time, she was making her first visit to the headquarters of the Howard Hughes Medical Institute (HHMI)*—twenty manicured acres nestled in Chevy Chase, Maryland, just outside Washington, DC. On the faculty at Yale University, Doudna was a newly minted investigator of the nonprofit institute, selected after a rigorous nationwide competition as one of the most talented young scientists in the country.

Doudna had a stellar scientific pedigree, having trained with not one but two Nobel laureates. With the HHMI's hefty financial support of about $1 million a year (a blip on what was then a $10 billion endowment), Doudna could indulge her scientific curiosity. I congratulated her on her appointment over a glass of wine. She graciously told me about her research plans. I nodded along enthusiastically, hoping my lack of structural biology expertise wasn't too obvious.

Born in Washington, DC, Doudna was seven years old when her parents moved to Hilo, on the big island of Hawaii. Her father was an English professor at the University of Hawaii (UH) Hilo, while her mother taught history at the local community college. The Hilo scenery provided abundant plants and animals for Doudna to explore and ponder their evolution. When she was about twelve, Doudna's father left a book on her bed—a dog-eared paperback copy of *The Double Helix*, Jim Watson's riveting personal tale of the discovery of the structure of DNA. She ignored it at first, assuming it was a detective novel—which, in a sense, it was.

The Double Helix remains an astonishing story of naked scientific ambition and fierce rivalries. Watson was widely criticized for the sexist manner in which he portrayed Rosalind Franklin. It was Franklin's unpublished X-ray image of DNA fibers—photograph 51—that inspired Crick and Watson to construct their classic model. Watson literally pieced together the final pieces of the three-dimensional puzzle, showing how the four

* HHMI was set up as a nonprofit medical institute by the businessman-investor-aviator Howard Hughes. It was initially little more than a tax shelter but today HHMI spends a portion of its $20-billion endowment funding hundreds of researchers in the life sciences. Doudna has been an investigator for more than twenty years, and looks set for another twenty. Several other leading CRISPR researchers, including Feng Zhang and Luciano Marraffini, are also HHMI investigators.

DNA bases fit together, adenine (A) always pairs with thymine (T), while cytosine (C) partners with guanine (G). The report was published in *Nature*, eight hundred words of pure gold, "tight as a sonnet" in the words of Colin Tudge.[2] Franklin died in 1958 and thus was denied a share of the Nobel Prize, which was awarded to Crick, Watson, and her former colleague, Maurice Wilkins, in 1962.

The Double Helix captured Doudna's imagination, as it has countless young scientists, revealing how biologists could solve the secrets of life by probing the atomic structure of biomolecules such as DNA. Outside the classroom, Doudna experienced isolation and ostracization; she was a minority, referred to by the locals as a haole (a disparaging term for non-native). She found refuge in the library and the lab, delightedly spooling translucent DNA fibers around a glass rod. A family friend, Don Hemmes, let Doudna spend a summer working in his UH lab, playing with an electron microscope while studying worms and mushrooms. When her high school counselor told her "girls don't do science," Doudna's determination only intensified.

Doudna graduated in chemistry from Pomona College in California after briefly flirting with the idea of switching to French, supported by her mentor and undergraduate advisor, Sharon Panasenko.[3] She moved to Boston in 1985 for her PhD at Harvard Medical School, working with Jack Szostak, a brilliant RNA biochemist who went on to win the Nobel Prize. She published a major paper with him in 1989 revealing a surprising enzymatic function for certain RNA molecules. Doudna became smitten with RNA, not to be fashionable but to learn more about the versatility and functionality of ribonucleic acid.

In the central dogma of molecular biology first articulated by Francis Crick—DNA >>> RNA >>> protein—RNA was sometimes overlooked. It was considered by some a disposable copy of the genome, lacking the majestic symmetry of the double helix or the exquisite three-dimensional complexity and diversity of proteins, which carry out the essential functions in life. In 1960, Crick and Sydney Brenner postulated the existence of messenger RNA (mRNA), a facsimile of DNA relaying the freshly copied instructions in the book of life beyond the cell nucleus to the protein-manufacturing factories, the ribosomes.

While Doudna was focusing on RNA, she was acutely aware of colleagues who were breaking new ground in other branches of genetics. One was Jim Gusella, who wanted to help patients with the incurable Huntington's disease. Like a darts player throwing a lucky bullseye, Gusella had fortuitously pinpointed the location of the Huntington's gene on chromosome 4 in one of his first experiments in 1983. Fatefully, it would take a decade of toil before his team identified the gene itself.

Doudna's next move was to join another Nobel laureate, spending three years in Colorado working on RNA enzymes with Tom Cech, who won the Nobel Prize in 1989. "I think Jennifer had something to do with both of those [Nobel Prizes]," says her colleague Barbara Meyer, although that would require Doudna mastering in time travel. She opened her own lab at Yale University, and immediately made an impression. Vic Myer, a leading gene editor, recalls, "she was bright, energetic, insightful, and asked very good questions. Fundamentally she is driven by the science, excited by the science, and a great person."[4]

In 2002, Doudna moved her lab across the country from Yale to Berkeley, California, to be closer to home, family, and a major synchrotron source for her structural biology studies. Her husband, Jamie Cate, whom she met in the Cech lab, set up his own lab next door. The big question scientifically at the time centered on the RNA World hypothesis, the idea that life began on earth in the form of RNA molecules.

Jill Banfield, a microbiologist at Berkeley, characterizes new species of bacteria and Archaea, expanding our understanding of the evolutionary tree of life. Like an archeologist searching for rare living microbes, her research takes her around the world to a range of environments, some more exotic than others: salt lakes in her native Australia, mine shafts in Colorado, geysers in Yellowstone National Park, as well as simple groundwater wells. Banfield's team has characterized literally hundreds of new microbial species from these extreme locales, the winners in a billion-year-old game of *Survivor*.

In 2006, Banfield was stumped. Samples of the same species collected from the same location should have identical sequences, she reasoned, but to her surprise no two DNA traces had the same sequence. "It was shocking," she recalls. She had stumbled on the fast-evolving CRISPR region. The latest news on that front had come in a paper from Kira Makarova and Eugene Koonin at the NIH that included a rare sighting in science: a confession. The duo had previously suggested that the function of the CRISPR-associated genes was in DNA repair. They were wrong. Abandoning that hypothesis, they now proposed that CRISPR was a genetic defense system that targeted viruses via a mechanism called RNA interference (RNAi), which would earn its discoverers, Stanford's Andy Fire at Stanford and Craig Mello at the University of Massachusetts, a Nobel Prize.[5]

Banfield typed in "RNAi" and "UC Berkeley" into her search engine. The first name that popped up was Doudna's. She decided to give her colleague a call. "You're doing the type of research that I think could be very interesting for something that I've stumbled across in my own work," she said.[6] And she introduced Doudna to a new piece of scientific jargon: CRISPR. Now it was Doudna's turn to google Banfield. A few days later, the geomicrobiologist and the RNA biochemist met at the Free Speech Movement Café, a popular central meeting spot in the heart of the Berkeley campus. Seated at an outdoor table, Banfield excitedly told Doudna about her work sequencing bacterial genomes and the clusters of strange palindromic repeat sequences contained in some of her newly discovered organisms.

Banfield took her notepad and sketched a circular bacterial genome and then, magnifying a section of the DNA, drew a series of symbols:

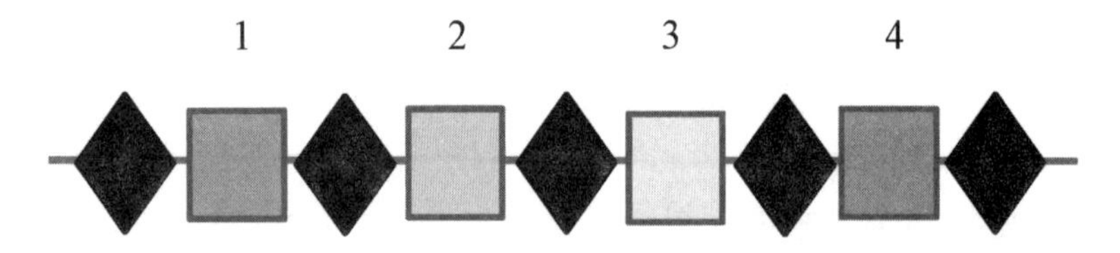

Array of Light: In her first meeting with Doudna, Jill Banfield sketched the CRISPR array showing the alternating pattern of repeats (diamonds) and spacers (squares) derived from viral DNA.

This was the CRISPR array, a series of identical motifs about thirty bases in length interspersed with other sequences—Banfield drew squares, giving each a number—that apparently had nothing in common with each other. Banfield knew the spacers were derived from viruses and that they evolved at a faster clip than other parts of the bacterial genome. And now there was the suggestion that CRISPR was the blueprint for an antiviral defense system involving RNAi. In Doudna, one of the world's experts on RNA structure and function, she'd found the perfect ally.

Banfield's expertise in cataloguing the vast diversity of microbial life—"the weight of evolutionary history" as she calls it—promises to unearth many new tools for the CRISPR toolbox. In 2013, Kim Seed and colleagues at Tufts University made a remarkable discovery: a phage that has pilfered CRISPR repeats from bacteria—like a hostage grabbing an assailant's weapon—and turned it on the host cell.[7] Banfield's team has also discovered so-called jumbo phages—phages with giant genomes larger than some bacteria, blurring the boundary between life and death—lurking in the gut microbiome of people in Bangladesh with a non-Western diet. The CRISPR machinery not only helps the viruses evade the bacterial defense mechanisms but may also thwart competing viruses.[8]

It would actually be several years before the two women, both leading large groups with different research interests and areas of expertise, managed to collaborate on a research paper, but Banfield had piqued Doudna's interest. Looking back at her pivotal role in the CRISPR drama, Banfield, one of the most accomplished microbiologists in the world, can only laugh. "One thing they'll write on my tombstone is: 'Told Jennifer Doudna about CRISPR-Cas.' Like, that will be the sum of my life!"[9]

In the first few months of 2007, two new postdocs joined the Doudna lab. Blake Wiedenheft introduced CRISPR to the laboratory; Martin Jínek ensured the lab's legacy.

Jínek hails from Třinec, a city on the border between the Czech Republic and Poland. One hundred miles to the west is Brno, the birthplace of genetics. Surprisingly, Jínek didn't pay a visit Gregor Mendel's monastery until a few years ago, when he was invited to give a lecture. At sixteen, Jínek won a scholarship to a private boarding school in England. He then spent four years at Cambridge University studying chemistry, but always had an inclination towards biology, especially RNA, thanks in no small part of Doudna's string of successes with Szostak and Cech. "It's such a versatile molecule," he told me. "It can do catalysis, fold into 3D structures, and it's a carrier of information. It's an all-rounder"—a term, I surmise, he picked up watching cricket at school.

After completing his PhD in Germany, Jínek wanted to study RNAi. The Doudna lab had just published the structure of an important RNA processing enzyme called dicer. Doudna likened it to a "molecular ruler" that measures RNA sequences prior to cutting them into precise lengths. RNAi was attracting huge biotech interest, underpinning new companies such as Alnylam Pharmaceuticals and Moderna. Jínek arrived in Berkeley just before publication of the Barrangou et al. *Science* paper, which elevated CRISPR to prime time. Doudna's group discussed that paper in a journal club meeting. "Everybody was quite excited about it," Jínek recalls. "We decided that this was going to be an RNA-guided mechanism, like RNA interference. There was going to be some kind of connection."

A short time later, Wiedenheft, a swarthy scientist from Montana, visited for an interview with Doudna and talked about studying CRISPR. "We were all primed for that by the journal club—Jennifer included," Jínek said. The man from Montana got the job and performed the first CRISPR experiments in Doudna's lab. Wiedenheft was interested in the pathways by which phages infect bacteria and conversely, how bacteria ward off phage infection. Like Banfield, he had collected microbial samples in Yellowstone and other extreme locations. "Without Blake being in the trenches, I don't know if the Doudna lab would've had a meaningful thrust in that direction," said Ross Wilson, who joined the lab two years later.[10]

Jínek wasn't working on CRISPR himself but he'd regularly talk science with Wiedenheft. The Czech's first involvement was helping his friend learn about protein crystal structures for a first look at a DNA-cutting enzyme called Cas1, published in 2009.[11] Some afternoons, the two men would take a break to go biking in the Berkeley hills. Wilson shakes his head recalling the sweaty duo returning to the lab in their spandex shorts.

Just as Doudna's interest in CRISPR was getting off the ground, her head was turned by an irresistible offer in 2009 from one of the most famous names in biotech. She admits she was undergoing a mini midlife crisis, looking for a new scientific challenge. "Am I going to get to the end of my career and feel like I did some cool stuff, had some fun, published some papers we're proud of, but did I really solve any problems?" she told journalist Lisa Jarvis.[12]

The offer came from Richard Scheller, Doudna's former colleague at HHMI before he became head of R&D at Genentech, one of the most successful and coolest biotech companies in the world.[13] Doudna would have the chance to apply her RNA expertise in the search for novel drugs and therapies. Her appointment as Genentech's new vice president of Discovery Research was announced in a press release, and she listed her change of address in an article with Jínek in *Nature*.[14] She hoped that most of her group would join her in South San Francisco, but industry wasn't appealing to Jínek. Plan B was to move down the hall and finish his postdoc in the lab of Doudna's husband.

However, Doudna soon had a change of heart. "I realized what I'm good at doing and what I really like. It all boiled down to creative, untethered science," she said. After a somewhat painful two months, she resigned and returned to Berkeley, where she reclaimed her HHMI professorship. Having given up most of her administrative obligations, she was free to pursue "crazy, creative projects" that might not be clinically relevant but she considered cool science. And the craziest project in the lab was CRISPR. "Had I not made the foray to Genentech and then back to Berkeley, I might not have done any of the CRISPR work," she acknowledged.[15]

In a 2010 lab photo, Doudna is pictured with her troop of CRISPR devotees—Wiedenheft, Jínek, graduate student Rachel Haurwitz, and longtime lab manager Kaihong Zhou. There is an air of innocence in the group, blissfully unaware of how their lives were about to change. Jínek had finally run out of independent funding after four years but Doudna was happy to keep him around. In 2011, he began searching for faculty positions back in Europe, but wanted to enjoy a last hurrah to end his Californian adventure. Jínek fancied taking a closer look at the little-known type II CRISPR system. Doudna soon offered him a gilt-edged opportunity.

It wasn't just the Doudna lab that was struck by the Danisco CRISPR story. On May 23, 2007, just two days after the *Science* paper came out in print, Virginijus Šikšnys emailed Horvath to offer a collaboration. In the Lithuanian's world, the discovery of a new antiviral defense system called CRISPR was almost as momentous as the collapse of the Soviet Union. For years, Šikšnys had been trying to understand how bacteria defend themselves against viruses. Suddenly, scientists from a yogurt company had described an entirely new bacterial immune system. It was the start of an important partnership that would have a major impact on the development of genome editing.

Vilnius, the capital of Lithuania, is not exactly the Mecca of molecular biology. When I first visited in the Spring of 2017, the main cultural attraction was the former KGB headquarters.* The city center is a curious mix of narrow medieval streets in the old town, where I washed down a pungent beaver stew with real mead, adjacent to rows of designer clothing boutiques.

The Vilnius Institute of Biotechnology is a few miles outside the city center. A chemist by training, Šikšnys built a reputation studying the 3D structure and properties of bacterial restriction enzymes (there are four thousand all told).** CRISPR offered an exciting new research opportunity. But first, he needed to transfer the CRISPR system into a bacterium other than *S. thermophilus*, for good reasons. "We don't know how to make cheese and yogurt!" Šikšnys joked.

---

* The Museum of Occupations and Freedom Fights.

** I was part of the first wave of PhD students in the 1980s who had the luxury of just ordering these enzymes from a catalogue. My predecessors recounted horror stories of spending hours camped in the cold room having to purify them from scratch.

The obvious choice was *E. coli*, the lab workhorse. But which CRISPR system? It turns out *S. thermophilus* has four different CRISPR systems—all have a cluster of spacers but differ in the architecture and adjacent Cas genes. Šikšnys chose the simplest (type II) system with the smallest number of Cas genes, including one called Cas9, known to be required for phage resistance. With his intimate knowledge of DNA-cutting enzymes, Šikšnys noticed a couple of spots in the Cas9 structure that resembled catalytic active sites he'd observed in restriction enzymes. It was a sign that Cas9 might have interesting DNA-cutting properties of its own.

The Lithuanian's entry into the CRISPR timeline was one of many new dots on the map. But it took five years to connect the dots from yogurt and pizza starter cultures to the brink of a universal gene-editing technology. The next steps were to answer some practical questions: How were the CRISPR spacers captured from the phage invaders and stitched into the bacterial genome? And how were they weaponized to target and destroy incoming phage?

Moineau's group first described a key recognition sequence,[16] the critical first touchpoint of Cas9 alighting on DNA. As phages mutated or "escaped" CRISPR-mediated destruction, Moineau catalogued a series of single-base mutations in the CRISPR repeats as well as this recognition motif, which Mojica dubbed the proto-spacer adjacent motif, or PAM.[17]

Over the next few years, researchers around the world began piecing together the molecular details of the CRISPR immune system. It was like the British children's party game pass the parcel: at each round the music stopped with a different participant peeling off another layer of wrapping before revealing the gift. After documenting CRISPR in her own microbial samples,[18] Banfield reached out to Barrangou to propose they organize a meeting of CRISPR disciples. That summer in 2008, about thirty scientists gathered at Berkeley's Stanley Hall. Members of Doudna's lab, located on the seventh floor of the same building, dropped in. Barrangou put the wine and beer costs on his corporate credit card.

The action next shifted to The Netherlands. Studying CRISPR in *E. coli*, John van der Oost, a microbiologist at Wageningen University, and Stan Brouns showed that the CRISPR spacers are first transcribed into a long contiguous RNA, which is then sliced into discrete CRISPR RNAs corresponding to individual spacers. This RNA then forms a complex with Cas proteins that targets the corresponding phage.[19] (The *E. coli* CRISPR is of the type I variety, in which the role of Cas9 is played by a complex of five proteins known as the Cascade complex.) The result was a CRISPR milestone, duly recognized a decade later when van der Oost shared Holland's most prestigious science award, the Spinoza Prize. "I don't think John's work will ever be forgotten," says Koonin, the unofficial master record-keeper of CRISPR gene evolution.[20]

In Chicago, Erik Sontheimer and his Argentine postdoc, Luciano Marraffini, designed some clever experiments using *Staphylococcus epidermis* to settle another big question: does CRISPR-Cas target the viral RNA—mimicking RNAi—or DNA? Marraffini suspected it would be more efficient for bacteria to dispose of viral infections if they cleaved DNA, taking a machete to the viral genome in one fell swoop. He was correct.[21] Practically speaking, they wrote, "the ability to direct the specific addressable destruction of DNA that contains any given 24-48 [base] target sequence could have considerable functional utility, especially if the system can function outside of its native bacterial context."

Sontheimer and Marraffini saw a glimmer of clinical relevance—the possibility "to impede the ever-worsening spread of antibiotic resistance genes and virulence factors in staphylococci and other bacterial pathogens." It was one small step on the road to precise CRISPR genome editing, albeit one that exclusively used the delete button. "We were the first to recognize and explicitly articulate the possibility that CRISPR could be repurposed for genome engineering," said Sontheimer.[22] But the celebration was short-lived: Sontheimer's patent application was denied for lack of experimental evidence, as was an ambitious grant application that was years ahead of its time. (As we shall see in chapter 12, it was the beginning of a contentious legal battle for inventorship.)

As interest in CRISPR grew, additional types of CRISPR systems were discovered. At the University of Georgia, Michael Terns showed that type III CRISPR systems target RNA rather than DNA. Meanwhile, Doudna published her first papers on CRISPR with Wiedenheft and Haurwitz.

In 2010, Moineau, who has been fascinated by phages since he first saw them under an electron microscope, took center stage. “If you go to the ocean and take water in the palm of your hands, you have more viruses in your hands than there are humans on this planet,” he says.[23] His forte was the phages that infect *S. thermophilus*—the key ingredient for yogurt and cheese. Moineau says we eat more than one sextillion (10^{21}) *S. thermophilus* cells per year.

Moineau’s interest is the ongoing arms race between phage and bacteria. Still collaborating with Horvath, Barrangou, and the Danisco team, Moineau’s lab demonstrated that CRISPR RNAs cleaved their DNA targets directly, literally cutting a circular DNA molecule (a plasmid inside a bacterium) in one spot into a linear fragment, producing clean blunt ends in the vicinity of the PAM sequence.[24] Moineau published the story in *Nature*, one of the highlights his career.

Meanwhile, researchers continued to expand the family of Cas genes by combing through the expanse of microbial life on earth, much of the credit belonging to Koonin and Makarova. The number of different CRISPR classes and subtypes has grown increasingly complex, with six classes known by 2020. Fortuitously, the *Strep* bacteria studied by Horvath and Barrangou have a type II system, the most rudimentary of CRISPR systems. And that was huge.

The first time I emailed Emmanuelle Charpentier, in early 2017, I received an immediate response. It was an “out of office” message that said:

> Due to my full schedule associated with attendance of prize ceremonies, I will not be able to reply to your email . . .

Well, that was a first. She wasn’t kidding. Charpentier has won dozens of prestigious awards since 2012. Asked about the distraction of being a semi-permanent fixture on the awards circuit, she told *Le Figaro*: “It’s very exotic! Let’s say that’s not the reason I did research. As a researcher, I like being isolated in my laboratory with my team."[25] Her PhD supervisor said she was so resourceful, she could start a lab on a desert island.[26] It didn't quite come to that, but after two decades of nomadic existence—moving from the United States to Austria to Sweden—she has found a home in Berlin. Time will tell if she can match the success she earned in 2012.

Charpentier may have a lower public profile in the United States than Doudna or Zhang, but it is a different story in Europe. A 2015 profile in *Le Monde* dubbed her "the charming little monster" of genetic engineering, with "the air of a mockingbird, perched in the forest of the best-rooted authorities of science in France, Europe, and America."[27] She is called "pugnacious and courageous," inspired by three muses—curiosity, daring, and freedom. The following year, the same newspaper, in an article titled "The new icons of biology," hailed Charpentier and Doudna as the "Thelma and Louise" of biomedical research, acquiring scientific honors and prizes like a squirrel harvests hazelnuts.[28]

In September 2018, just a few weeks after receiving the Kavli Prize in Oslo, Charpentier flew to New York to deliver a memorial lecture at Columbia University.[29] As usual, her appearance attracted a packed house, as hundreds of students and faculty crammed into the auditorium including the gangly figure of Nobel laureate Richard Axel. Arriving late, the eminent neuroscientist folded himself into a front row seat, as intrigued as everyone else to hear the petite Frenchwoman on course to follow his journey to Stockholm. After an interminable introduction, Charpentier politely chided her host for exaggerating her resume. "I did not publish that many papers!" she said. "I've always been more on the perfectionism side—I focus more on the quality than the quantity."

In her book, Doudna said her first impression of Charpentier was "soft-spoken and retiring." That was a fair reflection of Charpentier's lecture, a surprisingly subdued account of her CRISPR journey, framing her work in the context of legendary French molecular biologists like François Jacob and Jacques Monod. She dwells on a quote from Monod in 1970:

> Modern molecular genetics offers us no means whatsoever for acting upon the ancestral heritage so as to improve it with new

> features—to create a genetic "superman": on the contrary, it reveals the vanity of any such hope: the genome's microscopic proportions today and probably forever rule out manipulation of this sort.[30]

Charpentier seldom looks up at the audience and strikes me as guarded, reluctant to let loose and share stories. Axel fidgets in the front row. She wonders if her own mobility "helped me to understand the way bacteria defend themselves against mobile elements." It's more a statement of fact than an attempted joke. But if humor is in short supply, humility is not. Charpentier acknowledges a crucial slice of fortune. For years the pet organism in her lab was a bacterium called *Streptococcus pyogenes*, which can cause life-threatening infections. This was the source of the Cas9 protein. "Cas9 in *Strep pyogenes* is very efficient," Charpentier said. "We've tested other Cas9 proteins, some are close in efficiency, but if we'd identified mechanisms in another bacterial species, I'd not be here in front of you."

After her lecture, Charpentier fields questions from the audience: She bemoans a recent European court ruling on GMOs ("a big disappointment for scientists in Europe") and dismisses concerns about CRISPR's safety. "Delivery is a bigger bottleneck than the CRISPR mechanism itself," she says confidently. Students line up to take selfies with her; she is a bona fide scientific celebrity after all. I join the queue to invite Charpentier for an interview before she heads to dinner with some university VIPs.

The next morning, we rendezvous on Manhattan's Upper West Side. She arrives looking chic in blazer and jeans, two months' shy of her fiftieth birthday. By contrast, I am sporting fresh Central Park pigeon excrement down my laundered white shirt. "It is good luck, I think," she smiles, as we head to a nearby French bistro.

Charpentier was born fifteen miles south of Paris in 1968, six months after the student protests and civil unrest. From an early age, she was intent upon going to college, inspired by her older sister. Her father taught her the Latin names of plants, which might have inspired her to pursue biology. At age twelve, Charpentier came home from school one day and told her mother that she would eventually study at the Pasteur Institute. Ten years later, she made good on her promise. "I got my worst grade for [microbiology] and it became my specialty!" she laughs. She earned her PhD in microbiology in 1995, writing her thesis in a library overlooking the Cathedral of Notre-Dame.

Charpentier recognized that life as a research scientist "would fit the many aspects of my personality—my curiosity, intellectual drive for knowledge, enjoyment of communicating knowledge to others, and working as a team, and my desire to turn complex scientific discoveries into practical applications that would help society."[31] She spent the next six years in the United States, beginning at the Rockefeller University in New York. She studied bacteria responsible for skin infections in mice, searching for new biochemical pathways and drug targets. After working with *Streptococcus pneumoniae*, she turned to *S. pyogenes*, which became her favorite organism. Her experience in America introduced her to many talented researchers, many with a strong entrepreneurial spirit. That lesson, too, was not lost on her.

There were no openings at the Pasteur when she was ready to return to Europe, but she landed at the University of Vienna in laboratories named after the Nobel laureate Max Perutz, a contemporary of Crick and Watson.* "It was important to be independent and to not have anyone around me," she said. I'm intrigued by Charpentier's willingness to travel in search of freedom and funding. Without missing a beat, she says: "In twenty-seven years, I've worked in five countries, seven cities, ten institutions. Fourteen different offices, thirteen different departments, and fourteen apartments. It's a very big turnover!" Each move was motivated by an incentive or a better position.

Charpentier's first exposure to CRISPR came in 2006. Two of her students, Maria Eckert and Karine Gonzales, were performing computer

* In early 1953, Perutz shared a Medical Research Council report containing Rosalind Franklin's unpublished DNA crystallography data with Crick and Watson. That sneak peak proved critical in the assembly of the double helix.

searches for DNA sequence matches similar to the studies reported by Mojica one year earlier. This time CRISPR was not the bait but the prize. Eckert and Gonzales were scouring the *S. pyogenes* genome for traces of small RNAs (encoded by genes that produce RNA molecules rather than proteins). One of the most abundant hits was a novel trans-activating CRISPR RNA (tracrRNA). The gene encoding this tracrRNA sat in the vicinity of the CRISPR array, although the significance of that location wasn't immediately obvious. Charpentier's major interest was not CRISPR or bacterial immunity, but her tracrRNA had a certain *je ne sais quoi.* It would prove a critical piece of the genome editing puzzle.

In 2008, Charpentier decided to leave the history of Vienna for the hinterland of northern Sweden and Umea University. The weather was bleak but Charpentier warmed to the atmosphere and people at the Laboratory for Molecular Infection Medicine Sweden. She focused on tracrRNA, collaborating with German biochemist Jörg Vogel, planning experiments on the flights back and forth to Sweden.

The next year, her group established a link between the CRISPR-Cas9 system and tracrRNA. "It was a very simple experiment," she recalls. When the group knocked out tracrRNA, they found that the CRISPR RNA (crRNA) was not made, and vice versa. The logical conclusion was that Cas9 formed a physical complex with these two RNA molecules. When Charpentier's team compared tracrRNA sequences from a variety of bacteria, they found one thing in common: a sequence that forms a duplex with crRNA. While other groups were describing more complex types of CRISPR systems, the beauty of the type II system in *S. pyogenes* was that only a single gene, Cas9, was necessary for viral interference, along with crRNA. And then there were three. Charpentier and Vogel had brought the third component—tracrRNA—into the story. "Yes, it is an essential component, because Cas9 is an enzyme guided by two RNAs," she said.[32]

One month after Charpentier submitted her polished account of the discovery of tracrRNA to *Nature* in September 2010, she traveled to the Netherlands to give a talk at the annual CRISPR conference, which had ballooned to about two hundred CRISPR aficionados. Few members of the CRISPR club knew about the French scientist working in Sweden with colleagues in Austria and Germany. But now, Charpentier recalled with a smile, "they discovered the famous story of tracrRNA." It was the pinnacle of Charpentier's career to date. "The pioneers all came up to me and shook my hand and said, 'I think you got the story!'" One notable absentee was Doudna, whose research wasn't consumed by CRISPR just yet.

With her paper accepted by *Nature* in early 2011, Charpentier plotted the next step of her story, which she would set in motion at an upcoming conference. She needed an expert in RNA biochemistry and structural biology. "I had in mind to approach Jennifer and to ask her whether she would be interested in deciphering the structure of Cas9."[33]

CHAPTER 5
DNA SURGERY

In March 2011, Doudna and Charpentier met for the first time at a small conference at the InterContinental San Juan in Puerto Rico, hosted by the American Society for Microbiology. The theme was "Regulating with RNA in Bacteria." After both women had given their respective talks, Charpentier suggested they take a stroll through Old San Juan. In some respects, they were a study in opposites: one tall with fair hair, the other a shorter brunette. Charpentier was the junior scientist in several respects—five years younger, with a relatively modest publication record, working at a remote Swedish university. Doudna, by contrast, had trained under two Nobel laureates, was a full professor and approaching her fifteenth year as an HHMI investigator. She had also been an author on almost 20 research articles in the top three science journals, compared to just a couple for Charpentier.

As they ambled down cobblestone streets, Charpentier shared results from her upcoming *Nature* paper—her first as a group leader in the journal. The lab work was led by a Master's student, Elitza Deltcheva, and a Polish grad student, Krzysztof Chyliński.[1] After years of perseverance, she had identified the critical role of tracrRNA in the weaponizing of the CRISPR antivirus defense mechanism. Now she needed help in deducing the role of Cas9 in *S. pyogenes* and offered a collaboration. Charpentier had identified several different Cas9 proteins from different bacteria, but they always worked with a duplex of RNAs—the crRNA, corresponding to the spacer sequence, and tracrRNA. Sorting out the enzyme's 3D structure was crucial because "if one wanted to reduce the system into practice, then the structural biology might bring clues to shorten the proteins and do some protein engineering." There was no doubt this was Doudna's domain.

The two scientists hit it off. "I really liked Emmanuelle," Doudna said. "I liked her intensity. I can get that way, too, when I'm really focused on a problem. It made me feel that she was a like-minded person."[2] And Doudna had published her first papers on CRISPR by this time.

Back in Berkeley, Doudna persuaded Martin Jínek to work on the Cas9 project with Charpentier's lab. Jínek began skyping with Chyliński, the grad student, who had stayed in Vienna after Charpentier, who didn't have tenure, had taken a group leader position in Sweden. Although they could converse in something approximating Polish—Jínek learned the language watching Polish television as a child—the two scientists mostly communicated in English.

The Charpentier lab had tried to purify Cas9 "but it wasn't working out," said Jínek. After the Puerto Rico accord, the collaboration was sealed when Charpentier attended the 2011 CRISPR meeting at Berkeley. The team captured the union with a casual group photo on the steps of Stanley Hall on the eastern edge of the Berkeley campus, where Doudna's lab was located. Jínek stood in the center of the group, all clad in jeans, with Doudna and Charpentier on his right, Chyliński and Ines Fonfara (a Charpentier postdoc) to his left. They could have been a country music quintet posing for the cover of their soon-to-be-platinum debut album.[3]

The first task was to purify enough Cas9 protein to obtain crystals to analyze the structure using X-ray crystallography. Jínek's interest wasn't so much genome editing at this stage but understanding the molecular mechanism of how Cas9 works. "It's a system that uses guide RNAs but most likely targets DNA," he said. "There are parallels with RNA interference, but it's not RNA that's being targeted." This just added to Jínek's interest in demonstrating that Cas9 was acting as an RNA-guided DNA cutter. Perhaps this wasn't headline news but it should provide a stimulating swansong for his postdoctoral fellowship before heading back to Europe.

Working with a summer student, Jínek purified Cas9 from *S. pyogenes* samples shipped from Europe. Wilson remembers Jínek being quite secretive. He gave the precious Cas9 plasmid DNA an inscrutable lab name,

MJ923. "This is how he catalogued things until it was safe to talk about," said Wilson. Jínek's first experiments using crRNA alone to target DNA failed. That all changed when Jínek added the tracrRNA to the mix.

Jínek's breakthrough—although he would be loath to use such dramatic language—was fusing the crRNA and tracrRNA, both required for gene targeting, into a single chimeric RNA. This was actually a fairly standard procedure, Jínek reminds me. As a biochemist, he was always looking for the minimal requirements in any reaction or system, attempting to break down the components in a very reductionist fashion. If both RNAs were part of a duplex, then presumably the two ends must be in close proximity to each other. "Then you can stitch them together with a loop," Jínek said. Those sorts of permutations were not uncommon in the Doudna lab. For example, adding or removing a base from the end of an RNA molecule can dramatically alter the ability to form crystals. Jínek found he could trim the crRNA from one end and likewise truncate the tracrRNA. But he had to maintain a degree of base pairing between the two.

Jínek modestly describes the eureka moment thus: "We had a brainstorming session with Jennifer," but then clarifies that by "we," he really meant him. As he sketched the CRISPR parts list on the whiteboard in Doudna's office, they sensed that by synthetically fusing the two essential RNA molecules, forming a single guide RNA (sgRNA) molecule, they would in principle have the ability to preprogram any guide sequence. In other words, they could specify and target any gene of interest, not just naturally occurring viral sequences. All that was required was to design a custom RNA sequence of some twenty bases that matched the desired target sequence. Doudna remembers it being a "transformative" moment, although when she initially mentioned it to a few Berkeley colleagues, they couldn't see what all the fuss was about.[4] "Krzysztof and Emmanuelle were brought on board shortly after that," Jínek recalls.

It took Jínek a few weeks to make the chimeric RNAs in-house but he quickly demonstrated that the sgRNA system was able to cut DNA at a matching sequence. Suddenly CRISPR had the makings of another gene-editing technology, in the same toolbox as TALENs and ZFNs. Things moved quickly after that. Jínek presented his results at an internal lab meeting, a weekly gathering in which Doudna's students and postdocs took turns to present their latest results. Wilson doesn't recall too many fireworks when Jínek presented his sgRNA results, but he does remember asking if they could use this method for RNA interference. Doudna replied, "We could use this for something better, like genome editing."

In June 2012, Jínek and Chyliński presented their discovery at the annual CRISPR conference, which had returned to Berkeley. Šikšnys presented his own unpublished results, which also demonstrated that Cas9 was a DNA-cutting enzyme. The overall impact "wasn't revolutionary," Wilson says, probably because there was no overt mention of gene editing. But Jínek sensed growing excitement. The CRISPR field was poised to move from an obscure branch of molecular microbiology to cutting-edge biotech.

On June 8, 2012, Doudna submitted her paper to *Science*. Jínek and Chyliński were given joint top billing, while Doudna and Charpentier's names completed the author list (the standard convention in the natural sciences, reflecting the funders and directors of the work, like the opening credits of a film). Both were listed as co-corresponding authors—an even division of credit. The *Science* editors moved quickly, accepting the manuscript in a mere twelve days, and posting the article online twenty days after submission.[5]

The *Science* paper showed that the CRISPR-Cas9 system was customizable, able to cleave almost any given DNA sequence in a test tube on demand. The nifty sgRNA would enable any other researcher to adapt the system for their own purposes. Showing ample self-restraint, Doudna and Charpentier closed by stating that CRISPR showed "considerable potential for gene-targeting and genome-editing applications." It was the modern-day equivalent of Crick and Watson's teasing understatement in 1953: "It has not escaped our notice that the specific [base] pairing we have postulated immediately suggests a possible copying mechanism for the genetic material."

The authors chose their words carefully because these experiments were limited to bacterial DNA—the team hadn't shown that CRISPR-Cas9 would work in plant or animal cells, including human. Whether that was a formality—*E. coli* DNA is the same twisty inert molecule as *H. sapiens* DNA—or a massive technical challenge given the added biochemical complexity of the cell nucleus of humans and other higher organisms, was the $64,000 Question.

The UC Berkeley press office crafted a press release to tout the importance of Doudna's report. The release trumpeted "potentially big implications for advanced biofuels and therapeutic drugs, as genetically modified microorganisms, such as bacteria and fungi, are expected to play a key role in the green chemistry production of these and other valuable chemical products."[6] But tellingly there was no mention of any clinical applications in humans, which was far beyond the scope of the paper.

Asked to supply a quote on the invention of "programmable DNA scissors," Doudna's comments are revealing for their candor and reluctance to overhype the results. "We've discovered the mechanism behind the RNA-guided cleavage of double-stranded DNA that is central to the bacterial acquired immunity system," she said. "Our results could provide genetic engineers with a new and promising alternative to artificial enzymes for gene targeting and genome editing in bacteria and other cell types." And she added: "Although we've not yet demonstrated genome editing, given the mechanism we describe it is now a very real possibility."

Despite the media outreach, the paper didn't catch fire beyond the scientific community. The *New York Times* didn't see fit to publish an article on CRISPR until 2014.[7] Doudna's hometown *San Francisco Chronicle* also passed.[8] But for insiders, the colliding worlds of CRISPR and genetic engineering sparked genuine interest. Stan Brouns hailed Cas9 as "the Swiss Army knife of immunity."[9] Fyodor Urnov, who coined the term "genome editing," has no doubt where the Doudna-Charpentier paper ranks in the annals of scientific literature. "I will never forget reading the last paragraph of the . . ." he pauses, grasping for the right words like a nuclease embracing a guide RNA. "*Immortal* is a strong word so I'm going to use it carefully:

IMMORTAL *Science* paper, in which they describe [that] Cas9 can be directed."[10]

But the overriding question was: Could a bacterial enzyme, hundreds of millions of years old, make the massive evolutionary leap and find its DNA target in the alien surroundings of a eukaryote cell nucleus? Human DNA might have the same four-letter alphabet as bacterial or viral DNA, but in its natural habitat, the double helix in eukaryotic cells is wrapped, bundled, and looped like a garden hose around protein cores in a material called chromatin. Nobody knew for sure how Cas9 would fare with chromatin.

Two experts weighed in on this very issue. Barrangou said that the potential use of the CRISPR-Cas system for genome editing in human, plant, and other complex cells hinged on whether the molecular scissors could cleave chromatin. "Only the future will tell whether this programmable molecular scalpel can outcompete ZFN and TALEN DNA scissors for precise genomic surgery," he wrote.[11] He told me later: "2012 was not genome editing . . . It's the CRISPR-Cas9 technology, the single-guide technology."[12]

Dana Carroll, a pioneer of ZFNs, agreed. Unlike CRISPR, the known genome editors were derived from DNA-binding proteins active in eukaryotic cells. "There is no guarantee that Cas9 will work effectively on a chromatin target," Carroll wrote. "Only attempts to apply the system in eukaryotes will address these concerns."[13] In other words, the proof was in the pudding. Carroll concluded: "Whether the CRISPR system will provide the next-next generation of targetable cleavage reagents remains to be seen, but it is clearly well worth a try. Stay tuned."

Stay tuned is the kind of stock throwaway line that scientists have written hundreds of times in journal reviews and commentaries. This is how science works, by raising more questions than answers. The distinguished Utah biochemist could have no idea that his words would be dissected and parsed not only by fellow scientists but also by armies of patent attorneys sparring over the inventorship of CRISPR gene editing for years to come.

In science, the race to be first to publish a groundbreaking result means the world: acclaim, funding, promotion, tenure, prizes. Every day, researchers place their trust in the editors of the leading biomedical journals. At three of the leading journals—*Nature*, *Science*, and *Cell* (sometimes dubbed the CNS journals)—the editors are full-time professionals, not part-time academics. As Mojica, Vergnaud and many others can attest, unsympathetic, indecisive, or ill-informed editors and reviewers can make mistakes or cause excruciating delays in publishing decisions, often requiring authors to waste months in search of a suitable home for their findings. In early 2012, the publishing gods struck again.

For five years, Šikšnys, the Lithuanian biochemist, had been collaborating with Horvath and Barrangou. After successfully transferring the CRISPR system from *S. thermophilus* to the lab-friendly *E. coli*, he surprisingly found that it could still defend against invading DNA, despite the two bacteria species being very distantly related in evolutionary terms.[14] The next step was to sequentially strip away each of the four Cas genes adjacent to the CRISPR array and watch what happened next. Removing three of the four genes had no effect on phage defense, but inactivating Cas9 crippled the defense system, like disabling an alarm system. It was a sure sign that Cas9 was the key ingredient in the interference provided by CRISPR. One of Šikšnys' students, Giedrius Gasiunas, succeeded in isolating active Cas9 in a tube. "I still remember the excitement when he actually did the first experiment," Šikšnys recalled, cutting DNA in a programable fashion using the purified Cas9. He had published dozens of papers in good journals, but this was a big story worthy of a very big journal.

In 1974, a former *Nature* editor named Benjamin Lewin launched a journal called *Cell,* devoted to "the molecular biology of cells and their viruses." Whereas *Nature* and *Science* printed on gossamer-thin paper the consistency of a toilet roll, *Cell* looked and felt like a fashion magazine for science, with a thick cover, sensual glossy paper, and a heretical Helvetica logo and typeface. Whereas *Nature* editors would typically squeeze authors down to fewer than 1,000 words for their reports, Lewin gave scientists free rein to lay out and discuss their results without arbitrary page limits. And Lewin, who authored a string of successful textbooks on genetics, knew the science and befriended the scientists. He set up shop in Harvard Square, where top scientists from Harvard and MIT would often hand deliver their latest manuscripts. And although *Cell* was a biweekly publication, in the ultracompetitive world of molecular biology, Lewin routinely scooped his rivals. In 1990, someone composed a *Cell* parody[15] and distributed it by fax. It was called *Cool.*[*] Lewin sold his company Cell Press to the Dutch publishing giant Elsevier in 1999 for more than $100 million.

Šikšnys was proud of his Cas9 results, which he concluded "pave the way for the development of unique molecular tools for RNA-directed DNA surgery." He submitted the paper, with Barrangou and Horvath as co-authors, to *Cell* in March 2012, but soon regretted the decision. Within a week, the editor had rejected the paper without consulting any outside experts, doubtful that the story was of "sufficient general interest." Šikšnys was upset. "We thought that it was a big thing because we showed in this paper that in principle you can reprogram this Cas9 to [cut] any sequence." He resubmitted to *Cell's* sister journal, *Cell Reports*—a notch lower on the prestige pole—but *Cell*'s song remained the same. The clock was ticking.[16]

By now it was May: Šikšnys resubmitted to the *Proceedings of the National Academy of Sciences*—a famous journal but lacking the wow factor of the CNS glamour journals. He addressed the cover letter to a member of the editorial board whom he felt would have the most expertise to appraise his manuscript: one Jennifer Doudna. But by the time his revised paper was finally published[17] in September 2012, Šikšnys was yesterday's news. It was an important paper in its own right, but didn't have the single-guide RNA that Jínek had developed. His results were a day late and a dollar short, and while the experiments were conducted around the same time as (if not before) Jínek, the belated publication date made the study look merely like a confirmation of the Doudna-Charpentier breakthrough.

"These two papers, if you look side by side, look nearly identical," Šikšnys told me. "The only thing I think that they showed [beyond our paper]

* *Cool*'s scope of articles read: "*Cool* only publishes articles that it deems to be astonishingly cool beyond belief. Any dorky shit submitted will be returned immediately to the authors postage due; there is just too much cool shit submitted (mostly by Cool Dudes) to waste our precious thick glossy sexy stock on it."

is that they can use this single-guide RNA." Unfortunately, *PNAS* took more than three months to review and publish the report. The rejection by *Cell* in itself was not a problem, says Horvath. "The problem appeared when we saw the [Jínek et al.] *Science* paper!" Would history have been different if his paper had been published in *Cell*? Horvath shrugged: "Nobody knows," he said. "Nobody will ever know."

Over the past few years, the CRISPR narrative has been shaped around the dream team partnership between Charpentier and Doudna. As the profile and fame of the two women grew, Šikšnys was, except for CRISPR insiders, the forgotten man. That began to change following an insightful article in *WIRED* by Sarah Zhang, who spotlighted the Lithuanian's lament.[18] Šikšnys earned some belated international recognition when he shared the 2018 Kavli Prize with his two illustrious peers.

After their "immortal" 2012 study, the transatlantic collaboration between Charpentier and Doudna devolved naturally. Jínek was busy juggling his research with job interviews. Charpentier was relocating from Sweden to Berlin. Chyliński was writing up his PhD thesis. Charpentier joined the Berkeley team on a follow-up report in 2014, but all good things come to an end. "We didn't decide to end the collaboration," Jínek recalls. "I'd enjoyed it immensely. We made a really good team."

After almost six years with Doudna, Jínek took a faculty position in Zurich in early 2013, where he continues to study the finer details of Cas9 and other DNA-cutting enzymes. In 2016, he won a prestigious Swiss prize for young biochemists, named after the physician Friedrich Miescher, the father of DNA. In 1868, Miescher set up a lab in Tübingen Castle, where he studied white blood cells in pus isolated from surgical bandages collected from a nearby clinic. Miescher extracted a substance from the cell nuclei that he first described in a letter in February 1869. He dubbed the substance, which "[does] not belong to any known type of protein," nuclein.[19] Almost 150 years later, Jínek won the Miescher Prize for his seminal role in laying the groundwork for editing nuclein.

And yet, Jínek's vital contribution to the CRISPR story remains underappreciated. Jínek is humble to a fault, not one to draw attention to himself. When he was asked after a lecture in Germany what he considered the keys to his breakthrough, he offered three reasons. First was the old "we stand on the shoulders of giants" quote.* Nobody does science in a vacuum, he said. Second was stressing the value of curiosity, the fundamental importance of basic research, and the freedom to pursue the unexpected. Jínek's goal with Doudna was simply to understand how RNA-targeting enzymes work. "I trained as a structural biologist. I think very deeply about how molecules work," he said. "I get very excited when I see molecules like this!" he said, gesturing to a cartoon of Cas9 and drawing laughter.

Some politicians take pleasure in cherry-picking federal research grants ("sexual preferences in fruit flies," etc.) as evidence of wasteful government spending. But the CRISPR gene-editing discovery and the scientific, medical, and economic bounty it has delivered, would not have happened but for public funding of unfashionable research conducted by diehard microbiologists, evolutionary biologists, biochemists, and structural biologists studying CRISPR purely for the thrill of discovery, not the lure of money or prizes. Fundamental and applied research are not "two spigots that can be operated independently," says Stuart Firestein, a professor at Columbia University. "They are one pipeline, and our job is to keep it flowing."[20]

Jínek's third key to success was serendipity—being in the right place at the right time. "I was very privileged to be in a great academic environment, where we were encouraged to think about our projects, to ask the right questions, to think outside the box, where we had the freedom to explore ideas, to work with other people, exchange ideas, and be quite pragmatic about the way we worked." The rest, he said, was just hard work: 10 percent inspiration, 90 percent perspiration.

In December 2018, Jínek was invited back to his hometown to give a TED-style talk in his native language.[21] He spoke about the rapid progress

* Scientists do love to trot out Isaac Newton's famous quote about "standing on the shoulders of giants" to acknowledge the scientific contributions of their colleagues and predecessors, although there is a school of thought that says Newton was actually trolling his diminutive rival, Robert Hooke.

in DNA sequencing, the $1,000 genome (a shout-out would have been nice!) and the diagnosis of genetic diseases. Now, thanks in part to his work, we have the ability to rewrite genes and transform medicine. To illustrate gene editing, he used a famous Czech nursery rhyme, *Skakal pes pres oves* ("the dog was hopping over the wheat"). To explain how bacteria capture the signatures of viruses in the CRISPR array, he showed a photo of an *ockovaci prukaz*—an international vaccination booklet containing stamps for various immunizations. As for germline editing, regulation had to be a global decision, not nations or researchers alone but the entire human race.

The Doudna-Charpentier *Science* paper was the impetus for several groups to demonstrate CRISPR-Cas gene editing in living cells. Doudna was at an initial disadvantage. Her specialty was RNA and structural biology, not cell biology and human genetics. Other groups clearly had the edge in expertise and materials. But Jínek was optimistic. After all, researchers had succeeded in using bacterial DNA-cutting enzymes at the heart of the ZFN and TALEN systems. Some engineering would definitely be required—adding a nuclear localization signal to ferry the Cas9 complex into the nucleus, and making some subtle changes to the DNA sequence (a process called codon optimization) to better fit a mammalian cell. But a priori, Jínek reckoned there was "nothing fundamentally different or any impediment to getting it to work in mammalian cells."

On October 3, Doudna's inbox chimed with a message that supported Jínek's belief. The sender was Jin-Soo Kim, a leading molecular biologist in South Korea. His lab had been pursuing CRISPR editing since Doudna and Charpentier's "seminal paper" and was preparing to submit a new report on "Genome editing in mammalian cells." Kim generously asked if Doudna (and Charpentier) would be interested in publishing together. "I do not wish to scoop you because your *Science* paper prompted us to start this project," Kim wrote. But he wasn't interested in getting scooped, either.[22]

Six weeks later, Church likewise emailed Doudna and Charpentier "a quick note to say how inspiring and helpful" he had found their CRISPR paper. "I'm sure you have received similar appreciative comments from other labs."[23] Church divulged that he was trying to get CRISPR working in human stem cells, only adding to the stiffening competition. Doudna felt the pressure of growing demands and responsibilities. "My inbox was exploding and journal editors were calling me," she recalled. "It was just crazy. You could see this tidal wave coming toward you."[24]

Jínek worked diligently in his final few months in Doudna's lab to prove his hunch. On December 15, 2012, Doudna and Jínek sent their latest manuscript to the online journal *eLife*, funded in part by HHMI. The rhetorical question posed in the opening of their paper—could the bacterial Cas9 system work in eukaryotic cells?—had been answered in the affirmative. Given the evident competition from Church, Kim, and who knows who else, Doudna was hoping for a quick decision. She got it. On January 3, 2013, the *eLife* editor emailed her to say her paper had been accepted. Both referees, one of whom revealed himself to be Dana Carroll, loved the work. The editor called the manuscript "excellent" and concluded that the ease of programmability meant that CRISPR would probably supplant ZFNs and TALENs for genome editing. Doudna's results would "likely have a transformative impact in the field of genome engineering for human and many other species with complex genomes."[25]

But Doudna's delight was short-lived. That afternoon she received another email:[26]

> Greetings from Boston and happy new year!
>
> I am an assistant professor at MIT and have been working on developing applications based on the CRISPR system. I met you briefly during my graduate school interview at Berkeley back in 2004 and have been very inspired by your work since then.
>
> Our group in collaboration with Luciano Marraffini at Rockefeller recently completed a set of studies applying the type II CRISPR system to carry out genome editing of mammalian cells. The study was recently accepted by *Science* and it will be publishing [sic] online tomorrow. I have attached a copy of our paper for your review.

> The Cas9 system is very powerful and I would love to talk with you sometime. I am sure we have a lot of synergy and perhaps there are things that would be good collaborate [sic] on in the future!
>
> Very best wishes,
>
> Feng

The Boston bombshell came from Feng Zhang at the Broad Institute. Just like the double helix, the story of CRISPR genome editing has a complementary strand, the twists and turns of which are still being unraveled.

CHAPTER 6

FIELD OF DREAMS

One could say that Feng Zhang's immigrant story—from China to Iowa, Harvard to Stanford, back to Harvard, to MIT, *60 Minutes* and scientific celebrity, fame, and a fortune in the making—is the quintessential American Dream. In 1993, Shujun Zhou, a Chinese computer engineer, brought her eleven-year-old son from Shijiazhuang, a heaving city of 10 million people about 150 miles southwest of Beijing, to the United States. They settled in Des Moines, Iowa—a world apart from his bustling birthplace. Driving around the Hawkeye state, Zhang would see a relentless landscape of cornfields and cattle, and highways lined with religious billboards with uplifting signs such as "After you die, you will meet God." The car radio might have picked up the "Bible Bus," a twenty-four-hour Christian station. His new home was indeed "a lot more 'Zen-like,' a lot more calm," Zhang jokes. Compared to his home, Des Moines felt "empty."

Zhang enrolled in the Callanan Middle School and soon became fluent in English. He got his first taste of molecular biology in eighth grade during a Saturday afternoon enrichment program. He joined a group of fellow nerdy middle school students called STING—Science and Technology in the Next Generation, inspired by *Star Trek*, organized by Ed Pilkington. The first documentary he watched, he jokes, was Steven Spielberg's *Jurassic Park*, which for all the revisionist criticism of Michael Crichton's dinosaur-DNA-locked-in-amber premise, inspired Zhang to imagine the possibilities of engineering and programming biological systems.

Zhang didn't know it, but Crichton's storyline gave rise to one of the first documented accounts of genome editing. In his original novel, Crichton included an excerpt of DNA sequence purported to represent dinosaur DNA. But when bioinformatician Mark Boguski idly typed the sequence into his computer, he was crushed to find that Crichton's dino-DNA actually belonged to a bacterium. After Boguski busted Crichton in a short essay,[1] Crichton contacted Boguski offering both an apology and an invitation to supply a more fitting sample for the sequel. Boguski chose a segment of chicken DNA, representing a living dinosaur descendant, which Crichton included in *The Lost World*. What the novelist didn't know was that Boguski had left an easter egg in the sequence: when the genetic code was translated into its corresponding amino-acid code,* there were four small insertions that, taken together, read:

MARK WAS HERE NIH

In his sophomore year, Zhang got his first intoxicating taste of scientific research, volunteering in a gene therapy lab at the Iowa Methodist Medical Center. He used a virus to ferry a jellyfish gene encoding a green fluorescent protein into human cancer cells. The magical glow that emanated from those cells was early proof that Zhang had green fingers at the bench. His mentor, John Levy, gave Zhang some words to live by: "Try to do something on the sexy side of practical."[2] From this point on, Zhang's future in biological research was pretty much assured. Pilkington observed: "Not everybody is able to fail twenty experiments and then get up and then come up with something brilliant."[3]

At nineteen, Zhang placed third in the prestigious Intel Science Talent Search competition, winning a $50,000 scholarship. His mother accompanied him through to the finals, exhorting her son, "*Zhan zhi!*" ("Stand up straight!"). He excelled at Harvard, matriculating the year before Mark Zuckerberg, earning a degree in chemistry ("the central science") and physics.

* In the single letter code, each of the twenty amino acids, encoded by a three-letter codon of DNA, is represented by a single letter. For example, M-A-R-K stands for methionine-alanine-arginine-lysine.

Zhang wanted a grounding in the fundamentals that he could apply as he saw fit. Eager to experience Silicon Valley, Zhang moved to Stanford for his PhD. His first choice supervisor, Nobel laureate Steven Chu, had moved on; camped out in Chu's former office was a new faculty member, Karl Deisseroth.

A psychiatrist by training, Deisseroth had treated many patients with schizophrenia and depression, but was frustrated at how poorly we understand those diseases. He was developing a new technique called optogenetics for studying neuronal activity and neurological diseases. Deisseroth's brainstorm was to introduce a light-sensitive protein called an opsin into a rodent neuron, affording him the ability to trigger and control its activity.*

Zhang's initial task, building on his familiarity with gene therapy, was to introduce the opsin gene, which was derived from a pond-dwelling green algae, into single rat neurons on a tissue culture plate using a recombinant virus. When triggered by light, the newly transfected neurons would emit an electrical signal. Over the next few years, Deisseroth, Zhang, and another grad student, Ed Boyden, progressed from neurons in a dish to rats in a maze. When a *New York Times* reporter visited the lab one Sunday, Zhang had prepared a show-stopping experiment: inside a white plastic tub was a solitary transgenic brown mouse expressing the opsin gene in a particular type of neuron. Zhang inserted a small metal tube into the mouse's head, into which he threaded a tiny fiber optic cable. When he flicked on the light switch, the mouse suddenly started spinning in circles; switched off, the mouse stopped moving.[4] Optogenetics was ready for the big time: a tool to study brain function far more precisely than sticking electrodes into the brain or taking blurry MRI pictures.

"This is one of those things that only comes around every five or ten years," Deisseroth told his student.[5] And he was right: Deisseroth won the Breakthrough Prize in 2015 and helped lay the foundation for President Barack Obama's $300-million BRAIN Initiative. Deisseroth credited

---

* This idea was initially proposed by Francis Crick, who spent the final three decades of his life at the Salk Institute studying the brain and consciousness. Crick called for a new technique that could probe brain function at the single-cell level, and even suggested that this level of control could be achieved through light.

Zhang's skills as "absolutely essential to the creation of optogenetics."[6] Not too many PhD students get to see their handiwork featured in the *Times* or win a share of a major scientific prize. It was a sign of things to come.

Still in his twenties, Zhang returned to Boston in 2010 and joined the lab of George Church as a Harvard junior fellow. For a gifted scientist intent on developing new tools for studying genetics and the brain, there could be no more inspiring training ground. For twenty-five years, Church had been a driving force in genomics, pursuing the kind of bold, fearless research that appealed to a talent like Zhang. Church and his dozens of students developed new sequencing technology for reading DNA but was increasingly interested in writing entire genomes. In collaboration with stem cell researcher Paola Arlotta, Zhang and a new PhD student, Le Cong, threw themselves into the emerging field of gene editing. The idea was to develop a method to improve the creation of sequence-specific TALENs to target genes and modulate gene activity.[7] Joining them in a corner of the Wyss Institute were two other postdocs, Prashant Mali and Kevin Esvelt.

Born in Beijing, Le Cong shared Zhang's love of engineering. He enjoyed tinkering with radio sets and designing computer games as a child. While studying electronic engineering at Tsinghua University, Le Cong's interests turned to medicine after the death of some close relatives. "We've made so much progress in modern medicine, but there's so much we don't know," he told me.[8] Even simple diseases such as type 1 diabetes can take people's lives.

Arriving at Harvard on a fellowship, Le Cong quickly bonded with Zhang and was swept up by the infectious lab chatter about the genomics revolution. His wanted to develop tools to engineer the human genome to model brain diseases such as autism, schizophrenia, and bipolar disorder. Although officially Church's student, Le Cong says that in January 2010, "Feng became my advisor and mentor. We were the only folks interested in working on new tools in gene editing." It was difficult work: the techniques to synthesize bacterial chromosomes were still rudimentary, and extrapolating to larger mammalian genomes would be even harder.

Across the Charles River in Cambridge, Bob Desimone, the director of the McGovern Institute for Brain Research at MIT, was looking for new faculty. The institute was established by the late Pat McGovern, who had dropped out of MIT to start a publishing company in his basement. That company evolved into IDG (International Data Group), the global publisher of *Computerworld*, *MacWorld*, and *Bio-IT World*. McGovern became a billionaire but ran his enterprise with the warm touch of a family-run business.* McGovern and his wife Lore proudly donated $350 million to create the eponymous institute.

The faculty search wasn't going particularly well until Zhang's name came up. His research credentials were impeccable, the only question was whether his flourishing interest in genome manipulation would be the right fit. Desimone asked MIT's top stem cell researcher, Rudy Jaenisch, to assess the candidate. Jaenisch's verdict: "He's certainly very clever. If he can do 10 percent of what he's proposing, he's going to be a star."

Zhang joined the McGovern Institute in January 2011 with a joint appointment at the nearby Broad Institute. His goal was to continue developing TALENs and other systems to engineer genome mutations to model and devise treatments for autism, Alzheimer's disease, and schizophrenia. Zhang concedes he might have had a slight case of Imposter Syndrome, but it didn't linger. Le Cong decided to hitch a ride as well. "We'd been working colleagues, it was wonderful to continue our relationship," he said. Even though he was still technically Church's graduate student, they shared a taxi over the Harvard Bridge to start their new adventure at MIT.

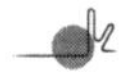

In early February, 2011, Zhang paid a return visit to Harvard Medical School. The Broad Institute's annual Board of Scientific Counselors meeting was at the Joseph Martin Auditorium, a block from Zhang's old lab. A lecture from a most unlikely source was about to change his life.

---

* Every Christmas, McGovern toured every IDG office location, meeting with each employee in person to hand them a cash gift and thank them for their specific contributions to the company, which he knew cold.

Michael Gilmore is an expert on antibiotic resistance in bacteria. In 2007, while at a microbiology conference in Pisa, Italy, he was impressed by a poster presented by Danisco on the link between phage immunity and DNA repeats with a weird name: CRISPR. Gilmore wanted one of his new postdocs to study CRISPR. One promising candidate, Luciano Marraffini, opted instead for a position in Chicago. In stepped Kelli Palmer who, with colleagues at the Broad Institute, began comparing the sequences of multidrug-resistant strains of bacteria with older strains. She made a striking observation: genomes from multidrug-resistant clinical strains dating back to the 1970s were larger and generally lacked CRISPR. By comparison, the other isolates still possessed CRISPR.

Gilmore's deduction was that "the repeated introduction of various antibiotics since the '40s not only selected for resistance to those antibiotics, but it selected for the ability to acquire [drug] resistance with enhanced facility." In other words, bacteria were gaining a selective advantage by losing their CRISPR defense system. That might leave them vulnerable to phages, but it would make it easier for them to acquire new genetic elements that could confer antibiotic resistance (via a mechanism called horizontal gene transfer). The overuse of antibiotics has resulted in bacteria effectively lowering their genome defenses to make it easier to acquire resistance from other microbes.[9] The genomes of clinical isolates of some bacteria might be 25 percent larger than those of commensal strains.[10]

Zhang's ears pricked up as Gilmore casually mentioned that bacteria contain CRISPR and its attendant nucleases. The acronym had appeared in several top journals by this time, but Zhang had to google the term to learn more about it. The next day, Zhang flew to Miami to attend a conference,* but hunkered down in his hotel room instead, devouring the CRISPR literature. The more he read, the harder it was to contain his excitement. He read a CRISPR review article by Horvath and Barrangou in *Science*[11] and the "amazing" 2010 *Nature* paper by Moineau's group that showed that the type II CRISPR-Cas system cleaves phage

* The Miami 2011 Winter Symposium, "Epigenetics in Development and Disease." co-organized with Nature Publishing Group.

DNA.[12] But Zhang wasn't interested in bacterial immunity or how to make a cheesier pizza. His focus was genome editing—a technique that could work in animal models and ultimately in humans. If, as Moineau's paper suggested, you could use RNA and CRISPR to target DNA, this would be much easier than either ZFNs or TALENs.

On Saturday February 5, Zhang fired off an email to Le Cong: "Take a look at this," he wrote, including a link to the Horvath-Barrangou review. "Maybe we can test in mammalian system." Le Cong replied, "It should be very cool to test in mammalian systems." Two days later, Zhang emailed: "Hey let's keep this confidential. This can completely replace any kind of [zinc finger] system. I ordered the Cas genes for synthesis. We should be able to test them . . . I've done a patent search."

On February 13, Zhang filed a "memorandum of invention," an internal Broad Institute document that summarized his new invention: multiplexed genome engineering. The idea, which he said originated just nine days earlier at Gilmore's lecture, was a clear statement that Zhang thought CRISPR might complement, or even replace, the current methods of gene editing, ZFNs and TALENs.[13] Zhang judged that CRISPR had the makings of a programmable gene-editing technology that could be targeted to almost any DNA sequence.

In a short McGovern Institute video filmed at the end of 2011, Zhang discussed his research goals. Coming from an engineering background, he said, "I think about how to take things apart, put them together, and then try to fix it." Using the same approach, Zhang hoped "to understand disease mechanisms and be able to fix the brain." The main tools in his arsenal were the TALEN proteins.[14] He exuded a quiet fearlessness, as if nothing was beyond his reach. But there was no mention yet of CRISPR or the potential to edit human DNA.

Zhang and Le Cong's early efforts using Cas9 did not work as planned. Two key modifications were required to get Cas9 to work in human cells, as I mentioned earlier: one was codon optimization to make the gene appear less foreign to a human cell. The other was to add a nuclear localization signal, a motif that helps ferry DNA into the cell nucleus (obviously not an issue in bacteria as they lack a nucleus). But for some reason, the *S. thermophilus* Cas9 wasn't behaving nicely. Zhang needed a new system and a Cas expert. The man he was looking for—who had almost been a colleague of Gilmore's—was just down the road in New York City.

Shortly before 10:00 P.M. on January 2, 2012, Zhang sent a short email to Marraffini at the Rockefeller University. The Argentine had received flattering faculty offers from MIT and Yale, but the prospect of living in cosmopolitan New York was irresistible. And Rockefeller's rich tradition in microbiology and genetics proved a perfect fit. Zhang didn't beat about the bush:

> Dear Luciano,
>
> Happy new year! My name is Feng Zhang and I am a research [sic] at MIT. I read many of your papers on the Staphylococcus CRISPR system with great interest and I was wondering if you would be interested in collaborating to develop the CRISPR system for applications in mammalian cells.
>
> Would it be possible to schedule a phone call with you in the next few days?[15]

Marraffini had never heard of Zhang but a quick Google search left him suitably impressed. He replied ninety minutes later and said yes. "We have been working on a 'minimal' CRISPR system that could be useful . . . Happy 2012!" They sealed the collaboration in a phone call the next day. A week later, after receiving a nudge from Zhang, Marraffini emailed an eight-page document that highlighted DNA sequences and other useful information about CRISPR in *S. pyogenes*. He closed with a five-step plan to produce the "minimal" Cas system that he believed could be used to edit human genes. The relevant materials, including the gene template for Cas9 and the tracrRNA, followed by mail.

Zhang was a man in a hurry. He was working in the trenches with his team, his weapons of choice a row of personalized pipettors with FENG taped across each one. On January 12, he told Marraffini he'd already identified a pair of target sites in a human gene (*AAVS1*) and was preparing to express the bacterial genes in mammalian cells. Around this time, he was included in an $11 million NIH grant application filed by the Broad Institute's then deputy director, David Altshuler.[16] The main idea was to use genome editing to engineer stem cell models for type 2 diabetes and other diseases. Zhang proposed using the four components described by Charpentier—the CRISPR array, Cas9, the tracrRNA, together with another enzyme (RNase III)—to reconstitute the active CRISPR complex in human cells.

Within a few months, Zhang had enough data to show he could deliver the CRISPR-Cas9 machinery into the nucleus of mouse or human cells (by flanking Cas9 with nuclear localization signals) and target a gene of choice. Marraffini also had preliminary results showing that Cas9 could target a human gene sequence. They also found an isoform of the tracrRNA that was expressed in human cells. Zhang toyed with the idea of writing up these preliminary results—potentially the first demonstration of CRISPR gene editing in animal cells—but decided against it. "I [wanted] to wait until we have a paper that can make a significant difference, not just to be first with something," he said.[17]

In late June, Zhang saw the names of Doudna and Charpentier on a CRISPR paper in *Science*. The consensus was that this was a landmark paper, but Zhang was uncharacteristically dismissive. "I didn't feel anything," he told *WIRED*. "Our goal was to do genome editing, and this paper didn't do it."[18] But the report impacted Zhang's team in at least two important respects. First, it showed that competition in CRISPR was heating up. "We didn't think we got scooped, but we knew people would jump on the topic so we had to speed up," Le Cong told me.[19] Second, it introduced the idea of the single-guide RNA (sgRNA). Zhang and Cong didn't hesitate to put this idea into action with the help of a new recruit with something to prove.

Fei Ann Ran was born in Szechuan, China, and moved with her family to Pasadena, California, when she was ten years old. She traveled to the East Coast for college and enrolled for her PhD at Boston Children's Hospital. But more than halfway through, Ran suffered a devastating setback:

her supervisor, geneticist Laurie Jackson-Grusby, ran out of funding and closed her lab. "She left," Ran told me in a cafe on Longwood Avenue, the central artery of the Harvard Medical School campus. "She decided not to publish any of our work. I was in my fifth year of grad school. It was a pretty terrifying time."[20]

A member of Ran's PhD committee, Harvard chemistry professor Greg Verdine, suggested she contact Zhang, whom Verdine remembered as a star student. Ran didn't know much about genome editing or Zhang. A friend told her Zhang was famous for his work on TALENs. But she didn't know what TALENs were, either—she misheard this as "Talent." Luckily, she fit the bill. One of Zhang's postdocs, Neville Sanjana, was developing a method to treat an inherited brain disorder called Angelman syndrome* by building a TALEN activator to switch on a silent gene. Ran made one TALEN but soon switched to working on the easier CRISPR system.

In July, Zhang gave a public lecture at the Broad Institute on "Engineering the Brain" in which he discussed the potential of genome editing to understand the brain and potentially treat incurable brain disorders. There was still no mention of CRISPR. Meanwhile, Ran was devouring the CRISPR literature and most of all hoping to salvage her PhD. "I was so single-minded on getting this to work in mammalian cells," she said. "Feng and Le had been doing a lot of work in mammalian cells. There was no question it couldn't work but how could it work robustly?" Le Cong and Ran were keen to try the new sgRNA approach, but it didn't perform as well as adding the crRNA and tracrRNA separately. By incrementally extending the tail of the tracrRNA, they were able to boost the gene-editing efficiency.

Ran found the atmosphere in the lab exhilarating despite the long hours, the group working around the clock almost in shifts. Le Cong still technically belonged to Church, but had a special bond with Zhang, "almost a father-son relationship." The group would eat an early dinner and a late-night supper, frequently Chinese takeout, in the kitchen adjacent to Zhang's laboratory. From the Broad's tenth floor, Ran could gaze across

---

* Angelman syndrome is an example of a class of genetic diseases called imprinting disorders, in which mutations effectively silence one copy of a gene, depending on whether it was inherited from the mother or father.

the Charles River and admire the Boston skyline—the twin peaks of the Hancock Building and the Prudential Center, the sparkling floodlights of Fenway Park. But the only nightlife that mattered was in the lab. "I was the later crew, but Feng was the omni crew," she said. "He'd show up before everybody and leave after everybody." His wife, Yufen Shi, whom Zhang met at Stanford and married in 2011, would often wait patiently for him in his office.

Occasionally there was time to relax outside the lab. Shortly after her arrival, Ran accompanied her colleagues on a summer camping trip to nearby Harbor Island. Most of the group slept in a twenty-person tent. In a lab photo from one of those group activities, Zhang's youthful appearance blends in with his students and postdocs. Anyone not knowing him would be hard pressed to pick out the professor. Zhang displayed a childlike excitability, desperate to see everyone's latest results. "It was like taking a kid to a candy store," Ran said. She kept improving the gene-editing efficiency and showed they could edit more than one gene at a time. The experiments worked routinely, unlike the frustrations with TALENs.

The final question was could they get CRISPR-Cas9 to edit genes inside human cells? Ran and Le Cong took Cas9, made a new guide RNA, and put everything into human cells growing in a petri dish. "And then we waited . . . and then we waited," she recalled.[21] A few days later, they sequenced the genome of those cells. "We could see the scars of DNA damage and repair—in other words, mutations—exactly where we thought they'd be. This was really exciting!"[22] Zhang was excited, too. "I want to see!" he said. He told her: "Isn't this cool that you're one of the only people in the world to see this?"

As noted earlier, many groups were desperately interested in applying CRISPR to genome editing, not least members of the Church Lab. Before Zhang left, Prashant Mali had been working on TALENs and dabbling with other gene-editing technologies. "CRISPR was just on our list of nucleases," Church told me. "But we were always looking for precision editing. And we wanted to do it in human cells."[23] Kevin Esvelt had joined the Church lab after a successful PhD with Harvard chemist David Liu. He'd got a rudimentary version of CRISPR working in the lab, which he was using like antivirus software to prevent infections from stray viruses. Intrigued by Esvelt's experience working with Cas9 in bacteria, Mali asked for his help getting CRISPR to work in mammalian cells. Esvelt had his doubts, assuming Doudna had an insurmountable lead. But Mali persisted. "This is going to be so big that if we discover one tiny little piece that the other groups miss, it will be worth it."[24]

As it turned out, that "one tiny piece" was demonstrating the need for the whole sgRNA—truncating it reduced activity. With help from another talented grad student named Luhan Yang, Church submitted their paper to *Science* in late October. It not only demonstrated "facile, robust, and multiplexable human genome engineering" but also predicted that more than 40 percent of human gene sequences were amenable to genome editing using CRISPR. As a courtesy, he emailed Doudna to let her know his group had extended her results to editing in normal human cells. As far as he knew, that was a first.

Of course, it wasn't. Three weeks earlier, Zhang had submitted his group's manuscript to the same journal. Le Cong and Ran were credited as co–first authors, and Marraffini and his student Wenyan Jiang were also included. Barrangou was one of the referees helping *Science* judge the merits of those and similar results from other teams. Barrangou judged that the reports from Church and Zhang demonstrating CRISPR editing in human cells were the most significant of the bunch. Of the two, Feng's paper was perhaps the better but "they've got to take both, back to back. The single-guide RNA technology is the tipping point. The real credit should go to George *and* Feng to show genome editing in human cells."[25]

Science published Zhang's article alongside Church's paper on January 3, 2013.[26,27] Jin-Soo Kim's demonstration of CRISPR gene editing in human cells came out four weeks later, in *Nature Biotechnology*.[28] Jínek and Doudna presented their human cell success in *eLife*. There were also reports from Keith Joung (Massachusetts General Hospital) working on zebrafish[29] and

Marraffini, squeezed out by *Science*, who published his studies on bacteria in March.[30]

Ran was on vacation when she got the news of her first major publication, which more or less secured her PhD. Six months of hard work at the cutting edge of science had made up for five years of fruitless labor and frustration. Her reaction was more relief than jubilation. She remained in the lab for another year before writing up her thesis. "I have this great suggestion," Zhang told her at one point. "Download Arno Pro. It's a great font!" Ran still has it on her computer. "That's how I got my PhD, with the help of Feng's font," she laughs. She didn't see much of him as she finished in the lab, and learned about the birth of his first child from an unusual source. While receiving the referee reports for a subsequent paper, the journal editor's email said, "Congratulations on Ingrid." Ran smiles: "We learned about it from a journal editor!"

Some would argue that Zhang and Church were merely confirming the obvious, extrapolating from Doudna and Charpentier's work in bacteria to engineer human DNA. After all, several teams reported success within a couple of months of each other. But getting the system to work in human cells was a major step forward in developing a new type of genetic therapy. It also set the stage for a massive legal dispute about the invention of genome editing (see Chapter 13). Any doubts that Cas9 could work in human and other animal cells had been well and truly answered.

One might ask what if *eLife*, located in Cambridge, England, had expedited review and publication of Doudna's follow-up paper and published it before the New Year. Shepherding reviewers over the festive break isn't easy, especially when Brits typically log off from Christmas to New Year. By the time Doudna's paper was published, few took notice. A few years later, Doudna was asked: Why did Zhang and Church demonstrate CRISPR gene editing in human cells before her? "They were absolutely set up to do that kind of experiment," she acknowledged. "They had all the tools, the cells growing, everything was there. For us, they were hard experiments to do because it's not the kind of science we do. What speaks to the ease of the system was that a lab like mine could even do it."[31]

The media still didn't fully grasp the implications of the genome editing papers from Zhang and Church.* One of the first to comment was author and columnist Matt Ridley, who marveled in the *Wall Street Journal* at scientists' ability to precisely edit a single base.[32] Two months later, *Forbes* science correspondent Matthew Herper surmised, "The [Cas9] protein could change biotech forever." Herper's story focused on Church, who had a much bigger profile than his former fellow, and said CRISPR was "spreading like wildfire." Not only was Zhang not quoted, but ironically, he also fell victim to an editing error: his name was misspelled "Zheng."[33]

Steadily, interest in CRISPR from scientists and media alike picked up. Geneticist Konrad Karczewski recapped the buzziest terms of 2013 in a game of catchphrase bingo popular at conferences. "CRISPR" was included along with nanopores, big data, and Myriad (for the blockbuster supreme court ruling banning gene patents).[34] A report from the Harvard labs of Chad Cowan and Kiran Musunuru compared CRISPR and TALENs side by side, with CRISPR winning convincingly.[35] "It was a surprisingly important paper," said T. J. Cradick,[36] the former head of genome editing at CRISPR Therapeutics, not least because it gave venture capitalists the green light to explore CRISPR's commercial potential.[37] The *Boston Globe* covered CRISPR for the first time in a story about the launch of local biotech company, Editas Medicine, featuring Zhang, Doudna, and Church among the cofounders. *Science* recognized CRISPR in its annual "Breakthrough of the Year" awards—albeit as a runner-up to cancer immunotherapy.[38] CRISPR achieved top billing two years later, having "matured into a molecular marvel."[39]

* Two industry publications ran reports based on press releases: *Genetic Engineering & Biotechnology News (GEN)* covered the Zhang paper, while Genomeweb highlighted the Church lab report.

CHAPTER 7
PRIZE FIGHT

In 1994, I attended a science conference in Philadelphia for my journal, *Nature Genetics*. During a coffee break, a geneticist named Dennis Drayna took me to one side and whispered that his team at Mercator Genetics had made a thrilling discovery: the gene mutated in patients with hemochromatosis, one of the most common genetic disorders in people of European descent. I said we'd be thrilled to review the paper and would handle it expeditiously. I sent copies by courier to four referees to ensure the results were thoroughly vetted. But a week later, as the faxed reviews landed on my desk, my worst fears were realized: a split decision, two reviewers loved the paper but the other two expressed concerns. In desperate need of a tiebreaker, I called the best man for the job: Eric Lander. He read the paper over a weekend, and transmitted his unequivocal endorsement, which we proudly published in 1995.*

Lander trained as a mathematician and taught economics at Harvard. He's a Rhodes Scholar and a MacArthur Genius Award recipient. In the late 1980s, he decided to apply his skills to biology—his brother is a leading neuroscientist—taking a fellowship at the Whitehead Institute in Cambridge, a new flagship research center on the fringe of the MIT campus. His

* My editorial accompanying the hemochromatosis gene paper was entitled *Definitely Maybe*, a clever nod I thought to Oasis's debut album and the line Kirstie Alley exhaled in *Cheers* after Ted Danson first plants a kiss on her. It was criminally edited to "A Definite Maybe." (I'm still upset 25 years later.)

stature grew rapidly in the 1990s, as he helped build the theoretical framework for human gene mapping before helping to lead the international genome project consortium's response to the Celera threat. Lander wrote the lion's share of the landmark draft human genome paper published in *Nature* in 2001.

After you've helped to orchestrate the successful effort to sequence the human genome, and quietly cofounded several successful biotech companies, what do you do for an encore? Lander set his sights on building a new biomedical institute (literally overshadowing the Whitehead) affiliated with both Harvard and MIT, anchored by a world-class genome center but venturing into cancer biology, neuroscience, cell biology, chemistry, and eventually CRISPR and genome editing. The philanthropists Eli Broad and his wife Edythe have committed $700 million to Lander's institute, which the billionaire art collector calls his greatest treasure. In 2008, President Obama named Lander to cochair the White House science council, praising Lander's work on the Human Genome Project as "one of the greatest scientific achievements in history."[1]

Over the years, like the general manager of a dynastic baseball team, Lander has assembled a team of all-star scientists including top Harvard chemists Stuart Schreiber and David Liu, former Harvard provost Steve Hyman, and Merck's former head of research, Ed Scolnick. He identified Zhang's potential before CRISPR became the biggest game in science and Lander could take some personal pride in having this rising phenom under his roof at the Broad.

As the competition for credit and awards heated up, one McGovern Institute executive, Charles Jennings, was concerned that Zhang was in danger of not receiving the recognition he deserved. The press was warming to the story of the Doudna-Charpentier alliance forged in Old San Juan. Jennings took it upon himself to nominate Zhang for his first prize. To be sure, the annual *Popular Science* magazine "Brilliant Ten" award for scientists and engineers may not rank as one of the world's most prestigious scientific awards, but Zhang happily joined the class of 2013.[2] Jennings's nomination praised Zhang for his work developing gene-editing tools and his willingness to share them widely with his fellow scientists. "These technologies are so fundamental, it's best to keep them as open as possible," Zhang said. "If someone had protected the HTML language for making

Web pages, then we wouldn't have the World Wide Web." Three years later, MIT awarded Zhang tenure. Desimone's nomination letter to the tenure committee began by saying this was probably the easiest decision the committee would ever have to make.

It is hard to overstate how much the CRISPR world exploded following Zhang's breakthrough paper in January 2013. Before then, a few dozen CRISPR-related papers were published each year. In the three years following that report, the number rocketed to 3,000. As researchers worldwide gleefully embraced the technology, the competition for prizes and patents intensified.

In January 2016, *Cell* published an extraordinary perspective by Lander. Titled "The Heroes of CRISPR," Lander framed this lengthy article as an educational essay paying tribute to the dedicated unsung heroes who had paved the way for the CRISPR gene-editing revolution.[3] He wrote eloquently, more in the style of a *New Yorker* piece than a typically turgid scientific review.

But some suspected that Lander was using this platform to spin the accomplishments of his protégé. The cues were apparent from the opening line of the abstract: "Three years ago [in 2013], scientists reported that CRISPR technology can enable precise and efficient genome editing in living eukaryotic cells." From the outset, Lander was charting the birth of the CRISPR revolution as Zhang's landmark paper. Lander appropriately paid tribute to many figures but downplayed the contributions of Charpentier and Doudna, devoting just a few paragraphs to their work compared to a page-and-a-half extolling the life and work of Zhang.

The article also included a map of the world, with the locations of the CRISPR pioneers marked in colored dots, from Japan to Lithuania, Germany to Spain, Boston to Berkeley. But there was something askew. On a closer look, the Atlantic Ocean had been magically compressed, while Greenland and Iceland had been erased completely.* As a result, the map conveniently centered on Cambridge, Massachusetts, home of the Broad.

* While Lander was playing at plate tectonics, he could have scratched off the UK as well. For a country that has contributed so much to our understanding of evolution and molecular biology—Darwin, Fleming, Crick, Rosalind Franklin, Sydney Brenner, Fred Sanger, Alec Jeffreys, Paul Nurse, and more—the origins of the CRISPR revolution surprisingly sidestepped the UK.

Readers also objected to the fact that Lander declared no conflicts of interest: he had no direct financial stake in Zhang's work but the Broad was embroiled in a fierce patent dispute with Doudna and Charpentier's respective institutions (see chapter 13). Cell Press, the publisher, said it did not require authors of commentaries to declare any conflicts of interest, although in the circumstances, it wouldn't have hurt either the editor or author to have included such a note.

The critics pounced. Michael Eisen, an outspoken geneticist and a friend of Doudna's, posted a scathing rebuttal, branding Lander "The Villain of CRISPR." "There is something mesmerizing about an evil genius at the height of their craft," Eisen wrote.[4] Lander's masterwork was "so evil and yet so brilliant that I find it hard not to stand in awe even as I picture him cackling loudly in his Kendall Square lair, giant laser weapon behind him poised to destroy Berkeley if we don't hand over our patents." He accused "the most powerful scientist on Earth" of scheming to help Zhang win a Nobel Prize and give the Broad Institute the inside track on "an insanely lucrative patent."

Such a clapback among prominent scientists is both rare and fascinating. Eisen's objection to Lander's history lesson was nothing personal against Zhang but a judgement of what he considered the most crucial discovery in the CRISPR timeline. If there was a pivotal step in bringing CRISPR to the genome editing party, Eisen said, it was Doudna and Charpentier's 2012 demonstration that CRISPR could be adapted into a molecular scissors following a decade of heroic foundational work. "Once you have that, the application to human cells, while not trivial, is obvious and straightforward," Eisen declared.

Piling on was science historian Nathaniel Comfort, who labeled Lander's essay a "Whig history" of the CRISPR saga—an effort to rationalize the status quo and spin the establishment's point of view. Comfort was pleased to see Mojica and others receive some overdue credit. "Too often the early players and the scientists at lesser-known universities become lost to history altogether. But we should also recognize how Lander uses those actors to create a crowd in which to bury Doudna and Charpentier."[5]

Not surprisingly, neither Doudna nor Charpentier were too thrilled with Lander's account. "The description of my lab's research and interactions with other investigators is factually incorrect, was not checked by the author and was not agreed to by me prior to publication," Doudna said.[6] Charpentier added that the description of her group's contributions was "incomplete and inaccurate."* By this time, however, the two women were receiving ample opportunities to give their own accounts of the story.

George Church wasn't thrilled with the media narrative, either. After all, his paper appeared in the same issue of *Science* as Zhang's, but as the Nobel Prize is awarded to a maximum of three recipients, there is a tendency during prize season to search for the holy trinity. In CRISPR circles, that trio was usually Charpentier, Doudna and Zhang. Church aired his frustration a couple of years later to *The Scientist.* He wasn't trying to take anything away from Doudna and Charpentier, pioneers who deserved credit for getting gene-cutting to work. "The spark that [they] had was that CRISPR would be a programmable cutting device." But getting it to do precision editing was another matter. Indeed, Church argued that his human cells were a more accurate system than the aberrant culture cells that Zhang's group had used.** In terms of credit, Church said, "you could say two and two. But to oversimplify that back down to three is like consciously omitting one."[7]

Church later told me it wasn't so much to cement his own place in history, whatever that matters, but the "egregious omission" of the postdocs who did the work—his and others. "I felt that Martin Jínek had been left out of the story, and Prashant Mali, and Luhan Yang, and Le Cong. You just never heard of them."[8]

Doudna did not make a habit of putting on her "out of office" email message but her travel schedule soon became packed with prize ceremonies, media

* Some commentators also accused Lander of diminishing the contributions of the women at the center of the CRISPR story in favor of their male rivals. This was silly; Lander has mentored many superb female scientists, including Stacey Gabriel, Jill Mesirov, Pardis Sabeti, Anne Carpenter, and Aviv Regev, who in 2020 was recruited to lead R&D at Genentech.

** Zhang used cultured 293 cells, a kidney cell line.

interviews, and keynote lecture invitations. Her talks were polished and accessible, generously crediting Charpentier and her colleagues. Despite her rapidly growing profile, she wasn't thinking about writing a book until she received a surprise invitation from Max Brockman, the son of a leading New York literary agent, John Brockman. Doudna's initial proposal, co-written with graduate student Samuel Sternberg, was a little dry, with references to Thomas Kuhn's *The Structure of Scientific Revolutions*. As Sternberg admitted later: "What kind of person on the street was going to read that?"[9]

But public interest was intensifying. That was driven home when Sternberg accepted an invitation to breakfast at a Mexican restaurant in Berkeley from a woman who asked if he would be interested in starting a company to deliver CRISPR gene editing to future parents. Sternberg had no interest in that particular venture, but it supported giving the proposal another shot. The result was *A Crack in Creation*, published in spring 2017, which tells Doudna's personal story, although she deftly sidestepped any commentary or controversy on the patent dispute.[10]

In various permutations, Charpentier, Doudna, and Zhang have hoovered up almost every major science prize, with two conspicuous exceptions: the Lasker Award, which is often referred to as America's Nobel Prize and the Nobel Prize. Those appear to be a sure thing, but to whom and for what is a topic of much speculation.

The two women have shared the "Nobel Prizes" of Japan, Spain, Israel, and Canada (with Zhang), to name a few. The most lucrative award was the Breakthrough Prize, created by Silicon Valley billionaires including Priscilla Chan and Mark Zuckerberg (Facebook), Sergey Brin (Google) and his ex-wife Anne Wojcicki (23andMe), and Dick Costolo (Twitter). At a black-tie awards ceremony in November 2014, Doudna and Charpentier received their awards from Hollywood actress Cameron Diaz. Charpentier flashed her Gallic humor on stage. "It's kind of surreal to receive the prize from Cameron," she said, then turned to Costolo: "Three powerful women . . . I was just wondering if you're Charlie?"

Two years later, Charpentier and Doudna joined Barrangou, Horvath, and Zhang for the annual Canada Gairdner Awards, the most prestigious

Canadian scientific honor. At the banquet dinner, it is a tradition for each awardee to choose their own walk-up music as they head to the stage to accept their award. Zhang naturally chose John Williams's stately theme from *Jurassic Park*. He thanked his parents for their sacrifices on his behalf and his wife for keeping him company in the lab on late nights and for the birth of their daughter. Horvath selected a jazzy rendition of the *Mission: Impossible* theme. He joked that his scientific career began working on sauerkraut and quoted a famous French proverb: "*Impossible n'est pas français.*" Barrangou chose "Happy" by Pharrell Williams, mugging shamelessly for the cameras as he shimmied up to the stage in his trademark cowboy boots. Charpentier, by contrast, selected a moody slice of French electronica by Daft Punk.

The most interesting speech was by Doudna, who selected Billie Holiday's "On the Sunny Side of the Street." She thanked her students, colleagues and mentors, as well as her two special guests, husband-and-wife Harvard Medical School professors George Church and Ting Wu, for inspiring her when she was a student at Harvard. She also paid tribute to Church's under-recognized work in CRISPR. "His work has had a huge impact on the gene editing field over the years, including adapting the CRISPR-Cas system for gene editing in mammalian cells."[11] (Some might wonder if she wasn't throwing some shade in the direction of Zhang and Lander, who was also a guest at the dinner.) Then she announced that she was donating her $100,000 award to the nonprofit organization for genomics education cofounded by Wu and Church.

In a photograph from that evening, Barrangou stands at the center of the CRISPR quintet, his boots putting him a head taller than his peers. (Also in the group was another awardee, Anthony Fauci, recognized for global health.) The Gairdner was the undoubted highlight of his career, recognition for a landmark study that fermented the CRISPR revolution. Doudna and Charpentier were deservedly recognized for developing the single-guide RNA technology—the tipping point as he calls it.[12] "Single-guide RNA is an invention—it's novel, not obvious, not natural. They didn't just recapitulate it like Virgis. They engineered it, they designed it." But genome editing is not until 2013, when "George and Feng and Luciano and Jin-Soo Kim and eventually Jennifer show that."

Horvath shared the Massry Prize with Doudna and Charpentier in 2015 and Harvard's Alpert Prize with Barrangou, Charpentier, Doudna, and Šikšnys. "I have mostly been in the shadow of Charpentier and Doudna, but not for the Bower," he told me. The 2018 Bower Award and Prize for Achievement in Science was perhaps his greatest honor. Created in 1824, prizes have been bestowed on more than 2,000 scientists and inventors, including Tesla, Edison, Einstein, Hawking, Church, and Bill Gates. Horvath's citation read:

> For the foundational discovery of the role of CRISPR-Cas as a microbial system of adaptive immunity that has been developed as a powerful tool for precise editing of diverse genomes.

Horvath said his awards had created "some stress" among his colleagues, but I also sense that he feels his contributions haven't been adequately recognized. "I recognize that the real interest is in the gene-editing aspect, and it's possible that we'll forget those who discovered the natural bacterial system and remember only the final developers of tools that allow this revolution," he told me. Horvath also spares a thought for Jínek and Chyliński, the bench scientists who led the CRISPR breakthrough in 2012. "Maybe their defect is not to be women," he said in a flash of political incorrectness. "Currently there is a demand for women in science. There is a positive discrimination for women in science. It's good," he says quickly, "but there might be some drawbacks to that. As soon as you have a woman who is at this level who gets the recognition, it's obvious."

A fun ritual each September is to predict the next group of Nobel laureates. For CRISPR, it is surely a matter of when, not if. A maximum of three people can share the award for each category. And you have to be alive.* Some might argue that Nobels have already been awarded for gene targeting, the forerunner of genome editing, shared by Mario Capecchi,

* There is an exception: if you die but nobody on the Nobel selection committee knows at the time of the announcement, you may still be awarded the Prize on a technicality. This happened to Ralph Steinman in 2011, who passed away three days before the winners were revealed.

Oliver Smithies and Martin Evans in 2007 (for "introducing specific gene modifications in mice"). Perhaps the prize will go to the disciples of genome editing "before CRISPR", whom we'll meet in the next chapter.

Barrangou thinks people are asking the wrong question: it's not when, or whom, or for which discovery. In other words, which committee? Chemistry or Medicine? If the award goes for chemistry, then the development of the sgRNA favors Doudna and Charpentier, but a strong case can be made for Šikšnys, who shared the Kavli Prize, or Jínek, who performed the signature sgRNA experiments. If it's for the discovery of Cas9, then maybe Moineau. If the award is given for physiology or medicine, then it must go for genome editing, most likely Zhang and Church. "George was right there!" says Barrangou. "He's been written off the books of history for no reason. You can't keep George out of that, that's crazy."[13] But for all its potential, CRISPR-Cas—indeed the entire field of genome editing—still has to prove itself as a game-changing, life-saving therapeutic.

Luciano Marraffini's key role in helping Zhang kick-start his CRISPR program in 2012 was omitted from Lander's "heroes" narrative. The affable Argentine's technical expertise was central to the gene-editing discovery but, with the exception of the 2017 Albany Prize, has largely fallen under the radar. At the Albany Prize ceremony, Marraffini shared the stage with Mojica, who was asked to reflect on life as the grandfather of CRISPR. It was like adopting a child. "You give it a nice name—CRISPR," he said. "You're very proud of this child. It feels like [someone] that belongs to you, even though it's not true. You try to look after them." After ten years, the child becomes "a very clever person," and then "a very important person." Overall, he said, "I feel full of joy, I feel happy, I feel proud."[14]

As for the Nobel speculation, Mojica wishes it could be put to rest. "If I get it, I will disappear from the planet," he says.[15] Of course, if he gets it, there is not the slightest chance of that happening.

While the developers of genome editing rack up accolades, the task of improving the original CRISPR system and expanding the CRISPR toolbox marches on. The potential of CRISPR has inspired thousands of new researchers around the world to study the fundamental biology of CRISPR and apply it in a host of settings including new forms of therapy.[16]

In CRISPR circles, the original Cas9 still commands the lion's share of the attention. But researchers including Banfield and Koonin are mining the astonishing diversity of microbial life on planet earth to identify unknown organisms and catalogue the diversity of CRISPR immune systems and Cas genes therein. Scientists have wasted no time in adapting some of these systems for new research and diagnostic purposes.

One early and profound addition to the toolbox was to take the molecular scissors and immediately blunt the blades to mute Cas9's DNA cutting function. That may sound counterintuitive, but the RNA programmability of Cas9 serves a multitude of purposes beyond cutting DNA. Cas9 can be used to ferry many kinds of molecules to a specific spot in the genome to modulate gene expression up or down (CRISPR activation or interference). This non-cutting Cas9, described by Stanley Qi and colleagues, is called "dead Cas9."[17] Its applications include base editing,[18] a new riff on CRISPR genome editing in which Cas9 is used not to cut DNA but to position different enzymes to nick the DNA and perform pinpoint chemistry on a specific base. (We'll return to this in chapter 22.)

The original Cas9 protein from *S. pyogenes* (SpCas9) is made up of 1,366 amino-acid building blocks, which makes for a tight squeeze when packing this molecule into the limited cargo space of the most popular gene therapy vector, the adeno-associated virus (AAV). There are many flavors of Cas9 derived from other microbes, some significantly smaller than SpCas9, others that recognize a different PAM site. Many of these provide new tools, increasing the options for the CRISPR engineer. Doudna's team also found a way to put Cas9 on pause, holding it in a locked formation with the molecular equivalent of a plastic zip tie, which can be snipped as required to release the nuclease.[19] This affords researchers more control in exactly where—or when—they unleash Cas9, reducing the chance of undesirable off-target effects.

In 2016, almost a decade after they first talked CRISPR over coffee, Doudna and Banfield unearthed a trove of new CRISPR tools from a metagenomic sampling of microbes that have not yet been cultured in the laboratory.[20] Among the highlights were two diminutive Cas nucleases, named CasX and CasY. CasX is only 60 percent the size of SpCas9. It cuts DNA in much the same manner, even though it bears no sequence similarity to SpCas9, suggesting it evolved quite independently. And unlike SpCas9, it is derived from a bacterial species that does not naturally infect humans, so in principle it would not have any of the potential immunogenicity concerns.

Another interesting Cas scalpel is Cas12 (formerly known as Cpf1), in particular Cas12a, discovered by the Zhang lab.[21] Unlike Cas9, this enzyme produces a staggered cut—slicing the two strands of the helix in different places, rather than a clean cut. It is a small protein and doesn't require a tracrRNA. While independently investigating the properties of Cas12a, members of the Zhang and Doudna labs were shocked to find that Cas12a did something quite different to single-stranded DNA—it didn't so much cut DNA as shred it.[22] Another new addition to the toolbox, Cas13, did much the same thing to RNA. If the detection of a specific DNA or RNA molecule could be coupled to some sort of chemical signal, the groups would have a simple diagnostics platform.

That's exactly what the two groups did. Two of Doudna's students, Janice Chen and Lucas Harrington, helped create Mammoth Biosciences to commercialize their diagnostics system, dubbed DETECTR. (Chen's brother is world champion figure skater Nathan Chen.) Meanwhile, two of Zhang's protégés, Omar Abudayyeh and Jonathan Gootenberg, joined Zhang, Pardis Sabeti, and the cofounders of Sherlock Biosciences; you don't need to be pipe-smoking detective to know what their system is called.*

Here's an example: let's say we want to program Cas12 to detect the SARS-CoV-2 virus responsible for the COVID-19 pandemic. A series of guide RNAs are designed to recognize certain sequences that have been amplified from the coronavirus genome. But once Cas12 recognizes that sequence, a new enzymatic property is switched on, such that it will cut (and keep cutting) any single-stranded DNA molecules in the vicinity. By adding

* SHERLOCK stands for Specific Hypersensitive Enzymatic Reporter unLOCKing.

reporter molecules that light up when cut, the presence of even trace amounts of the virus can be detected using a simple color assay on a paper strip.[23]

Similarly, the Cas13 family can be used to detect infections such as flu, dengue, and Zika, and of course COVID-19.[24] Once activated, Cas13 exhibits what Zhang calls "collateral RNase activity"—it keeps cutting RNA. By supplying a suitable quantity of chemically tagged RNA reporter molecules, his team has the basis of a simple, portable detection system that can work on urine, blood, or saliva. The presence of virus will switch on Cas13, cutting the RNA reporters and releasing a fluorescent marker that can be read on a simple paper strip much like a pregnancy test.

The reliability of simple, cheap, one-stop diagnostic tests have a bad reputation following the debacle of Theranos, the theatrically overhyped Silicon Valley unicorn launched by Elizabeth Holmes that crashed from a $9 billion valuation to bankruptcy following great investigative reporting by John Carreyrou.[25] Unlike Theranos, which only belatedly published a single peer-reviewed study,[26] the Doudna and Zhang teams have already laid out the science and technology behind their diagnostics discoveries in a series of top-tier publications.

The portability of Mammoth's and Sherlock's kits could find huge markets—at home for the flu, in hospitals for antibiotic resistance, and in the field where outbreaks of coronavirus and other viruses emerge. Chen's group has shown DETECTR can accurately detect HPV samples in a fraction of the time of a conventional test. Led by Harvard's Sabeti, the SHERLOCK test has already shown promising results in detecting cases of Lassa fever in Nigeria, dengue in Senegal, and Zika virus in Honduras. Both companies are actively adapting their platforms to detect the COVID-19 virus. Beyond diagnostics, applications beckon in areas from food security and agriculture to bioterrorism. The Zhang lab has already applied the SHERLOCK to gene detection in plants, pointing to an array of applications in detecting pathogens or pests.[27]

In Toronto, Joseph Bondy-Denomy "found something amazing that we never expected," said his PhD supervisor Alan Davidson.[28] He discovered anti-CRISPRs, a growing family of viral proteins that are able to disarm or neutralize bacterial CRISPR defenses—the rocks and paper to CRISPR's scissors. Erik Sontheimer described a means to use anti-CRISPRs to restrict genome editing to a tissue of choice. Harvard Medical School's Amit Choudhary is identifying small chemicals that can fine tune Cas enzyme activity. DARPA has launched a program called Safe Genes to fund research into anti-CRISPRs, and Bondy-Denomy wasted little time in cofounding a company, Acrigen Biosciences, to make gene editing safer and more efficient.

Other toolbox additions include Cas3, a DNA shredder to generate large deletions; a system called EvolvR to introduce mutations and evolve a specific target region; and systems that engineer programmed DNA insertions at a target site. Similar ideas were developed in parallel by Samuel Sternberg's group at Columbia University, and the Zhang lab. Sternberg adapted a CRISPR system from *Vibrio cholera* to develop a programmable system based on transposons (parasitic jumping genes) to insert a custom DNA sequence at a specific site in the genome.[29] The system offers an appealing alternative to genome engineering without breaking DNA or triggering the cellular DNA damage response.

Scientists are just starting to appreciate some of these new tools, but these are just the tip of the iceberg. Zhang says about 150,000 microbial genomes have been sequenced, but we only understand something about the defense systems in about one third of them. There are so many more secrets yet to be revealed from the sequences of our microbial ancestors, which have had a mere billion years to innovate and evolve.

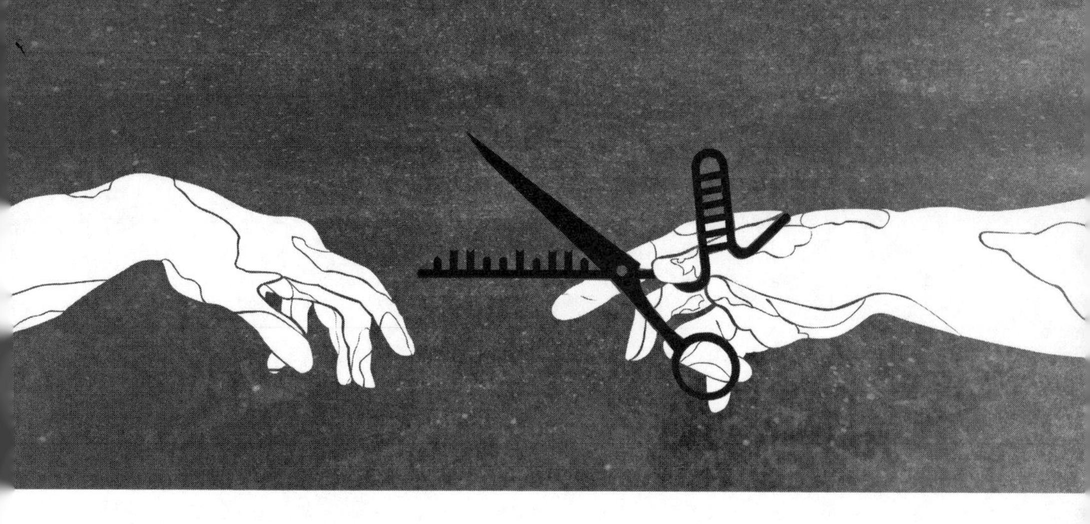

PART II

"I hold that while a man exists, it is his duty to improve not only his own condition, but to assist in ameliorating mankind."

—Abraham Lincoln

"The advance of genetic engineering makes it quite conceivable that we will begin to design our own evolutionary progress."

—Isaac Asimov

"Your scientists were so preoccupied with whether or not they could, they didn't stop to think if they should."

—Ian Malcolm, *Jurassic Park*

CHAPTER 8
GENOME EDITING B.C.

"If I had a ruble for every time I've heard about the promise of gene editing, I'd be an oligarch!" declares Fyodor Urnov. "What hypothetical promise? It's been in the clinic for nearly a decade!"[1] Urnov should know: for more than a decade, he was one of the molecular musketeers at a biotech company called Sangamo that took the lead in developing genome editing and brought it to the clinic, developing a therapy for HIV. It wasn't an unequivocal success by any means, but it opened the door for CRISPR and an avalanche of new therapies, some of which might turn into cures.

Now back in academia working with Doudna at the Innovative Genomics Institute, Urnov is all in on CRISPR, allied with the biggest name in the field. "I'm happy Jennifer Lopez is doing a TV show [on CRISPR], but what the other Jennifer is doing is a lot more interesting," he joked the first time I heard him give a lecture in 2018. He speaks fast and enunciates crisply in a vestigial Russian accent mellowed by more than two decades on the West Coast. If there is a guru in the world of genome editing, Urnov is the man. But before we consider what CRISPR means for humankind now and in the future, I first need to tell a bit more of the back story regarding CRISPR.

Genome editing did not burst onto the scene fully formed like Athena, with what Urnov termed the "immortal" Charpentier-Doudna CRISPR discovery in 2012. In fact, the year before, the journal *Nature Methods* declared gene editing its "Method of the Year" based on the promise of two forerunners of CRISPR—zinc finger nucleases (ZFNs) and TALENs.[2] Although expensive and difficult to deploy, the technologies entered the clinic years ahead of CRISPR—ZFNs developed commercially by Sangamo, and TALENs championed by Paris-based Cellectis.

If Doudna and Charpentier's teamwork in 2012 is the pillar for genome editing in the modern era, the New Testament if you will, then Urnov brands the era leading up to that moment as "Genome editing B.C."—*before* CRISPR.

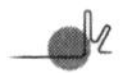

Few Nobel laureates have a more remarkable personal story than Mario Capecchi. Creativity and success in science requires "the abrasive juxtaposition of unique sets of life experiences that are too complex to preorchestrate."[3] Remarkably, Capecchi survived outrageous odds during World War II to become the first scientist to conduct a form of gene editing in mammalian cells. Capecchi was born in Verona in October 1937 as fascism flared across Italy. His father, an officer in the Italian air force, had an affair with a beautiful poet who lectured at the Sorbonne in Paris. After Capecchi's birth, "my mother wisely chose not to marry him."[4] As a bohemian, she staunchly opposed fascism and took her baby to the Italian Alps. But in 1941, Capecchi recalls the Gestapo arriving in Tyrol and arresting his mother, who was incarcerated in Dachau, Germany.

For a year, Capecchi lived with a neighboring family, living on homemade bread; he remembers jumping naked in barrels of freshly picked grapes. But when the money Capecchi's mother had provided ran out, he was left to fend for himself. Only four years old, Capecchi headed south, "sometimes living in the streets, sometimes joining gangs of other homeless children, sometimes living in orphanages and most of the time being hungry." Many memories of that period "are brutal beyond description." After the liberation of Dachau in 1945, Capecchi's mother returned to Italy to search for her son. Miraculously she found him in a hospital in Reggio Emilia, where he was being treated for malnourishment. In Rome, Capecchi had his first bath in six years. Later, the Capecchis sailed to

America. "I was expecting to see roads paved with gold," he wrote. "I found much more: an opportunity." Capecchi settled with his Uncle Edward, a physicist, just outside Philadelphia, He reveled in wrestling and still has the physique to prove it.

After graduating from Antioch College, Capecchi interviewed at Harvard with Jim Watson. When he asked Watson where he should conduct his PhD, Watson snorted: "You'd be crazy to go anywhere else." Capecchi joined the effort to defragment the genetic code. Capecchi admired Watson's bravado and stark honesty, as well as a sense of justice. "He taught us not to bother with small questions, for such pursuits were likely to produce small answers," he said. A few years later, Capecchi set up his own group at the University of Utah. By microinjecting DNA into the nucleus of living cells, he developed a method to swap a gene for a near-identical copy. In the early 1980s, an NIH panel rejected Capecchi's proposal, but he'd overcome tougher odds than that.

Capecchi's groundbreaking work, along with Oliver Smithies, a British geneticist then at the University of Wisconsin, and Martin Evans, provided researchers with a means to "knock out" a mouse gene using homologous recombination. By inactivating a gene in embryonic stem cells and then injecting those modified cells to create a chimeric embryo, scientists could do in small, furry mammals what they'd been able to do routinely in yeast and bacteria for decades. The technique was demanding, inefficient, and took months to perform, but the ability to create an animal model lacking a key gene was a godsend for geneticists and developmental biologists. Like a genetics gold rush, the journals were flooded with papers reporting what happened when one mouse gene after another was muted, many providing critical models of human genetic diseases. And it earned Capecchi, Evans, and Smithies the Nobel Prize in Physiology or Medicine in 2007.

Although a major advance for biology, the problem with targeted, or homologous recombination was its very low efficiency in mammalian cells—only about 0.01 percent. But Maria Jasin, a molecular biologist at Memorial Sloan Kettering Cancer Center in New York, knew that rates were much higher in yeast. In 1994, her group tested the idea that introducing a break in both strands of DNA could trigger the cell to repair the break. Using a restriction enzyme called I-*Sce*I, she found higher rates of recombination when she cut the DNA. Moreover, those breaks could be repaired with an exogenous piece of DNA. Jasin detected two different forms of DNA repair: homologous recombination and non-homologous end-joining (NHEJ). The former produced a clean repair, the latter a series of short deletions flanking the target site. Jasin's demonstration that a double-strand break was "editogenic" was a major landmark, arguably the first gene-editing experiment, even though its significance was not truly appreciated for a decade. In the Old Testament of Genome Editing, Urnov calls this the "gospel according to Jasin."[5]

"I am a textbook example of 'right place, right time.' My entire life!"[6] I've managed to corral Urnov for an hour in a hotel lobby in Florence, Italy, to reflect on the Old Testament of genome editing.

Urnov was born in the former Soviet Union during the calm of the 1970s. Socialism more or less worked. Moscow was a cultural Mecca. Friends and families argued about the meaning of life at the kitchen table. His father was a literary professor, his mother a linguist. Both were published biographers. His grandfather had edited the works of Charles Dickens. Urnov grew up on the works of Lewis Carroll, Dickens and Twain, while becoming an obsessive Beatles fan. Although not a scientific household, Urnov's father was friends with the family of the great Russian molecular biologist Vladimir Engelhardt, who lent him a copy of *The Double Helix*. Watson's book had a profound effect on the fourteen-year-old Urnov, especially the "incomparable taste of having discovered a secret." After one reading, any other career plans were canceled. He was hooked on DNA.

Urnov enrolled at Moscow State University in 1985, studying biology the year that Mikhail Gorbachev came to power. Glasnost was instituted in Urnov's freshman year, perestroika in his sophomore year. The Chernobyl disaster in 1986 also had a major effect on him. By the time he graduated, the Soviet Union had essentially fallen apart. Going west would no longer mean being smuggled in a suitcase or emigrating via Israel. With his parents' support, Urnov enrolled at Brown University. His PhD advisor,

Sue Gerby, patiently domesticated the ex-Soviet into her lab, where he spent six years studying how genes turned on and off. At a conference, a "starstruck" Urnov asked Alan Wolffe if he could join his lab as a postdoc. Wolffe was dynamic, charismatic, and the youngest institute director ever at the NIH. He said yes.

Urnov was catapulted into the premier league of genetics research. Wolffe was an expert in the emerging field of epigenetics, the study of chemical modifications to DNA that regulate the gene activity. Joining Wolffe's lab reminded Urnov of the Red Queen's race in *Through the Looking-Glass*, as Alice says, when we run this fast, we generally get somewhere else. Everybody in the lab was top-notch, working around the clock. He had to up his game and be ready when Wolffe ambled up to his bench and asked: "Ah, Dr. Urnov, what have you discovered?"

Alfred Hershey, one of the founders of molecular biology, once described "Hershey Heaven" as coming to the lab, running an experiment, and having it work every day. Urnov says he was "truly in Hershey Heaven." But then in 2000, in a surprising move, Wolffe accepted an offer to head research at a young biotech company in California. At age forty, Wolffe was ready for a new challenge. Urnov readily agreed to go west. It was the first time he heard the name Sangamo.

The home office of Edward Lanphier, the retired founding CEO of Sangamo BioSciences, is a carriage house in Marin County, about ten miles north of San Francisco. I stop to admire the mementoes of a successful biotech career. A framed front page of the *San Francisco Examiner* from 1981 has the headline: GENENTECH JOGS WALL STREET. A bookshelf houses the obligatory vanity license plate ("SANGAMO") that Lanphier belatedly detached from his car. A framed photo shows Lanphier and his daughter in New York outside the Nasdaq stock exchange when Sangamo went public, the neon sign flashing his name.

I don't have to look far for the origin of the company's name: a Sangamo Electric meter has been converted into the base of a table lamp. In the

1890s, Lanphier's great-grandfather, a Yale-educated electrical engineer, Robert Lanphier, cofounded a company in Sangamon County, Illinois. Lanphier designed and patented the watt-hour meter, with its familiar rotating wheel.[7] Sangamo Electric became a public company before being acquired by Schlumberger in 1975. Lanphier asked his father if he could borrow the name and the logo for his own engineering start-up in the mid-'90s.

Lanphier got his start in the pharmaceutical industry in the early 1980s at Eli Lilly, which had just licensed recombinant human insulin from Genentech. In 1992, he joined Somatix, a "first-generation" gene therapy company. Three years later, Lanphier launched Sangamo, and soon became enamored with the potential of a class of gene regulators for gene therapy.

Zinc finger proteins (ZFPs) are an abundant* class of gene activators that were discovered a decade earlier by Nobel laureate Aaron Klug, a Lithuanian Jew who emigrated to England for his PhD and worked with Rosalind Franklin shortly before her death in 1958. These transcription factors had an unusual structure—a series of digit-like projections that make direct contact with the DNA. Each digit consisted of some thirty amino acids, anchored at the knuckle by a zinc atom binding to a quartet of amino acids. "Zinc structural domain" didn't have much of a ring to it, so Klug coined the term "zinc finger" for each module. Further work showed each finger recognizes a specific three-base sequence of DNA, like a blind person reading braille. Thus, a DNA-binding protein containing three "zinc fingers" can recognize a specific nine-base stretch of DNA.

Sangamo's initial goal was to use zinc finger proteins to switch certain genes on (or off). Lanphier approached the leaders in the field, including Carl Pabo at MIT. In London, he sat outside the office of Klug, who was president of the Royal Society, like a schoolboy waiting for the headmaster, before sealing a partnership over a three-hour lunch. Klug became a key advisor to Sangamo after selling his own company, Gendaq, to Lanphier in 2001. Sangamo also brought some biotech muscle onto

* There are about seven hundred genes—3 percent of the total—encoding zinc finger proteins in the human genome.

the board, including Bill Rutter and Herb Boyer, cofounders of Chiron and Genentech respectively.

In 2000, Lanphier decided to ride the wave of irrational exuberance in the markets and go public, raising $150 million at the peak of the biotech bubble. To launch this new chapter, he lured Wolffe to become head of research. "It was an enormous coup," Lanphier recalls. "Alan was just a frickin' rock star" with an encyclopedic knowledge. "One of the most brilliant men I've ever known. Every brilliant, twentysomething alpha-male postdoc wanted to work for Alan."[8] Joining Urnov on the expedition was another Russian expat, Dmitry Guschin. Together with zinc finger designer Ed Rebar (from Pabo's lab), Michael Holmes (from Bob Tijan's lab at UCSF), Jeffrey Miller, and Andrew Jamieson, Sangamo assembled a boiler room of fearless young talent. Their mission was to take Sangamo, Lanphier said, "from the concept of a steam engine to an internal combustion engine to a freakin' Ferrari." The most critical hire was an English postdoc named Philip Gregory. "Amongst a group of unbelievably talented people, Philip organically rose to be the first amongst equals," Lanphier said.

Wolffe's expertise was crucial in creating artificial transcription factors—proteins that bind to specific DNA motifs to switch genes on and off—that can drive cells to particular developmental fates. Pabo had shown that by mixing and matching individual zinc finger units, researchers could design a new hybrid transcription factor to recognize a specific DNA sequence. By the time Urnov arrived, Sangamo had designer zinc finger proteins to show this approach was feasible, setting the stage to test their efficacy in human cells.

Then in May 2001, tragedy struck. While attending a conference in Rio de Janeiro, Wolffe was struck by a bus and killed while out running one morning. "Alan was the hub of the spokes . . . and then he's gone." Pabo, the chair of Sangamo's scientific advisory board, acted as the research team leader until Gregory assumed the role two years later.

Just as Sangamo was preparing to enter the clinic, using a novel ZFP to treat nerve damage in diabetes patients by switching on the *VEGF* gene, news arrived of another setback. In the summer of 2002, just months after Alain Fischer and colleagues had published exciting gene therapy trial results,[9] reports emerged that one of his patients had developed leukemia. The virus carrying the therapeutic gene had inserted itself into the patient's genome, causing cancer. Following the 1999 Jesse Gelsinger tragedy, gene therapy trials were put on hold or abandoned.

By this time, however, a glimmer of light for zinc fingers had emerged, offering the possibility of not merely replacing a disease gene with a healthy copy, or switching on a silent gene, but actually editing and fixing the mistake at the DNA level. A team at Johns Hopkins had devised a means to modify zinc fingers to target them to genes of interest.

In 1978, Hamilton "Ham" Smith, a 6' 5" biochemist at Johns Hopkins University, received the call almost every scientist dreams about: he had won the Nobel Prize. Smith, never comfortable in social situations, survived the scrum of press photographers and well-wishers, and a mild case of Imposter Syndrome. Even his mother was surprised: when she heard the news on the car radio, she turned to her husband and said: "I didn't know there was another Hamilton Smith at Hopkins."[10]

Smith's Nobel was awarded for his serendipitous discovery of restriction endonucleases, a large family of bacterial enzymes that recognize and cut specific DNA sequences or motifs. Genetic engineers turned these enzymes into the catalysts of the recombinant DNA revolution. "Everything about modern biology, from the idea of determining a DNA sequence to the idea of recombinant DNA to DNA fingerprinting, it all starts with restriction enzymes," said geneticist David Botstein.[11] By the mid-1990s, thousands of restriction enzymes had been catalogued, shipped commercially around the world in polystyrene buckets of dry ice. But their usefulness as a scalpel for precision gene editing was limited. These enzymes cut DNA at very short recognition sites, typically only four to six base pairs. While those motifs might only occur a few times in the tiny genome of a virus, they crop up thousands of times scattered across the human genome.

Smith often discussed the idea of engineering artificial enzymes that could be more selective in cutting DNA with his students, including in

1986 a visiting chemist named Srinivasan "Chandra" Chandrasegaran. Years later, Chandra set out to engineer a chimeric restriction enzyme, a new kind of nuclease. Flicking through the enzyme catalogue offered by New England Biolabs, Chandra and his colleague Jeremy Berg settled on the amusingly named *Fok*I from *Flavobacterium okeanokoites.* Like a *Star Wars* TIE fighter seeking the thermal exhaust vent, *Fok*I scans the DNA in search of a specific landmark—a sequence of five bases, GGATG. But once it settles on the DNA, the actual cutting is carried out by a different domain of enzyme about ten bases downstream. As the two domains were separate, Chandra reckoned he could alter the target parameters by tethering a different DNA recognition domain to the cutting site.

Chandra published his "hybrid restriction enzyme" breakthrough in 1996.[12] His team fused the DNA-cutting domain of *Fok*I with zinc-finger domains that supplied the specificity. "In theory," Chandra wrote, "one can design a zinc finger for each of the sixty-four possible triplet codons, and, using a combination of these fingers, one could design a protein for sequence-specific recognition of any segment of DNA." These zinc-finger nucleases (ZFNs) could be programmed to latch onto any DNA sequence that would serve all manner of applications. Interestingly, Chandra's choice has stood the test of time. "Like the fact that a [soccer] match lasts ninety minutes or the QUERTY keyboard starts with the letter Q, it is widely accepted," says Urnov. "People haven't seen the need to evolve beyond that."

Chandra was in no doubt that his chimeric nucleases—"a new type of molecular scissors"—could transform gene therapy: in 1999 he said his goal was to excise a gene mutation and replace it neatly with its normal counterpart. Ethical issues aside, he wrote, "gene therapy will be routinely used in clinical practice, signifying a paradigm shift in the treatment of human disease."[13]

Chandra teamed up with Dana Carroll, a biochemist at the University of Utah, who wanted to customize a ZFN to engineer a mutation in a classic animal model such as the fruit fly. If done right, the *Drosophila* cells would turn brown to yellow. Carroll's colleague saw the yellow bristles down the microscope.[14] "If I were you, I'd be pretty excited," he told his boss. By 2002,[15] Carroll's group had demonstrated the ability to engineer DNA in living organisms, the first use of ZFNs not merely to modulate the expression of certain genes, but actually to change their DNA sequence. Carroll's development of ZFNs coupled with the editogenic insights from Jasin and colleagues laid the foundation for genome editing in humans.[16]

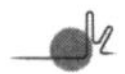

From 1997, Sangamo's headquarters was in a building in Point Richmond, shared with Pixar, the animation studio behind *Toy Story* and *A Bug's Life* later acquired by Disney. When Pixar moved to a larger headquarters in Emeryville, Sangamo expanded into the space. For three years, Urnov and Holmes shared an office that was formerly Pixar's screening room. Urnov says their partnership was akin to Lennon and McCartney, before conceding that might be a bit of a stretch. Assisting the Sangamo team was Matthew Porteus, a physician-scientist at Stanford who had trained with David Baltimore. He'd also been inspired by Carroll's ZFN papers and wanted to get them to work in human cells. Porteus developed an assay using the green fluorescent protein that could report successful gene targeting using ZFNs.[17]

Sangamo's young musketeers were on a mission and there was no time for failure. "Nothing creates a sense of urgency like being on Nasdaq," says Urnov. Over the next few years, Sangamo figured out how to turn good ZFNs into effective gene editors. There were multiple disease targets—sickle cell disease, hemophilia, and severe combined immunodeficiency (SCID). (Urnov and Holmes even dabbled with editing the CCR5 gene.) With the French gene therapy setback in everyone's minds, Sangamo began looked to repair the genetic glitch in those SCID patients—a mutation in the gene for the interleukin-2 gamma receptor (IL2Rζ).

One day, Urnov was reading the results skipping off a lab instrument called a phosphoimager. It looked like "we'd achieved a one-in-five efficiency of gene editing. Efficiency like this happens spontaneously in about one in a million cells." Urnov shared the results with Holmes. "If this is real, we've just entered into a new era!" Holmes concurred, already planning the next experiments to pressure test the result while an exhilarated

Urnov paced around the room. "Extraordinary claims demand extraordinary evidence. We both knew nobody would believe us!" Urnov recalls. They kept the results to themselves, while secretly running every control experiment that they could think of. Sangamo's expertise was starting to pay off. "We'd finally built the fast engine in the car with the superb tires, a super-aerodynamic frame, and a super-flat racing track."

Urnov and Holmes had glimpsed gene correction—the ZFN created a DNA break, replaced by a piece of genetic information. They looped in Gregory, Rebar, and Miller to make a circle of five, and came up with a definitive experiment to rule out the possibility of an artefact caused by sample contamination. Urnov selected a cell that was homozygous for the gene being edited. If there was contamination, there would always be two forms of the gene. But if the only signal was the sequence being introduced, that would indicate that the native gene had been replaced—edited—by the external sequence. Over one long weekend, Urnov tested a group of edited cells. The first few were normal, unchanged, boring. The next was a heterozygote—one normal variant, one altered. This continued until—"Ode to Joy!"—he found a cell in which both copies of the gene had been changed. Urnov dashed off an email to Holmes: "A HOMOZYGOTE!!!" That was soon followed by "ANOTHER homozygote!!!"

There's a Russian proverb that says: "If you grab the rope, don't complain that the cart is too heavy." It was time to open the curtain and put the finishing touches on the all-important scientific report. Urnov ran the most important experiment of his life—a wondrously low-tech experiment devised by Ed Southern in the 1970s, in which DNA fragments are ingeniously sucked out of a gel and transferred onto a nylon membrane with the absorbent assistance of a 4" stack of paper towels (the Southern blot). Holmes showed they could reverse the edited change, while the experiments in cancer cells were repeated in clinically relevant white blood cells.

Before submitting the paper to *Nature,* one of Sangamo's key advisors, Sir Aaron Klug, proposed Urnov and colleagues use the term "high-efficiency gene *correction*" rather than modification. (The manuscript copy bearing Klug's handwritten comments remains one of Urnov's most prized possessions.) After two rounds of review, *Nature* published the report that rewrote the gene therapy playbook in April 2005.[18] Sangamo had demonstrated the feasibility of correcting human genetic mutations. Moreover, the method avoided the problem of insertional mutagenesis that had marred the French gene therapy trial. "The 'hit and run' mechanism of ZFN action uncouples the therapeutically beneficial changes made to the genome from any need to integrate exogenous DNA, while still generating a permanently modified cell," Urnov wrote.

When the *Nature* editors asked for ideas for a cover headline, Urnov suggested "genome editing." (His father had just become the editor in chief of a Russian journal of literary criticism.) Five years after the completion of the first draft of the human genome, scientists had demonstrated the feasibility of rewriting the language of life to fix a genetic disease.

WIRED magazine's Sam Jaffe reported on the landmark "nano-surgery" technique with a headline that hopefully earned the copy editor a bonus: "Giving Genetic Disease the Finger."[19] Jaffe quoted David Baltimore: "This doesn't just deliver a foreign gene into the cell. It actually deletes the miscoded portion and fixes the problem." The potential to target any gene in the genome was plain to see. Chandra's review of the paper for *Nature Biotechnology* was entitled: "Magic scissors for genome surgery."[20]

The next step was to move toward treating SCID patients, which required performing gene editing in stem cells. But to the team's despair, all they found were small DNA sequence insertions and deletions, gene knockouts not precision repair. "This was, putting it mildly, not the droid we were looking for," says Urnov. The impasse was broken in style by the company's new chief medical officer, Dale Ando.

"I know exactly what to do," Ando said. "And I know what gene, and what disease. We're not going to do bubble boy disease. We're going to do HIV."

"Um, okay," Urnov said

"We're going to do *CCR5* in T cells"

"Okay."

"And we're going to collaborate with Carl June."

"Who's that?"

Ando started laughing. Not a bad way to make an impression on your first day in the job.

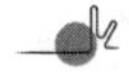

Few areas of medical research were more urgent or competitive in the mid-1990s than HIV, which was first described as an acquired immune deficiency syndrome (AIDS) in a handful of patients in 1981. As the epidemic spread, scientists in the Bay Area observed that some people possessed a natural immunity to the virus. Meanwhile, several groups identified the protein receptor footholds—CD4 and a co-receptor called CXCR4—that enable HIV to gain entry into white blood cells.

In June 1996, five separate reports incriminating another membrane protein, CCR5 (C-C chemokine receptor-5), as a second co-receptor were rushed into print by the top three journals, all within a week of each other. If HIV was a blimp that is snagged by the Empire State Building (CD4), then CCR5 was the cable car ferrying passengers—the HIV genetic material—to the ground. Like tabloid newspapers, premier science journals can get competitive trying to be the first to publish a research breakthrough. Alas, one of those CCR5 reports[21] in *Cell* was pushed into production so hastily that several pages ended up being printed upside down.

One of the senior authors of that report was Marc Parmentier, a Belgian physician-scientist, who had a hunch that abnormalities in CCR5 might explain the slow disease progression in some people exposed to HIV. Parmentier's team took samples from three such individuals and found a glaring thirty-two-base gap (Δ32 or "delta 32") in the middle of the *CCR5* gene.[22] The size of this deletion left little doubt that the function of the truncated protein was compromised. After testing hundreds of samples and volunteers, he found that the Δ32 variant was surprisingly common in Europeans—a carrier frequency (meaning one copy) of about 10 percent—but not a single HIV patient carried two copies of the Δ32 variant. A colleague showed that white blood cells with the Δ32 gene were resistant to HIV infection. By the time Parmentier submitted the report to *Nature* in July,[23] another group had found the same results on a larger cohort of patients.

In the 1980s, Stephen O'Brien, a lab chief at the NIH, embarked on a search for genetic factors that influence HIV susceptibility and progression.

O'Brien and geneticist Michael Dean began systematically screening candidate genes in their HIV population. After twelve years, the NIH team had examined more than one hundred candidate mutations in thousands of HIV patients without success. But the glut of CCR5 papers revealed one of the best candidates in years.[24] On July 4, while O'Brien was at the cinema watching the premiere of *Independence Day*, his team was furiously sequencing samples. They too uncovered the Δ32 variant, but didn't observe any Δ32 homozygotes in more than 1,300 HIV patients. About 1–2 percent of the American population is a Δ32 homozygote, but HIV patients almost never are. Without a portal into the white blood cell, HIV can land but it can't infect.*

The geographic distribution of *CCR5* Δ32 is interesting: it is most common in northern Europeans at a frequency of 5–15 percent. But as you travel farther south and east, the frequency drops—Δ32 is almost nonexistent in Africans and Asians. This pattern suggests that it must have been positively selected for a reason that has nothing to do with HIV, which didn't cross over to humans until the early 20th century. O'Brien felt the only reasonable explanation was "a mysterious, but breathtaking, fatal infectious disease outbreak which, like AIDS, exerted a huge mortality, and from which *CCR5* Δ32 carriers were resistant." The prime candidate is the Black Death, which ravaged Europe throughout the Middle Ages. Perhaps the Δ32 variant arose in Scandinavia in response to an earlier plague.

One year before the wave of CCR5 discoveries, a Seattle man named Timothy Ray Brown was diagnosed with HIV while studying in Berlin. He staved off the disease using the antiretroviral drug cocktail, but in 2006, after attending a wedding in New York, fell ill upon his return home. His doctor diagnosed anemia, but a painful biopsy revealed that Brown had acute myelogenous leukemia. The only treatment was a bone marrow transplant: fortunately, a search for potential donors turned up more than 250 matches. Brown's hematologist had the idea to select a donor who had

* Subsequent studies revealed that some Δ32 homozygotes are infected by a different strain of HIV, which gains entry via the CXCR4 co-receptor.

the defective *CCR5* gene. On the list of prospects, donor number sixty-one possessed the Δ32 variant.

Brown didn't want to be a guinea pig,[25] but he signed up for a transplant. Three months after the operation in February 2007, the virus was undetectable in Brown's blood and he ceased taking his HIV medications. After his leukemia returned, Brown had a second operation in February 2008. Doctors eventually declared him HIV-free with a normal T cell count.[26] "My name is Timothy Ray Brown and I am the first person in the world to be cured of HIV," the Berlin Patient proudly wrote.

Brown's experience gave Sangamo a lot to chew on. Mutation repair was still the Holy Grail for many diseases but inactivating key genes could have important medical benefits in certain situations, including the one Ando was advocating. The goal was to attempt "to recreate this HIV-protective genotype in the cells of HIV-positive individuals, in the hopes of essentially creating a compartment of the immune system that is protected from HIV infection," Urnov recalled.[27]

With Holmes leading the HIV program, Sangamo finally entered the clinic in 2009, in collaboration (as Ando had advocated) with Carl June, a leading gene therapy physician at the University of Pennsylvania. Five years later, Sangamo reported results on the first dozen HIV patients who had been treated with their own *CCR5*-edited T cells.[28] The results were mixed: the therapy was safe and there was some evidence of an antiviral effect, enabling some subjects to remain off standard antiretroviral therapy. In 2015, Sangamo received approval from the U.S. Food and Drug Administration (FDA) to extend the concept from T cells to stem cells, with the goal of protecting other cellular compartments of the immune system from harboring HIV. To date, Sangamo has treated more than one hundred HIV patients.

While Sangamo is known as the zinc finger specialist, a French company has taken the TALEN gene-editing technology to the clinic. The CEO of Cellectis, André Choulika, thought CRISPR was "super cool" when he first heard about it, but decided to stick with TALENs, mostly for immunotherapies. "We found them to be more accurate, precise, and powerful, and we thought they would be safer for patients," he says.[29]

Lanphier retired from Sangamo in June 2016 after twenty-one years, partly for health reasons but also because he felt it was time "to bring in the real pros." On CNBC's *Mad Money*, Jim Cramer asked Lanphier about CRISPR and Sangamo's faith in the ZFN platform. "The key to human therapeutics is specificity—the ability to target exactly the gene you want and only that gene," Lanphier replied. "That's where zinc finger nucleases have a complete monopoly."[30] Years later, I asked Lanphier if he still felt the same way. "CRISPR is bacterial. It's nonspecific. It's immunogenic," he said. "It's a great research tool. It's going to give a lot of visibility to genome editing. And when people actually want to use it therapeutically, that's where they'll end up talking to us." It must have been a wrench to remove that vanity license plate. "Nobody can do it the way Sangamo does it, on this scale, with the kind of precision," he said.

Before retiring, Lanphier launched therapeutic programs in blood disorders, including sickle-cell disease, beta thalassemia, and hemophilia. Ed Rebar came up with a clever strategy to switch on genes in the liver. Albumin, the most abundant protein in human blood, is produced by a gene that is extremely active in the liver. Rebar reckoned: what was to stop Sangamo from smuggling in a gene like a Trojan horse and taking advantage of this powerful albumin gene promoter? The method, dubbed *in vivo* protein replacement, or "invisible mending", involved snipping the albumin gene in the first intron, plugging in a transgene into this "safe harbor," and using the constitutive power of the albumin promoter to fire up the gene of interest. That fueled programs to target rare inherited disorders of genes normally expressed in the liver such as mucopolysaccharidosis (MPS) types I and II (also known as Hurler and Hunter syndromes, respectively) and hemophilia B.

On November 13, 2017, forty-four-year-old Brian Madeux climbed onto a bed in Room 1037—the Infusion Room—of the UCSF Benioff Children's Hospital in Oakland. Dressed in a gray sweatshirt and khaki shorts, Madeux nervously watched a nurse hook up an IV. He was no stranger to hospitals, having endured more than two dozen surgeries for hernias, bunions, spinal, eye, and ear problems resulting from Hunter syndrome. Surrounded by doctors, nurses, and a film crew, he was about to become the first patient to receive a direct infusion of a gene-editing drug. "It's kind of humbling," he told the Associated Press.[31]

Madeux's infusion took place on the centenary of the first description of his disease. Charles Hunter, a Scottish physician who had emigrated to Winnipeg, Canada, published a case report of two brothers, ages ten and eight, with a syndrome of physical abnormalities that would later bear his name.[32] The brothers had several common features—undersized, large head, short neck, broad chest, easily winded. We now know the disease is caused by a deficiency of an enzyme called iduronate-2-sulphatase. Hunter syndrome patients are unable to break down two particular carbohydrates, which consequently accumulate in various tissues. Enzyme replacement treatments involve weekly infusions can cost more than $100,000 per year. Madeux's doctors hoped his treatment would stem the progression of his disease and serve as an inspiration for other patients. For the first few days, he felt weak and dizzy, later he suffered a partially collapsed lung (probably unrelated to the therapy). Encouragingly, his liver appeared to be functioning normally. More patients were enrolled, some receiving higher doses. But initial results were equivocal.

Lanphier's successor, Sandy Macrae, is a Scottish physician who trained in the 1990s as a molecular biologist with the great Sydney Brenner. "My wife said it would never be of any use to me, and then this job came up," Macrae jokes. After revising the name of the company to Sangamo Therapeutics, he began inking deals for different disease targets with big pharma partners. Sangamo wasn't going to discard decades of expertise on ZFNs, but it is no longer just a zinc finger company. "If I was back doing my postdoc, I'd be using CRISPR," Macrae admitted.

Rarely does a biotech CEO acknowledge mistakes or failures, but Macrae has done both. Success in clinical genome editing comes down to three things: editing, delivery, and biology. The Hunter syndrome story showed that the albumin promoter strategy works beautifully in cells and animal models, and appears safe in human patients. Any complications in the trials were due to the AAV vector that was used. The trial was a "remarkably unremarkable event," Macrae said.[33] But the boost in enzyme levels only proved significant in a patient who received the highest dose. He then developed a side effect called transaminitis, which shut down production of the enzyme. "We succeeded in the editing, but it wasn't good enough for the biology," Macrae said. A new effort in the clinic with improved vectors is underway.

As for HIV, results could have been better. "We didn't understand enough about the biology. It was not the dramatic cure we hoped for," Macrae said. "We're *not* an HIV company." More promising data, albeit so far only in mice, suggests that a zinc finger approach can distinguish and shut down the faulty expanded version of the Huntington's disease gene from the normal counterpart.[34] But providing these medicines at an affordable cost to patients will be a challenge for the entire industry. Macrae says a typical Sangamo gene-editing drug costs about $300 million to move from idea to clinical trials to FDA approval.

Sometimes the pioneers are not the ones who reap the rewards. But veteran Ed Rebar, who briefly headed the Sangamo R&D team before joining Sana Biotechnology in 2020, remains a staunch believer in the power of zinc. "For therapeutic applications, ZFNs can do everything we need them to do," he told a crowd of genome engineers.[35] "Precision, any base, high levels of specificity." CRISPR is a great tool for basic research and has enjoyed widespread adoption. "But therapy is a different type of application."

Rebar wasn't exactly preaching to the choir. The answer to most genome editing applications in the clinic is to be found in the New Testament of CRISPR-based therapies. But ultimately, patients and their families won't care which technology is used if it answers their prayers and delivers a cure.

CHAPTER 9
DELIVERANCE OR DISASTER

The conceptual seeds of genetic engineering date back deep into the 20th century, two decades before the double helix and more than a decade before the demonstration that DNA, not protein, was the genetic material.

In 1932, some five hundred scientists traveled to Ithaca, New York, for the Sixth International Congress of Genetics. The registration fee was $10, a room in a hall of residence $1.75. Delegates could go for a day trip to Niagara Falls, attend a group picnic, or listen to an organ recital in Sage Chapel on the Cornell campus. The scientific program was dominated by the rock stars of the era: Thomas Hunt Morgan and his colleagues—Hermann Muller, A. H. Sturtevant, and Curt Stern—from the famous "fly room" at Columbia University. In a lab that smelled of rotten bananas, Morgan's group anointed the fruit fly as the ideal model organism to establish the "chromosome theory of heredity." Morgan's momentous discoveries were accepted as universal truths: His group built the first genetic maps of chromosomes and demonstrated that X-rays cause gene mutations. Morgan won the Nobel Prize the following year. But the answer to the existential question: "What is the gene?" would only emerge twenty-one years later, courtesy of Crick, Watson, and Rosalind Franklin.

Relegated to a Saturday breakout session, Hubert Goodale, the chief geneticist at the Mount Hope Farm in the northwestern corner of Massachusetts, didn't have the horsepower of a Morgan; instead of a fly room, he had a "mouse house" and a good story about applying genetic principles to animal breeding. Mount Hope was a leading genetics center in the United States: Goodale kept meticulous breeding records of poultry, cattle, pigs, and other animals, producing marked improvements in egg size, milk, and pork production. The farm's prize bull was named Satisfaction, but not for the reason you might think: an average Mount Hope cow sired by Satisfaction produced three times as much milk as a typical dairy cow.[1] Goodale's talk, entitled "Genetical Engineering," was perhaps the first public conceptualization of genetic engineering.[2]

The year 1932 was also when Aldous Huxley published *Brave New World.*[3] Almost two decades later, genetic engineering made its science fiction debut. In his 1951 novel *Dragon's Island*, Jack Williamson wrote:

> Man may now become his own maker. He can remove the flaws in his own imperfect species, before the stream of life flows on to leave him stranded on the banks of time with the dinosaurs and trilobites—if he will only accept the new science of genetic engineering.

Written two years before the double helix, Williamson understandably took some pleasure in his foresight—only to learn that the Oxford English Dictionary had unearthed a previous use of the phrase in 1949. "Everybody is famous, if only for fifteen minutes," he said.[4]

On March 19, 1953, Francis Crick wrote a long letter to his twelve-year-old son, Michael, at boarding school. As sneak peeks go, it was pretty special. "My dear Michael," Crick wrote, "Jim Watson and I have made the most remarkable discovery. We have solved the structure of deoxyribosenucleic acid (D.N.A.) . . ." On the next page, Crick sketched the double helix, showing the pairing of the four bases—C with G, A with T. After several more pages of near textbook detail, Crick invited his son to view the model during his half-term break. He signed off, "Lots of love, daddy."[5] Showing maturity beyond his years, Michael held onto his father's letter. It was a wise decision: sixty years later, Crick's letter fetched a world-record price at auction—$6.3 million. Half the proceeds went to the Salk Institute in San Diego, where Crick spent his final years.*

One week earlier, Watson had written a similar letter to one of his scientific idols, Caltech virologist Max Delbrück. Watson shared that he and Crick had fashioned a model of DNA with intertwining strands glued by interlocking base pairs running through its core, and would shortly submit a report to *Nature*. In a rare moment of humility, Watson conceded that (as had happened before) their model might be wrong. Then again, "If by chance, it is right, then I suspect we may be making a slight dent into the manner in which DNA can reproduce itself."[6]

Crick and Watson's historic success was made possible by data collected by Rosalind Franklin at King's College London. Franklin, working with her student Raymond Gosling, was a brilliant experimentalist, but uninterested in using her X-ray photographs of DNA crystals in trivial pursuit of building models. Crick was the mathematical brains in the Cambridge partnership, but as the late Brenda Maddox, Franklin's biographer, observed, he was unlikely "to have reached the goal without the pushing and prodding of the gauche young man from Chicago."[7]

In early 1953, Maurice Wilkins showed Watson a pristine, unpublished X-ray image of DNA—photograph 51—taken by Gosling six months earlier. To a trained eye, the trademark "X" pattern visible on "Photograph 51" could only mean that DNA was a helix. Watson pieced the final parts of the puzzle together. Working with cut-out representations of the four constituent bases of DNA, Watson's pairings—adenine (A) with thymine (T), cytosine (C) with guanine (G)—completed the structure of the molecule of life, two months shy of his twenty-fifth birthday.

A few weeks later, on April 25, 1953, the world—or at least *Nature* subscribers—got their first glimpse of the double helix. It was a family

---

* Perhaps inspired by Crick's auction windfall, Watson auctioned off his Nobel medal. It was bought by Russian oligarch (and part-owner of Arsenal) Alisher Usmanov. After meeting Watson in Moscow and learning that he wanted to use the proceeds for charity, Usmanov agreed to loan the medal back to Watson. It was returned to Cold Spring Harbor in an armored truck.

affair: the eight-hundred-word report was typed up by Watson's sister Elizabeth, while the double helix was elegantly sketched by Crick's wife, Odile. The report began with an immortal English understatement:

> We wish to suggest a structure for the salt of deoxyribonucleic acid (DNA). This structure has novel features which are of considerable biological interest.[8]

Word traveled slowly in those days. It took the *New York Times* six weeks before it saw fit to print a front-page story on the double helix. Watson and Crick published a follow-up paper in which they proposed that "the precise sequence of the bases is the code which carries the genetical information." For the next decade, the smartest minds in life sciences set about deciphering the code and figuring out how it was broken in genetic diseases. Only then could they contemplate how to fix it.

Watson has been justifiably criticized for his sexist portrayal of Franklin ("terrible Rosie") in *The Double Helix*, which was published over Crick's objections in 1968. In subsequent editions and other venues, he has acknowledged the importance of her scientific contributions. But Maddox, Franklin's biographer, defended Watson. "If it weren't for Watson, no one would have heard of Rosalind Franklin. He is deservedly in the top rank of writers of the 20th century."[9] Franklin died of ovarian cancer in 1958, denying her a thoroughly deserved share of the Nobel Prize. Today her contributions are widely recognized. In the 2015 West End production of the play *Photograph 51*, Nicole Kidman starred as Franklin.

Six months after Crick and Watson picked up their Nobel Prizes, Salvador Luria declared: "If knowledge is power, the science of genetics has placed in the hands of man an impressive amount of power in the last few decades."[10] But a new question loomed large: "Does the new knowledge of the genetic material and of its function open the door for a more direct attack on human heredity?"[11] Drawing an analogy with the physicists who split the atom and developed the atomic bomb, Luria was certain that geneticists would soon gain the power to contemplate "a direct attack on the human germ plasm."

A big name in bacterial circles, Rollin Hotchkiss was one of the first scientists to articulate concerns about the dangers of human genetic engineering. After excelling in high school in the 1920s and completing his PhD at Yale in organic chemistry in just three years, Hotchkiss turned to microbiology, discovering antibiotics and the first chemical modification to DNA.[12] Hotchkiss thought it was natural to feel "instinctive revulsion" at the thought of meddling with human nature but it would surely be done. "The pathway will, like that leading to all of man's enterprise and mischief, be built from a combination of altruism, private profit, and ignorance," he said.[13] Humans have long sought to improve on nature—seeking shelter, foraging for food, and defeating disease, whether modifying the diet of a baby diagnosed with phenylketonuria or administering chemotherapy to interfere with DNA replication in a cancer patient. Human genetic manipulation was on the horizon, Hotchkiss warned, and "we are going to yield when the opportunity presents itself." It was not too soon "to diminish the dangers to which this course will expose us."

Robert Sinsheimer, who passed away in 2018, is known as one of the architects of the Human Genome Project. In May 1985, while chancellor of the University of California Santa Cruz, he hosted a workshop to discuss a "big science" initiative to sequence the human genome (and put Santa Cruz on the map).* Sinsheimer's role in catalyzing the inception of the genome project crowned more than four decades in molecular biology. In 1953, he embarked on a six-month visit to Delbrück's lab to learn about phages. (Ironically, the greatest minds of the era were studying phages and bacteria, but oblivious to CRISPR.)

Sinsheimer studied a phage named ΦX174, the smallest phage known. His work helped lay the foundation for the sequencing of the very first complete genome, by Fred Sanger in Cambridge in 1977. Along the way,

* Sinsheimer wanted to leave his mark on UCSC but he was also hoping to save a $36 million pledge from the Hoffman Foundation for a space telescope that had been fully funded by the Keck Foundation. Sinsheimer didn't get his genome institute but UCSC would go on to play a major role in the completion of the first draft of the human genome.

Sinsheimer demonstrated that the viral genome was merely a single strand of DNA, a stunning result that overturned six years of double helix dogma, like "finding a unicorn in the ruminant section of the zoo," Sinsheimer said. He followed that with another heretical result: the ΦX174 DNA wasn't even a linear molecule, but a ring. The enzyme that closed the loop—DNA ligase—proved to be the missing link in the ability to replicate the virus in a test tube. In 1967, Sinsheimer collaborated with Nobel laureate Arthur Kornberg to successfully replicate ΦX174 that could infect bacteria. The result was even picked up by President Lyndon Johnson, who said on television, "Some geniuses at Stanford University have created life in the test tube!"

By this time, Sinsheimer was thinking hard about far-reaching implications of genetic advancement. The year before, he delivered a talk on the future of molecular biology at an event to celebrate the 75th anniversary of Caltech. He spent months preparing his lecture, mulling over the ramifications of humanity acquiring the keys to its own inheritance. On October 26, 1966, sporting a bow tie, Sinsheimer walked to the podium to warm applause, and addressed his "fellow prophets" in the audience. The title of his lecture was "The End of the Beginning."

He began by recalling his travels through the breathtaking canyons and landscapes of Arizona and Utah, where the sands of time formed layers of rock visible in cross section along the river gorges, revealing a billion years of geologic history. "On that immense scale," Sinsheimer said, "a foot represents the passage of perhaps 100,000 years. All of man's recorded history took place as an inch was deposited. All of organized science a millimeter. All we know of genetics, a few tens of microns. If we remember that timescale, then what vision can seem too long?"[14] Then he said this:

> The dramatic advances of the past few decades have led to the discovery of DNA and to the decipherment of the universal hereditary code, the age-old language of the living cell. And with this understanding will come control of processes that have known only the mindless discipline of natural selection for two billion years. And now the impact of science will strike straight

> home, for the biological world includes us. We will surely come to the time when man will have the power to alter—specifically and consciously—his very genes. This will be a new event in the universe. The prospect is to me awesome in its potential for deliverance or equally, for disaster.

Sinsheimer's mesmerizing words envisioning a future of human genetic modification predated the recombinant DNA revolution and genetic engineering, let alone the invention of DNA sequencing and the Human Genome Project. How might we change our genes, he asked rhetorically? Might we "alter the uneasy balance of our emotions. Could we be less warlike, more self-confident, more serene?" After two billion years, he said, "this is, in a sense, the end of the beginning."

Sinsheimer followed his speech with a powerful essay in *American Scientist* on "The Prospect of Designed Genetic Change."[15] "There is much talk about the possibility of human genetic modification—of designed genetic change," he wrote. "A new eugenics" was potentially "one of the most important concepts to arise in the history of mankind. I can think of none with greater long-range implications for the future of our species." *Star Trek* had just debuted on television in 1966, but no fancy hyperdrives or teleports were needed to conjure up visions of mankind boldly going where no one had gone before.

One hopeful idea was to treat diabetes by reanimating the insulin gene that, except for a few specialized cells in the pancreas, lies dormant in the human body. Viruses could be used to deliver the insulin gene to the necessary cells once scientists had sequenced and resynthesized it. Sinsheimer wasn't advocating for a utopian super race but for equality of opportunity. He wasn't pushing for Galtonian state-sponsored coercion but rather a voluntary improvement of the cognitively disadvantaged, such as 50 million Americans with an IQ of 90 or less. Should we "continue to accept the innumerable, individual tragedies inherent in the outcome of this mindless, age-old throw of dice," or instead "shoulder the responsibility for intelligent genetic intervention"? The stakes, Sinsheimer argued, were little short of astronomical:

> Copernicus and Darwin demoted man from his bright glory at the focal point of the universe to be merely the current head of the animal line on an insignificant planet. In the mirror of our newer knowledge, we can begin to see that in truth we are far more than another ephemeral form in the chain of evolution. Rather we are an historic innovation. We can be the agent of transition to a wholly new path of evolution. This is a cosmic event.

Sinsheimer's vision of "genetic change, specifically of mankind," was fueled by the successful elucidation of the universal genetic code. The beauty of the double helix had immediately suggested how DNA could replicate itself, each strand unzipped becoming the template for a new daughter strand. Kornberg won the Nobel Prize for identifying the key enzyme, DNA polymerase. But with the genetic material now reduced to atomic detail, the big question in biology became: what is the code that governs how the instructions inscribed in DNA are communicated and translated into proteins?

During the course of the 1950s, the work of Crick, Watson, Sydney Brenner, and others established the central dogma. A messenger RNA facsimile ferries instructions from the cell's data center (the nucleus) to the protein-manufacturing sites in the heartland (the cytoplasm). But what about the code itself? Proteins are made up of twenty different amino acids, whereas the DNA alphabet only has four letters. A two-base code would only yield a maximum of sixteen building blocks (4x4), whereas a triplet code (4x4x4) could in principle give rise to as many as sixty-four building blocks.

In 1959, Marshall Nirenberg, a biochemist at the NIH, developed a cell-free system to synthesize proteins in a test tube by mixing the raw ingredients—DNA, RNA, enzymes, and radioactively labeled amino acids. His colleague Bruce Ames felt the project was "suicidal." With Nirenberg traveling in California, his German student, Heinrich Matthaei, found himself alone in the lab after midnight on a Saturday morning (May 27,

1961). Thirty-six hours earlier, President John Kennedy, inspired by Alan Shepard's achievement in becoming the first American in space, asked Congress to commit to "landing a man on the moon and returning him safely to the earth."

Here in the late-night tranquility of an empty lab, Matthaei was poised to crack the first clue in the genetic code that governs life on earth, propelling the field of genetic engineering into orbit. He pipetted a synthetic strand of RNA made up entirely of just one base (uracil, U) into his cell-free solution. The resulting peptide was composed entirely of one amino acid—phenylalanine. Clearly some combination of U's provided the necessary code for phenylalanine. The first square in the 64-square genetic code bingo card—UUU—had been filled. Soon they had a second letter: CCC corresponded to proline.

That summer, Nirenberg delivered a lecture at a major conference in Moscow. His initial talk was attended by only a smattering of scientists, but Crick arranged for Nirenberg to give an encore performance in a plenary session. Nirenberg was heartily congratulated by Crick and other scientific legends afterwards and felt a bit like a rock star. An American literature student, who had spent the day touring art museums, was electrified hearing about Nirenberg's results from his roommate. That student, one Harold Varmus, would later win the Nobel Prize for cancer research and become the director of the NIH.

The following year, Crick and Watson received their Nobel Prizes. By this time, Crick had proven that the genetic code was indeed made up of 64 triplets. "We are coming to the end of an era in molecular biology," Crick said in his Nobel address. "If the DNA structure was the end of the beginning, the discovery of Nirenberg and Matthaei is the beginning of the end."

In August 1967, Nirenberg wrote a guest editorial for *Science* magazine, entitled "Will society be prepared?" The implications of the revolution in biochemical genetics, as he called it, and the prospect of "genetic surgery" were weighing heavily on him. Nirenberg believed that scientists were going to be able to reprogram cells—initially microbes, but eventually humans. And that made him nervous. He wrote:

> [M]an may be able to program his own cells with synthetic information long before he will be able to assess adequately the long-term consequences of such alterations . . . and long before he can resolve the ethical and moral problems which will be raised. When man becomes capable of instructing his own cells, he must refrain from doing so until he has sufficient wisdom to use this knowledge for the benefit of mankind. I state this problem well in advance of the need to resolve it, because decisions concerning the application of this knowledge must ultimately be made by society, and only an informed society can make such decisions wisely.[16]

The following year, it was Nirenberg's turn to win an all-expenses paid trip to Sweden. His students, one of whom was an ambitious physician named William French Anderson, hung a banner in his lab that read "UUU are great Marshall." Back home in Germany, however, Matthaei could only reflect on the Stockholm snub. Unlike Rosalind Franklin, he was still alive and eligible when the call came. But Nirenberg shared the stage with two others, and in Nobel math, four into three doesn't go.

By this time, some scientists were seeing a different side of genetic engineering—the concept of gene therapy. One of the first to do so was yet another Nobel laureate, Joshua Lederberg. The son of a rabbi, Lederberg graduated from Stuyvesant High School in New York at the age of fifteen. He was barely twice that age when, in 1958, he won the Nobel Prize, for discovering the transmission of genetic material between bacteria, including the process of transduction, involving phages. Sharing the prize that year were Lederberg's former supervisor, Edward Tatum, and George Beadle. (There was no mention of Lederberg's wife, Esther, who performed many of the crucial experiments and coauthored papers with her husband.) Lederberg went on to become the president of the Rockefeller University and a NASA consultant who coined the term "exobiology." Some believe he was the model for the hero in Michael Crichton's debut novel, *The Andromeda Strain*.

At a symposium on "The Future of Man" in London in 1962, Lederberg expressed sympathy with the "noble aims" of eugenics while noting it had been "perverted to justify unthinkable inhumanity." Advances in biology ultimately "could diagnose, then specify, the actual DNA composition of ideal man." But Lederberg proposed a new term, "euphenics," meaning the developmental engineering of organs as opposed to genetic engineering of the germline.[17]

In 1966, Tatum predicted that viruses could be used in "genetic therapy" via the introduction of new genes into defective cells of particular organs. He went on to describe what we now call *ex vivo* gene therapy. "The first successful genetic engineering will be done with the patient's own cells," he declared. The desired new gene would be taken from a healthy donor and transferred into the patient's cells. "The rare cell with the desired change will then be selected, grown into a mass culture, and re-implanted in the patient's liver."[18]

In a commentary for the *Washington Post* in January 1968, Lederberg launched a trial balloon for his idea of gene therapy—using viruses for vaccination. Drawing inspiration from Kornberg's demonstration of DNA replication in a test tube, Lederberg suggested that by screening enough natural viruses, it might be possible to isolate a virus that had naturally captured a medically important human gene, such as insulin or the gene encoding the missing enzyme in phenylketonuria.[19] He even considered "extracting DNA molecules that code, say, for insulin and chemically grafting these to the DNA of an existing tempered virus," forming the basis for virogenic therapy in man. Lederberg thought his idea of somatic gene therapy was more practical and palatable than genetic engineering, or "direct tackling of the germ line."[20]

In 1968, Nirenberg's medical student French Anderson, decided to add his name to the chorus of those advocating for gene therapy. "In order to insert a correct gene into cells containing a mutation, it will first be necessary to isolate the desired gene from a normal chromosome. Then this gene will probably have to be duplicated to provide many copies. And, finally, it will be necessary to incorporate the correct copy into the genome of the defective cell."[21] Anderson's ideas were too fanciful for the *New England Journal of Medicine*, which rejected the manuscript after some lively internal debate. One editorial member judged Anderson's proposal to be "a worthwhile adventure in pure speculation."

CHAPTER 10
THE RISE AND FALL OF GENE THERAPY

It took more than twenty years for French Anderson's worthwhile speculation to become a clinical reality. But the first tentative steps, albeit misguided, began shortly after his spurned manifesto. Stanfield Rogers, a physician at the Oak Ridge National Laboratory in Tennessee, had long been advocating the use of viruses to transmit genetic information. He had found that researchers handling the Shope rabbit papillomavirus* had lower levels of arginine than normal people, suggesting they were picking up the virus and supplemental activity of the viral enzyme called arginase, which breaks down the amino acid. Rogers had reported high levels of arginase in warts on the skin of rabbits infected with the virus, and speculated that the virus was a therapeutic agent in search of a disease. "The possibility of tying specific synthetic DNA information on to the genome of passenger viruses, thereby using viruses as a vector, could prove to be a useful technique," he suggested.[1]

Rogers got his chance after reading a report in the *Lancet* about a pair of young, mentally retarded German sisters who had a rare inherited disease called argininemia—excess arginine in the blood caused by arginase deficiency. Believing he could supplement the missing enzyme using the Shope virus, Rogers persuaded the girls' pediatrician to let him try a ludicrously

* The virus was named after Richard Shope, a Rockefeller University pathologist who, studying a flu outbreak in pigs in 1918, helped prove that influenza was caused by a virus, not a bacterium. In 1933, Shope injected himself with the eponymous virus.

premature experimental procedure—the first human genetic engineering experiment. In 1970, Rogers flew to Germany and injected small doses of the virus into the two girls, hoping to boost levels of the enzyme. There was no response. Later a third sibling received the virus, only to develop an allergic reaction. His reckless gene therapy adventure over, Rogers went back to studying plant viruses.

Two years later, Theodore (Ted) Friedmann and Richard Roblin published a commentary in *Science* entitled "Gene Therapy for Genetic Disease?"[2] Friedmann, a physician, is widely credited with coining the term "gene therapy." Friedmann was born in Vienna but fled with his family to the United States in 1938 to escape the Nazis. At the University of Pennsylvania, he attended lectures by Colin MacLeod, who with Oswald Avery had proven in 1944 that DNA was the genetic material. He later trained with Fred Sanger in Cambridge before joining the NIH.

Friedmann worked on Lesch-Nyhan syndrome, a debilitating, sex-linked genetic disorder in which affected boys suffer retardation, abnormal movements, and self-mutilation. Friedmann was able to correct cells from Lesch-Nyhan patients using gene transfer by replacing the DNA that codes for the key enzyme. The experiment was terribly inefficient—only about one cell in a million was corrected—because Friedmann was using a full genome's worth of DNA (this was years before the ability to isolate specific genes).

Friedmann admired the work of Renato Dulbecco, who had just discovered that a tumor virus did exactly what gene therapists wanted to do, "taking a foreign piece of genetic information, a foreign DNA, and inserting it into a cell and forever changing that cell."[3] Viruses could indeed be used to ferry normal copies of genes into cells carrying a broken version of the same gene. Friedmann helped popularize the concept of using modified viruses for gene therapy, while warning of the ethical dangers of pushing ahead too quickly. "Gene therapy may ameliorate some human genetic diseases in the future," he wrote. The idea of gene replacement therapy using viral vectors had just received a major shot in the arm.

This approach was exciting but why be content to just add a healthy gene, papering over the cracks in the genome as it were, rather than actually trying to repair the broken sequence? In 1978, the same year as he won the

Nobel Prize, David Baltimore offered one approach to this medical milestone. A patient with a blood disease like hemophilia or sickle-cell could be treated by transferring a normal gene into the patient's bone marrow stem cells that ultimately give rise to blood cells. This would allow a normal protein to replace (or be made alongside) the faulty protein and cure the patient's disease. "It is likely to be the first type of genetic engineering tried on human beings, and might be tried within the next five years."[4]

"The concept of repairing a defective gene such as the sickle-globin gene is appealing," wrote one physician, "however, existing technology cannot direct a site-specific recombinational event. Therefore, the concept of gene-repair in a genome as complex as that of man is for the moment impractical."[5] The author of those words was a UCLA hematologist named Martin Cline.

In 1979, Cline proposed treating patients with beta thalassemia with gene therapy, but a UCLA review committee insisted on additional animal experiments. Frustrated, Cline looked overseas, and in June treated two young women—a twenty-one-year-old at Hadassah Hospital in Jerusalem and a sixteen-year-old in Naples, Italy, a few days later. The process involved extracting some of the patient's bone marrow, transfecting the cells with the beta-globin gene, and infusing about 1 billion treated cells back to the patient following irradiation of their femur. Cline told the women that the chances of success were slim, but he felt compelled to try. "When do you consider animal experiments adequate?" Cline asked. "When do you feel ready for a transition [to man]? Here's a patient who has a life-threatening disease with a limited life expectancy and no options with modern treatment. Is now the time to try an experimental treatment?"[6]

In the opinion of the NIH and most of Cline's peers, the answer was emphatically no. Cline had taken it upon himself to conduct the first recombinant genetic engineering experiments on humans. Following censure by the NIH, UCLA's dean of medicine accepted Cline's resignation as chief of the oncology department in February 1981. He chastised Cline for conducting an unprecedented experiment on two patients without the necessary institutional approval. Although no medical harm had been done, he continued, "the freedom to conduct experiments of benefit to mankind is jeopardized by failure to act in accord with the relevant regulations."[7] Hematologist Ernest Beutler called the Cline episode tragic "because it interrupted the efforts of a highly talented, productive scientist who was in too much of a hurry to see patients benefit from the marvels of modern molecular biology."[8] Beutler softened his criticism when he judged that the patients were probably more at risk from the three hundred rads of ionizing radiation they received than the therapy itself.

Two years after the Cline affair, Bob Williamson, a prominent gene hunter in London at the time (and my PhD supervisor) argued in *Nature* that gene therapy, while not on the immediate horizon, "will be possible in the future, and it should be considered now, before the headlines break on us all."[9] But whereas Williamson preached caution, Anderson felt it would be unethical *not* to embark on human trials once issues of safety had been met. In 1984, he wrote:

> [A]rguments that genetic engineering might someday be misused do not justify the needless perpetuation of human suffering that would result from an unnecessary delay in the clinical application of this potentially powerful therapeutic procedure.[10]

Despite increasingly vocal opposition from anti-biotech campaigners led by Jeremy Rifkin, the realization of gene therapy had turned from a matter of *if* to *when*. "Egos and expertise will clash like cymbals as the technology of gene splicing keeps racing along so fast that it laps ethical debates about what it all means," wrote Jeff Lyon and Peter Gorner in *Altered Fates*.[11]

The year 1990 was an annus mirabilis for human genetics, notably the launch of a fifteen-year, $3 billion enterprise known as the Human Genome Project under the leadership of Jim Watson to build the definitive roadmap to identify the locations and identities of all genes, including those underlying the most devastating genetic diseases. In October 1990, I savored a taste of things to come as thousands of scientists traveled to Cincinnati,

Ohio, for the annual conference of human geneticists. Late one evening, with the World Series featuring the hometown Cincinnati Reds blaring in every hotel bar, UC Berkeley's Mary-Claire King hit a walk-off home run. In front of a standing-room-only crowd, she announced that she had mapped an errant gene, *BRCA1*, that when mutated increased a woman's risk of breast cancer.*

Identifying human disease genes was big news, and the genetic detectives commanding those search expeditions like King and Collins and Lander became scientific celebrities. Identifying the genes mutated in CF or DMD or hereditary breast cancer marked a transformation in medical diagnostics and immediately raised hopes for a successful drug or gene therapy. The genomics gold rush spread to Wall Street, but identifying a disease gene is just the start: it takes on average a decade and $1.3 billion to bring a drug to market, and even then success is not guaranteed.[12] The molecular basis of sickle-cell disease was identified in the 1950s but the disease is still incurable sixty years later. It took more than twenty-five years since the discovery of the CF gene in 1989 for Vertex Pharmaceuticals to develop a drug that successfully treated a subset of patients with a particular gene mutation.

Shortly before King's walk off in Cincinnati, a group of clinicians at the NIH took a major step on the path to gene therapy by treating a young girl with a rare genetic disease. If a disease is caused by a typo in the genetic code, then the most logical way to treat that disease is to introduce a normal copy of the same genetic sequence. If the disease was caused by a faulty nonfunctional gene, then why not just replace the gene? A gene transplant, if you will. After two decades of false starts and deliberation, gene therapy was about to get real. But as generations of gene therapists can attest, it is anything but easy.

---

* As I described in my first book, *Breakthrough*, King's quest to isolate *BRCA1* was thwarted by Myriad Genetics. Twenty years later, I served as the technical advisor for a film called *Decoding Annie Parker*, based on the true story of the first woman in North America to undergo *BRCA1* genetic testing. Helen Hunt played Mary-Claire King. Sadly, few of my suggestions were incorporated—the writers told me we weren't filming a *Nova* documentary. My name is buried at the end of the closing credits, right after Aaron Paul's guitar coach.

After three years of fierce debate, French Anderson finally won approval to start the first official gene therapy trial in the United States. Doctors and nurses gathered in the pediatric intensive care unit of the NIH Warren Grant Magnuson Clinical Center. It took all of twenty-eight minutes for the first history-making infusion to take place after the final paperwork was completed earlier that morning.

On September 14, 1990—12:52 P.M. to be precise—Kenneth Culver took a small syringe and injected some liquid into the left arm of four-year-old Ashanthi de Silva from Cleveland. Wearing a white top and turquoise trousers, Ashanthi was a model patient, distractedly sticking cartoon stickers on her doctors' lab coats. About one billion genetically modified T cells slowly flowed into her body. "She was wonderful, a lot calmer than I was," said lead investigator Michael Blaese.[13] Ashanthi suffered from a rare form of severe combined immune deficiency syndrome (SCID), caused by a deficiency of the enzyme adenine deaminase (ADA). Only about a dozen children are born with this recessive disorder each year in the United States. Ashanthi had been sick for most of her life—about as long as the bureaucratic wrangling over the trial.

This was Anderson's medical Mount Everest, a scientific summit he had wanted to conquer since his audacious manifesto was dismissed by the medical establishment two decades earlier. As a student, inspired by the double helix and Roger Bannister's four-minute mile, Anderson had two goals in life: "I was going to be in the Olympics and I was going to cure defective molecules."[14] His first goal had already come true: he was a team physician for the U.S. Olympic team in Seoul in 1988.[15] As Anderson, Blaese, and Culver exhaled after Ashanthi's treatment, Anderson could reflect on twenty-five years of hopes and dreams. "At long last, the great adventure has started," he said.[16]

Four months later, Anderson's team began treating another pioneer, ten-year-old Cynthia Cutshall. In order not to treat the two girls as guinea pigs, Blaese and Anderson continued to administer the standard treatment, PEG-ADA. Did the gene therapy work? Well, yes and no. A twelve-year follow-up on Ashanthi declared that about 20 percent of her T cells were producing the ADA enzyme.[17] More than twenty-five years after the trial began, Blaese said the amount of ADA produced in the treated cells was only about 15 percent of what had been expected. But the two girls are "both beautiful young ladies"[18] who happily invited Blaese to their weddings. Anderson was elated. "I eat and sleep and breathe gene therapy 24 hours a day," he told a *New York Times* reporter.

There was a giddy euphoria in gene therapy circles in the years following the NIH trial as more disease genes were identified, many amenable to genetic therapy. As the founding editor of *Nature Genetics*, I was down for the ride. More and more manuscripts arrived in our Washington, DC, office touting advances in gene therapy. Ted Friedmann, whom Horace Freeland Judson called "gene therapy's most ardent advocate," wrote a review I entitled "A Brief History of Gene Therapy." Friedmann declared that the first phase of human gene therapy—the emergence and acceptance of the general concept—was over. "We are now in an explosive second phase—one of technical implementation."[19]

In March 1994, I took the train up to Philadelphia to attend a press conference led by a rising star in the field. James Wilson of the University of Pennsylvania walked onto a makeshift stage in his pristine lab coat, flanked by two colleagues, to discuss his latest study that we were publishing the next day. Wilson's team had treated a patient with an inherited form of coronary artery disease by removing part of her liver, treating it with a recombinant virus, and restoring the liver. The woman's LDL cholesterol levels fell by 25 percent. Among the reporters in attendance was *New York Times* Pulitzer Prize–winner Natalie Angier. Her front-page story said Wilson's study was "the first to report any therapeutic benefits of human gene therapy."[20] It happened to be April Fool's Day.

The swashbuckling pace with which investigators were publishing results made some observers nervous. Harold Varmus, then director of the NIH, commissioned a report on gene therapy. "Today, the announcement of a [disease] gene being discovered is tantamount to the belief that gene therapy exists for the condition. We've seen an extrapolation from hope to hype. In the long run, this will be destructive to basic clinical science," said Varmus sternly. The report criticized the rush to clinical trials without sufficient understanding of disease pathology, the low frequency of gene transfer, and overselling of results leading to "the mistaken and widespread perception that gene therapy is further developed and more successful than it actually is."[21]

Nevertheless, by 1999, Friedmann and his fellow gene therapists were euphoric about the future. Despite the efforts of critics who had "fomented mistrust and misunderstanding of the goals and techniques of gene therapy," the inevitability of the science had been established "long before it was able to provide truly believable clinical benefits."[22] Friedmann continued: "The success of the concept of gene therapy has been phenomenal and represents a truly epochal new direction for medicine."

Even before the ink was dry on Friedmann's essay, those words rang hollow.

In 1984, a few months shy of his third birthday, Jesse Gelsinger fell into a coma while watching Saturday morning cartoons on television. After a lengthy hospital stay, he was diagnosed with a rare X-linked genetic disorder, ornithine transcarbamylase (OTC) deficiency. A missing enzyme results in an inability to process nitrogen (found in all proteins and other biomolecules) and a toxic buildup of ammonia. About fifty OTC deficiency babies are born in the United States each year; only half will live past the age of five.

Jesse's father, Paul, and his partner put Jesse on a low-protein diet to manage the disease. Fortunately he had a mild form of the disease, but if he forgot to take his pills (up to fifty per day), he could fall into another coma. In September 1998, Paul heard about an OTC deficiency clinical trial being conducted at Penn. But that Christmas, Jesse fell into a coma and almost died. Shortly after graduating from high school, Jesse volunteered to take part in the clinical trial as soon as he was eligible. The following June, the Gelsinger family flew to Philadelphia on Jesse's eighteenth birthday. They visited the usual tourist landmarks, with Jesse posing next to the statue of Rocky, arms aloft in a Minnesota Twins T-shirt.

The Penn doctors explained that the purpose of a phase I trial was to test safety, not to expect any clinical benefit. There would likely be side effects such as flu-like symptoms as Jesse's immune system ramped up to attack the virus. The doctors also explained that while OTC was rare, affecting one in 80,000 people, there were at least two dozen similar metabolic liver disorders, in total affecting one in five hundred people. Jesse would be a pioneer for thousands of other patients. Three months later, Paul took Jesse to the airport for his flight from Arizona to Philadelphia. "Words cannot express how proud I was of this kid," Paul said. "Just eighteen, he was going off to help the world."[23]

The OTC trial was directed by James Wilson, the founding director of Penn's Institute for Human Gene Therapy. Because Wilson had founded a biotech company, Genovo, he was not allowed to have direct contact with the patients. On Monday September 13, 1999, Jesse received his first infusion of a recombinant adenovirus containing a normal copy of the OTC gene. As planned, he received the highest dose among the eighteen volunteers. He soon developed a fever but that was not unexpected. Jesse spoke to his father by phone that evening. It was the last conversation they would have. The following morning, Jesse had developed jaundice and his ammonia levels were spiking. They informed Paul and Wilson. Over the next two days, Gelsinger's kidney and liver started to fail. He was put on an artificial lung to help his breathing. When Paul finally got to Jesse's bedside, he didn't recognize his son because Jesse's face was swollen beyond recognition. He had suffered irreparable brain damage.

With seven of his siblings and spouses in attendance, Paul held a brief bedside ceremony for his son. At 2:30 P.M. on September 17, physician Steve Raper shut off the ventilator and pronounced Jesse dead. "Goodbye Jesse, we'll figure this out," he said. The grim news reached the press two weeks later. "Teen Dies Undergoing Experimental Gene Therapy" was the *Washington Post* headline.[24] Ironically, Wilson had never met Paul or Jesse Gelsinger.

In early November, Paul led about two dozen mourners on a hike to one of Jesse's favorite places—Mount Wrightson, close to the Mexican border. After Paul shared some memories of his son, Raper read a poem by Thomas Gray:

> Here rests his head upon the lap of earth
> A youth to fortune and to fame unknown.
> Fair Science frowned not on his humble birth,
> And Melancholy marked him for her own.

Moments later, Paul and other mourners scattered Jesse's ashes into the Arizona air.[25]

A short time later, Wilson flew out to meet Paul Gelsinger for the first time. Sitting on Gelsinger's back porch, Wilson shared Jesse's autopsy results and told Paul that he was just an unpaid consultant to Genovo. Initially supportive, Gelsinger learned that the Penn team had seen fatalities in animals (albeit receiving much higher doses of virus) and adverse events in some patients before Jesse's fateful enrollment. Gelsinger eventually filed a lawsuit against Wilson and his two senior colleagues that was settled out of court. Wilson's gene therapy center was disbanded, and he was forbidden from running any clinical studies until 2010. In the formal scientific report published by Raper, Wilson, and colleagues, much of the blame was cast on the viral vectors that triggered Jesse's fatal cytokine storm.[26]

Writing shortly after the Gelsinger tragedy, molecular biologist Peter Little reached a grim diagnosis:

> I suspect the judgment will be that we were arrogant and came extremely close to using humans as experimental animals; we knew too little and expected too much, and the expectation of success was used to roll back objections.[27]

In his book *The Gene,* Siddhartha Mukherjee condemned the OTC trial as "nothing short of ugly—hurriedly designed, poorly planned, badly monitored, and abysmally delivered. It was made twice as hideous by the financial conflicts involved; the prophets were in it for profits."[28] Wilson, when interviewed for *The Gene* documentary, said: "I'll think about [the tragedy] frequently until the day I die. I don't know what else to say."[29]

The Gelsinger tragedy was soon compounded by news from across the pond. In 2000, French physician Alain Fischer of the Necker Hospital in Paris held a press conference to mark the preliminary success of gene therapy in two infants with an X-linked form of SCID. Fischer was using a retrovirus vector to shuttle the healthy gene into the patients' blood stem cells. But two years later, two boys were diagnosed with a form of leukemia that was traced back to the therapy itself, and one eventually died. Retroviruses work by integrating directly into the host genome, like hiding the joker in a pack of cards. In most cases, the integration event is harmless, but in rare instances it can trigger a cancerous event. Fischer's viral vector was finding a comfortable landing spot in the host DNA, inadvertently switching on an adjacent oncogene with devastating results. In 2003, the FDA responded by halting the use of retroviruses in the United States.

As Judson observed, despite hundreds of millions of dollars lavished on hundreds of gene therapy trials involving thousands of patients and volunteers, "new hopes cyclically turned to ashes, dramatic claims to sad farce."[30] Gene therapy itself was on life support.

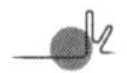

With some sober reflection, many of the setbacks could be understood. After all, viruses did not evolve simply to be used at our beck and call as delivery drones. As one gene therapy expert said: "We underestimated the fact that it took billions of years for the viruses to learn to live in us—and we were hoping to do it in a five-year grant cycle."[31] There was also the complication of our immune system, which is designed to combat foreign agents such as viruses. The human body isn't going to automatically give billions of recombinant viruses a pass just because they mean well.

It has been a long haul back to respectability and success for gene augmentation therapy. The roller-coaster ride follows the Gartner Hype Cycle: the inflated expectations of the 1990s, the trough—or abyss—of disillusionment at the turn of the century; followed by the slope of enlightenment.

What's been holding up the field is not a lack of suitable targets—we have an encyclopedic catalogue of thousands of eligible Mendelian genetic diseases—but the capability of delivering the therapeutic gene safely and effectively. Researchers had to go back to the drawing board, focusing on viral delivery and safety. Two new candidates emerged as reliable, adaptable, and effective delivery vehicles for a swath of gene therapy (and genome editing) indications: adeno-associated viruses (AAV) and lentiviruses.

AAV was discovered by accident in the mid-1960s as a contaminant of an adenovirus preparation. The attraction of AAV comes from its barebones structure—it is the tiny house of viruses, a protein shell that can carry a small gene cargo. The virus is very safe—about 90 percent of humans have been exposed and infected by AAV without knowing it. Wilson's group, clutching a financial lifeline from GlaxoSmithKline, got to work. There were only six known varieties when Guangping Gao in Wilson's team set out.[32] But at the end of 2001, he presented Wilson with a bounty of novel AAVs he had isolated from monkeys. More than one hundred forms of AAV are currently known. "Penn's viral vector center became the Amazon of AAV," observed science journalist Ryan Cross.[33]

Why so much interest in this tiny virus? AAV naturally carries just two genes, REP and CAP, which encode proteins that make up an icosahedral capsid coat. Fully clothed, the virus is just twenty-five nanometers in diameter, holding a payload of single-stranded DNA about 5,000 bases in length—enough for a small therapeutic gene. And unlike retroviruses, AAV does not integrate into the host genome; that means it will dilute out over time as the cells it infects divide.

Despite their popularity, there's room for improvement, says UC Berkeley's David Schaffer. "We need better viruses. Viruses did not evolve in nature to be used as human therapies."[34] Treating spinal muscular atrophy (SMA) patients, for example, has required the highest dose of AAV used in a human to date. Schaffer's team is evolving in the lab the amino acids on the AAV surface to create novel vehicles with improved targeting properties to the appropriate cells, such as the retina.[35] Jean Bennett was able to overcome the limitations of AAV2, which doesn't travel through the vitreous of the eye, by delivering it directly through the retina. Schaffer's group has evolved a new AAV that can penetrate the full surface of the retina by injecting into the vitreous.

Lentiviruses, the other emerging virus class, form a sub-family of retroviruses and includes HIV. The tenacity with which HIV can lurk in the T cells of a patient illustrates their potential value as a modified gene delivery vehicle. Lentiviruses can infect both dividing and non-dividing cells and have a cargo hold double that of AAV. The first clinical trial using a lentiviral vector was conducted in 2005.

A decade after Gelsinger's death, an editorial in *Nature* signaled a renewed sense of optimism in gene therapy circles. "The pervading sense of disillusionment is misplaced," *Nature* stated.[36] It was time for researchers and biotech "to consider its successes with as much intensity as its setbacks." Wilson reflected on the lessons he had learned. "With what I know now, I wouldn't have proceeded with the study," he said. "We were drawn into the simplicity of the concept. You just put the gene in."[37] Carl Zimmer penned a story for *WIRED* with a striking hero image of two viruses: on the left was the adenovirus that "laid waste" to Wilson's career. On the right, the AAV, the bright new hope of gene therapy that could bring Wilson "redemption."[38]

In 2015, Friedmann and Fischer were awarded the Japan Prize for their contributions to the gene therapy field. Friedmann opened his award lecture in Tokyo by showing a picture of the serpent-entwined Rod of Asclepius, the ancient Greek symbol of medicine. Next he substituted the snake with the double helix—the repository of all genetic information. "We would like to think that knowledge of this molecule is going to markedly change the way we understand disease and the way we treat disease," he said.[39] As an example, Friedmann paid tribute to a female physician in Philadelphia who was pioneering a method to deliver a gene therapy directly into the retina of patients with a rare genetic form of blindness. Those early results, Friedmann proclaimed, were of "biblical proportions."

Indeed, they were little short of miraculous.

CHAPTER 11
OVERNIGHT SUCCESS

The world got a prime time glimpse of the renaissance of gene therapy on a 2017 episode of the television show *America's Got Talent*. Christian Guardino, a genial sixteen-year-old from Long Island, New York, stunned Simon Cowell and the other judges with his rendition of the Jackson 5 hit, "Who's Lovin' You." While the audience cheered the teenager's amazing voice, the bigger story was out of sight. As an infant, Guardino was diagnosed with a form of Leber's congenital amaurosis (LCA type 2)—a genetic disorder that causes an inexorable degeneration of cells in the retina. A Fox News report made light of Guardino's medical ordeal. "When Christian Guardino was young, he learned that he would lose his sight. Fortunately, thanks to some gene therapy, he later regained the gift of sight. In the interim, he turned to music and stuck with it." The report irked geneticist Ricki Lewis, author of *The Forever Fix*. "Christian didn't just order up gene therapy like a side of fries," she grizzled.[1]

Indeed not. Christian was twelve years old when his parents heard about an experimental gene therapy being developed by Jean Bennett, a colleague of Wilson's at the University of Pennsylvania, and enrolled him. The results were almost literally off the charts. David Dobbs described the transformation: "Guardino could see. Everything that had posed an obstacle before—light and dark, steel and glass, the mobile and the immovable—now brought him pleasure. The world had opened before him."[2] That pleasure was shared with millions of TV viewers when

Guardino won something called the "Golden Buzzer." Bennett proudly shows the clip in her talks as Guardino disappears from view under a shower of confetti.

Bennett has dedicated her career to finding a therapy for LCA. In the early '80s, she'd spent time in the lab of French Anderson, who advised her to "go to Harvard Medical School, learn about the diseases you want to treat, go back to the lab and start working on them."[3] She followed his advice: at Harvard she met her future husband, Al Maguire, a retinal surgeon. In 1990, inspired by Anderson's historic gene therapy trial, Bennett and Maguire discussed the prospects of transferring genes to the retina. There was just one small problem: there were no known genes, no animal models, no natural history, no vectors, no surgical methods, and no outcome measures.

Bennett's naïveté proved to be a blessing. After moving to Penn in the early 1990s, she realized that "the retina is pretty cool and that it would actually be fun to look at retinal gene therapy."[4] Wilson and his colleagues were building new facilities and developing novel vectors. Eyes are "post-mitotic," meaning cells don't divide, which would stop the delivered transgene from becoming diluted. And there was little risk of triggering an immune response. (The eye is an "immune privileged organ," a mini isolation chamber with a blood barrier and no lymph system.[5]) The effects of the therapy could be tested in various ways and could be measured against the ideal control—the patient's untreated eye.

The pieces finally came together in 1997, when researchers identified the gene mutated in one of many forms of LCA. The defective gene, *RPE65,* was one of hundreds of mutated genes that constitute hereditary blindness. There are some 7 million blind people in the United States, including 700,000 children. With only about 1,000 LCA patients in the United States, tackling this disorder was a drop in the bucket. But Bennett had to start somewhere. She learned that the Penn veterinary college housed a breed of dogs with the same gene mutation. A blind four-year-old briard sheepdog mix named Lancelot joined the quest to cure LCA. (Lancelot's role was essential as the viral vector didn't work in mice.)

In dogs as in humans, the pace of the retinal degeneration is slow, giving the researchers time to assess the benefits of their treatment. Maguire injected the therapeutic *RPE65* gene, packaged in AAV, under the retina of Lancelot and two other dogs. Within a few weeks, Lancelot's demeanor changed: he began to see, watching and following the veterinary staff.[6] He became something of a canine celebrity, even visiting the United States Congress. He produced a large family with his sibling, Guinevere, and was eventually adopted by Bennett.

That early momentum stalled as the entire gene therapy community reeled from the Gelsinger tragedy. But in July 2005, Katherine High, a colleague at Children's Hospital of Philadelphia (CHOP), walked into Bennett's office and made her an offer she couldn't refuse: "How would you like to run a clinical trial at CHOP?" Five months later, Bennett and High addressed the Recombinant Advisory Committee (RAC) of the NIH. They were proposing to treat children—a controversial precedent for the gene therapy community. A pivotal moment came when patient advocates Betsy and David Brint recounted their daily struggle to help their youngest son, Alan, who has LCA, who needs the help of a dozen people to get through a day at school. Gene therapy, not to be overly dramatic, was Alan's only hope. The RAC voted unanimously to grant approval.

The first of a dozen patients in the phase I LCA2 trial were treated in Naples in October 2007, led by Francesca Simonelli. During the procedure, the retinal surgeon inserted a canula the width of an eyelash, through which the genetically modified virus was delivered. The procedure results in a localized retinal detachment, which typically flattens within a few hours.

Maguire warned his wife not to read too much into the early vision results. But one test gave her hope: she performed a pupillometry—a light-mediated reflex—on a patient one month after the injection. Bennett recalled: "When the retina is functioning, it sends a signal through the optic nerve to the brain, which then sends a signal back to the muscle that controls the iris and causes it to constrict. Nobody can constrict a pupil at will, so I couldn't contain myself when I saw this crystal-clear result."[7] She also called the test the bane of her existence, requiring hours measuring pupil diameters and analyzing spreadsheets strewn across her dining room table. But the results on the first three patients were unequivocal, and validated in the *New England Journal of Medicine.*[8]

Bennett's youngest patient was Corey Haas. He entered the hospital in 2008 walking with a cane, holding his parents' hands. After treatment, in a video that has done the rounds at medical conferences, Haas is asked to navigate a makeshift obstacle course coemprised of junk retrieved from Bennett's basement. With his treated eye patched, Corey couldn't stop bumping into obstacles. But when the patch was switched, he navigated the maze without difficulty. Soon he was riding a bike, playing video games, and throwing a baseball, just like any healthy nine-year-old boy. Bennett's other patients could suddenly see the moon, the stars, and their own faces. ("Mamma mia!" cried an Italian patient staying with Bennett on seeing her reflection for the first time.) Parents started clamoring for their child to have their other eye injected.

With no more dogs readily available, Bennett donated $10,000 to the lab of veterinary ophthalmologist Kristina Narfstrom to breed some new lines. Bennett picked up the puppies in person and adopted Venus and Mercury. Experiments showed little risk of immune response, allowing Bennett and Maguire to inject both eyes in patients in the Phase III trial. After twelve months, the control group was allowed to cross over to receive the drug as well. By all measures, the response a year later was as good as with the initial cohort of patients.[9]

In 2013, High cofounded Spark Therapeutics, and licensed Bennett and Maguire's original patent. Four years later, an FDA advisory committee unanimously recommended the approval of therapy, Luxturna. Final FDA approval for the first *in vivo* gene therapy drug* came four months after approval of Kymriah, Novartis's CAR-T cell therapy for acute lymphoblastic leukemia. In March 2018, hospitals in Boston, Miami, and Los Angeles administered Luxturna as a prescription drug for the first time. With a hefty list price of $425,000 per eye, it was no surprise that Spark should attract a big pharma suitor. Roche acquired Spark for $4.3 billion.

* Gene therapies can be classified as *in vivo* or *ex vivo*. For *in vivo*, the therapy is administered directly into the patient's body. For *ex vivo*, cells are removed from the body, treated in the lab, and then readministered.

Delivering a keynote lecture at a conference in 2018, Bennett ended her talk on a curious note: she had just received a phone call from French Anderson, the father of gene therapy. Anderson had just been released on parole after serving twelve years of a fourteen-year prison sentence, having been convicted of sexual molestation of the teenage daughter of his senior lab director.[10] Bennett insisted there was "abundant evidence of his innocence" and said he hoped to return to the field he'd helped launch almost three decades earlier. "I hope you'll welcome him back," she said.

That's unlikely, but the significance of Bennett's trailblazing work is not. Hundreds of patients have received gene therapy for ocular diseases since the LCA trial. Bennett hopes that her model will help to develop therapies for other blindness disorders, named after notable ophthalmologists such as Karl Stargardt, Friedrich Best, and Charles Usher. Bennett and Maguire continue their research at the new Center for Advanced Retinal and Ocular Therapeutics, or CAROT for short.

Luxturna was not the first gene therapy to make it through the roller-coaster ride to approval. The first drug was actually Glybera, which was approved in Europe for an ultra-rare disorder called lipoprotein lipase deficiency, resulting in patients essentially having heavy cream in their bloodstream. Manufactured by UniQure, the therapy earned the inglorious mantle of "world's most expensive drug." The European market could not support a $1.5 million price and it was subsequently withdrawn, having been administered to just one patient in Berlin.

But drugs like Strimvelis for severe combined immune deficiency (SCID) and Zolgensma for spinal muscular atrophy (SMA), coupled to success in developing CAR-T immunotherapies (Kymriah and Yescarta), signal a new era for gene and cell therapy. There were more than nine hundred registered gene therapies with the FDA at the beginning of 2020. UC Berkeley's David Schaffer put it nicely: "After twenty years, gene therapy is an overnight success."[11]

In late 2017, I invited Shakir Cannon, an African American patient advocate, to write a personal essay for the inaugural issue of a new science journal dedicated to all things CRISPR.* I had heard Cannon speak at the first CRISPRcon meeting at Berkeley about his struggles with sickle-cell disease (SCD) and his hopes that CRISPR might one day prove an effective therapy, if not a cure. His personal motto was, "Any day without pain is a good day." He accepted my invitation, signing his email "Thankful." After a few weeks, I reached out for an update. My emails went unanswered.

Then I heard the awful news. Shakir had died suddenly on December 5, 2017, of pneumonia, just thirty-four years of age. Shakir was one of 100,000 Americans and an estimated 20 million affected worldwide, primarily in Africa and parts of Asia. Together with other mutations in the beta-globin gene resulting in beta thalassemia, these are some of the most common genetic diseases on the planet. Each year 300,000 SCD babies are born.

SCD is a recessive disorder caused by the inheritance of a faulty gene from each parent. Hemoglobin, the protein that carries oxygen in our body, is made up of a quartet of peptide chains—a pair of alpha chains and a pair of beta chains. A tiny mutation, the switch of a T for an A in the beta-globin gene, produces a misformed protein that clumps together. This results in the red blood cells, normally a beautiful flexible biconcave shape, deforming to form rigid sickle-shaped cells that are prone to aggregate and block normal blood flow. Across Africa, SCD has various names—*Ahututuo, Chwecheechwe, Nuidudui, Nwiiwii*. Roughly translated, they mean "beaten up," "body biting," or "body chewing." About 30 percent adults with SCD experience debilitating pain every day, in some cases requiring heavy doses of prescription painkillers. "It's like having your hand slammed in a car door, but instead of it lasting for a few seconds, it lasts for weeks," said one patient.

Shakir's short life is a case in point. At age three, Shakir had a stroke, which he overcame with years of physical therapy. Once a month he skipped school to have a blood transfusion. Every night, he received a subcutaneous injection of a drug called Desferal. A portacath was implanted in his chest to help the injections (classmates joked it was his third nipple). He received

---

* *The CRISPR Journal* was launched in 2018, published by Mary Ann Liebert Inc., with Rodolphe Barrangou as chief editor.

growth hormone injections because of his short stature. While attending a basketball game with a friend, Shakir experienced a pain crisis so severe he could barely breathe or talk. His parents rushed him to the emergency room at Albany Medical Center, where he stayed for a week.[12] Despite this, Shakir cofounded the Minority Coalition for Precision Medicine and accepted an invitation from the Obama administration to present at the White House.[13]

While the average SCD patient lifespan in the United States is about forty, in Africa most sickle-cell children don't reach double digits. So why is this deadly disease so prevalent? Sickle-cell carriers (one mutant, one normal beta-globin gene) are inherently resistant to malaria, which kills 500,000 people in Africa each year. This "heterozygous advantage" provides a life-saving selective advantage like a built-in vaccine for SCD carriers and ensures that the sickle-cell gene continues to thrive in areas of the world prone to malaria.

"Blood is by far the most common cell in the body, so it's not surprising that critters want to feed on blood," says Merlin Crossley, a geneticist at the University of New South Wales in Sydney. There are about 100 trillion mosquitoes spread across most inhabited areas of the planet. Only a few species spread disease however, and of those, it is only the female mosquitoes that enjoy sucking human blood. In so doing, Crossley's critters, chiefly *Anopheles gambiae*, transmit parasites such as *Plasmodium falciparum*, which causes malaria.

Recent analysis suggests that the SCD mutation first arose in a newborn in West Africa some 7,300 years ago, during the African humid period.[14] That infant unknowingly possessed a stealthy superpower: (s)he would be resistant to malaria, greatly increasing the chances of reaching reproductive age—and a 50:50 chance of passing the same trait onto his or her children. Over the subsequent 250 or so generations to the present day, that single mutation has spread around the globe, especially in Africa, the Mediterranean and Asia, areas devastated by malaria.* Today it is estimated that

* There is good evidence to suggest that the SCD mutation has actually occurred spontaneously on four different occasions in different populations.

5 percent of the world population carries the sickle-cell trait or another mutation in the beta-globin gene.

We know more about SCD than almost any of the other 6,000 or more documented genetic disorders. The disease was first described by Chicago physician James Herrick in 1910.[15] Herrick reported "freakish" cells in a blood sample from "an intelligent negro of 20"—actually Walter Noel, a dental student from Grenada. Subsequent case reports with multiple affected siblings pointed to a genetic basis. The editors of the *Journal of the American Medical Association* made a striking pronouncement in 1947:

> The most significant feature of sickle cell anemia is not its characteristic bizarre deformation of erythrocytes but the fact that it is apparently the only known disease that is completely confined to a single race . . . [SCD] is independent of either geography or customs and habits. Its occurrence depends entirely on the presence of Negro blood, even though in extremely small amounts.[16]

Two years later, Nobel laureate Linus Pauling discovered that extracted red blood cell proteins from SCD patients ran differently in a gel than healthy controls, predicting (correctly) that the hemoglobin molecule in SCD patients (HbS) carried two additional positive charges. Pauling proposed that SCD was "a disease of the hemoglobin molecule"—the first molecular disease. His prediction was confirmed when a South African physician scientist, Anthony Allison, showed that sickle-cell carriers were resistant to the malaria parasite.[17]

Vernon Ingram* was a German national who immigrated to the United Kingdom as a teenager one year before World War II. In 1957, Ingram was able to zoom into the amino-acid sequence of the globin chains to pinpoint the molecular aberration in SCD predicted by Pauling—a single amino-acid alteration (glutamic acid to valine) in the beta globin chain. Ingram made his breakthrough at the Cavendish Laboratory in Cambridge, where Crick and Watson had assembled the double helix four years earlier,

* Ingram's given name was Werner Adolf Martin Immerwahr.

although Ingram's lab was a converted bicycle shed.[18] A decade later, Makio Murayama showed how the appearance of that rogue valine residue creates a hydrophobic surface that enables the sickle chains to clump together forming stiff polymers.

DNA sequencing was not invented for another twenty years, but once fellow Cavendish biochemist Fred Sanger developed his eponymous Nobel-worthy sequencing method, it was natural to work out the sequence of the HbS gene mutation. One year after the mutation was genetically spelled out in 1977, Yuet Kan and Andrée Dozy reported the use of a polymorphic DNA marker adjacent to the beta-globin gene to perform prenatal genetic diagnosis for pregnant women with a family history of SCD.

Despite knowing the molecular basis of SCD for more than sixty years, a treatment has remained elusive, dreams of a cure a mirage. That may be about to change: a Bay Area biotech called Global Blood Therapeutics in 2019 had a drug called Voxelotor approved by the FDA; it binds to the mutant hemoglobin and increases its oxygen affinity, although further studies are needed to ensure the drug significantly reduces pain crises.

Several gene therapy biotechs have set their sights on treating SCD and beta-thalassemia. With a large proportion of beta-thalassemia patients lacking any beta-globin production, there are two main strategies for genetic therapy. The most straightforward would be to replace the defective gene by restoring copies of the healthy beta globin gene. On Boxing Day, 2017, a twenty-eight-year-old African American, Jennelle Stephenson, arrived at the NIH clinical center for the beginning of a major clinical trial. Asked to describe the pain she experiences on a scale of 1–10, she said it went beyond a 10, a sharp stabbing pain affecting her shoulders, back, elbows, arms, cheekbones—her entire body.[19] Once she collapsed in a hospital emergency room, only to be accused by hospital staff of faking her distress to get narcotic drugs.

A team led by hematologist John Tisdale purified Stephenson's stem cells and inserted the correct version of the beta globin gene. To ferry the gene into her cells, Tisdale and his collaborators at Bluebird Bio in Kendall Square chose a modified lentivirus vector. After chemotherapy to cripple her immune system, Stephenson received an infusion of her modified stem cells. (Bluebird's first SCD patient, a French teenager, was treated with LentiGlobin three years earlier.)[20]

A few months later, Tisdale compared magnified images of Stephenson's blood. Before, the sickled blood cells are plainly visible like a biology textbook photo. Tisdale meticulously scans the new slide but comes up empty. "Her blood looks normal," he says. Stephenson is able to run, swim, and take judo classes, experiencing an endorphin high for the first time. NIH director Francis Collins, whose interest in the genetics of blood diseases traces back to meeting a sickle-cell patient as a young medical student in the 1970s, told *60 Minutes*: "I've got to be careful, but from every angle that I know how to size this up, this looks like a cure."[21]

Several other strategies are being tested to treat SCD patients, including tinkering with the regulatory switches of globin production. During pregnancy, fetuses produce a special form of hemoglobin, appropriately called fetal hemoglobin (HbF). This form has a higher affinity for oxygen than adult hemoglobin, all the better to pull oxygen from the maternal bloodstream. HbF is made up of four globin chains—a pair of alpha chains that also exist in adult hemoglobin, along with two gamma (γ-) chains. A few days after birth, production of γ-globin shuts down, replaced by beta globin. Reactivating the fetal form of hemoglobin in SCD and beta thalassemia patients offers a promising strategy. But where's the switch?

In a 1948 study, Brooklyn pediatrician Janet Watson observed fewer sickle cells in two hundred "Negro newborn infants" than in older patients. "Fetal hemoglobin thus appears to lack the sickling properties of adult hemoglobin," she concluded.[22] In the 1950s, physician Richard Perrine was puzzled that some SCD patients at the Aramco oil company in Saudi Arabia had low levels of anemia, with only mild pain episodes.[23] He too suspected an increase in levels of fetal hemoglobin compensated for the disease. Decades later, Collins began studying a benign genetic disorder called hereditary persistence of fetal hemoglobin (HPFH) in which, as the name suggests, fetal globin production curiously persists after birth into adulthood. Collins discovered a pair of mutations in HPFH patients located just in front, or upstream, of the start of the γ-globin gene. Collins had effectively located the site of a regulatory switch that tells the γ-globin gene to shut down. Disrupt that signal and γ-globin stays on, offering a lifeline to thalassemia and SCD patients.

It took more than twenty years to identify the genetic circuitry leading to the γ-globin switch. In 2008, researchers including Vijay Sankaran and Stuart Orkin at Boston Children's Hospital conducted a genome-wide association study to identify gene variants associated with high levels of HbF. One of the biggest hits incriminated a gene for a zinc finger transcription factor called BCL11A, which governs this "fetal switch" by shutting down HbF by clamping onto the DNA in the regions identified by Collins's sleuthing two decades earlier. Sankaran and Orkin nominated BCL11A as an attractive therapeutic target, suggesting that "directed down-regulation of BCL11A in patients would elevate HbF levels and ameliorate the severity of the major β-hemoglobin disorders."[24]

How would one go about suppressing BCL11A? Disrupting the gene itself was the simplest route but that won't work: BCL11A regulates many other genes including some that are expressed in the brain—patients with inherited mutations in BCL11A are on the autism spectrum. Orkin and Daniel Bauer came up with an alternative strategy. They identified a critical regulatory element that enhances BCL11A activity; if disrupted, BCL11A is switched off specifically in red blood cell precursors. By crippling the enhancer sequence, Orkin's team would silence BCL11A selectively in the cells that give rise to mature red blood cells. In turn, this would remove the brake on HbF production while simultaneously dialing down sickle globin production.

In May 2018, twenty-one-year-old Emmanuel "Manny" Johnson Jr. became the first patient treated in a clinical trial conducted at Dana Farber Cancer Institute, led by Orkin's colleagues David Williams and Erica Esrick. The Boston team isolated stem cells from Manny's blood and, using a modified lentivirus, engineered the genes so that when the hematopoietic stem cell (HSC) becomes a red blood cell, the gene switch is automatically turned on. Williams's team uses RNA interference, a Nobel Prize–winning technology. Following chemotherapy to allow the replacement cells to take hold, Manny's modified cells were infused intravenously, setting up shop in his bone marrow, pumping out healthy red blood cells.

After seventeen years of monthly blood infusions—Manny suffered a stroke at the age of four—Manny is hoping for a cure not just for himself but also his younger brother Aiden, who has the same disease. "I'm doing this so that my brother might not need all the years of treatment I've had to go through," he says. Six months later, Williams shows Manny before-and-after photographs of his magnified blood showing a frame of healthy round blood cells. "Oh wow, I've never seen this before, this is fantastic."[25] Manny hasn't needed a blood transfusion in the nine months since the therapy. Williams hopes for similar results in his next patients, including twenty-six-year-old Brunel Etienne Jr., who started chemotherapy shortly after attending the Super Bowl. The tickets were a gift from New England Patriots star Devin McCourty, who has a relative with SCD.

Early success stories like these from medical centers in Boston or Stanford or the NIH are wonderful, but have little relevance for the millions of sufferers in Africa, where the impact of SCD is felt the hardest. In October 2019, the NIH and the Gates Foundation announced a $200 million program over four years to produce affordable therapies for patients in Africa. The goal, Collins says, is "to make sure everybody, everywhere has the opportunity to be cured, not just those in high-income countries." But he readily admits: "This is a bold goal."[26]

In her final year at college, Shani Cohen gave birth to her first child, a beautiful baby girl named Eliana ("God has answered me" in Hebrew). Around eight months of age, Shani noticed that her daughter couldn't stand up in her crib like other infants her age. Eliana was eventually diagnosed with SMA type 2, often called "floppy baby," caused by a mutation in a gene called *SMN*. Shani was devastated, the only silver lining being that type 2 is not the most severe form of SMA. The motor neurons die, leaving patients basically paralyzed, many dying of respiratory failure, often before their first birthday. "Think of it as ALS [amyotrophic lateral sclerosis] but in infants," said Sean Nolan, the CEO at the time of AveXis.[27]

For Jerry Mendell, success treating patients with SMA and other forms of muscular dystrophy finally came after two decades of persistence. Following the Gelsinger tragedy, he built a group at Nationwide Children's Hospital in Columbus, Ohio, and set about developing new AAV vectors. In 2008, his colleague Brian Kaspar reported success with a newly engineered virus called AAV9 that had traversed the blood-brain barrier in mice.[28] Following more encouraging results in monkeys, Mendell believed they were ready to begin a clinical trial, but no drug companies wanted to take the risk. Kaspar cofounded his own company called BioLife (later renamed AveXis), and licensed the rights to Mendell's SMA program. But the high AAV doses required to ensure delivery to all the affected tissues, including the diaphragm and the heart, had experts worried. One even warned Kaspar: "You're going to kill someone—this is going to be Jesse Gelsinger all over again." Mendell wasn't about to stop now. "I'm sick of watching kids die," was his justification.

Mendell's trial launched in 2014 with fifteen children; within a few weeks, he saw an unwelcome spike in levels of liver enzymes in the first patient—warning signs of inflammation or liver damage. After consulting the FDA, Mendell changed the trial design, administering steroids before the gene therapy. Most complications disappeared. As reported in 2017, all patients showed rapid increases in motor function attributed to elevated levels of the SMN protein.[29] Most of the children can now sit unassisted, an unprecedented result. "What used to be called a science experiment, gene therapy, is becoming a reality," Nolan said, narrating a video of an SMA child sporting a Spider-Man backpack skipping out of hospital unaided. One of Mendell's first patients was Evelyn Villarreal, whose younger sister Josephine died before the therapy was available. At age three, Evelyn is dancing, doing pushups, and enthralling Senators and staffers on Capitol Hill.

The SMA success story proved irresistible to the Swiss pharma giant Novartis, which acquired AveXis for $8.7 billion in 2018. Zolgensma became the second FDA-approved gene therapy drug in May 2019, but how would Novartis price a potential one-time therapy? CEO Vas Narasimhan had hinted the price could be as high as $5 million. In that light, the final "responsible" list price—$2.1 million—seemed almost reasonable, but Zolgensma still claimed the dubious title of "the world's most expensive drug" in history. Even the SMA trade paper seemed taken aback: "At $2.125 million, a 60-minute intravenous infusion of Zolgensma costs more than a 2,000-square-foot apartment in Paris with a view of the Eiffel Tower, a brand-new 2019 Aston Martin One-77—among the fastest cars ever made—or a Cirrus Vision SF50 private jet."[30] Novartis allows patients to defray payments over five years, and will offer some sort of money-back guarantee if, for example, the patient dies.

Most experts however felt the exorbitant price was justified: it fell within the guidelines set by the Institute for Clinical and Economic Review, a drug pricing nonprofit, and it could work out cheaper than the full course of a rival drug, Biogen's Spinraza, over five to ten years. In Novartis's case, the price had to recoup not only the cost of developing the therapy—an estimated $550 million—but the billions of dollars to acquire AveXis. While there are 10,000–20,000 SMA patients in the United States, there are only about seven hundred children under two years of age, the cutoff for Zolgensma approval.

SMA patient groups welcomed the drug's approval as priceless. Nathan Yates, an economics professor who suffers from SMA, declared there should not be a price tag on life. He thought of his parents receiving the devastating news that their child has SMA and that there is no cure. "The price of Zolgensma seems insignificant now, don't you think?"[31] It wasn't to the parents of Eliana Cohen. Shani and her husband Ariel despaired as their health insurer denied coverage for Zolgensma and spurned their appeals. With Eliana's second birthday just a week away, the Cohens launched a desperate crowdsource campaign. Miraculously, they raised $2 million in just five days.

On Maundy Thursday 2019, the *New England Journal of Medicine* published a study in which a team at the St. Jude Children's Research Hospital in Memphis, Tennessee, successfully treated eight infant boys with "bubble boy" disease (X-linked SCID), the same disorder that Alain Fischer had treated two decades earlier.[32] "We believe that the patients are cured," said team leader Ewelina Mamcarz. "They're living normal lives, and they have normal, functional immune systems." They had returned home, were starting daycare, and making antibodies in response to vaccines.[33] The results were hailed as a complete fix for these patients, although investigators would be monitoring to ensure there were no adverse effects. "The data are extraordinary for every single patient," said Manny Lichtman, CEO of Mustang Bio, which licensed the rights to the therapy. But how will patients' families afford MB-107, which is designed to be a one-time treatment? Lichtman suggests a deferred payment scheme over a period of about ten years (assuming the therapy still works).

Gene therapies are expensive to develop and to manufacture, and the biotech companies that take the risks to develop these drugs deserve to recoup their investment. Skyrocketing prices are happening across all areas of the pharmaceutical industry, not just gene therapy. To address the soaring cost of drugs for rare diseases, some economists have floated the idea of healthcare loans, or "mortgages for medicines."[34] It is not clear whether current economics will sustain gene therapies and genome editing treatments. The arrival of new, improved therapies should lead to stiffer competition and reduced prices, but in practice, precision medicines and reformulated generics often justify higher prices. "Certainly, there must be a price that is too high," said philanthropist John Arnold. $5 million? $20 million? $100 million? "How should society answer that question?"

"I don't want my legacy to be the most expensive drugs in history," George Church told me. "We've brought down the price of the genome from $3 billion now to 'zero dollars.' That I'm proud of. I'm much more excited about that than I am about my contribution to expensive therapy."[35] This is not scalable. Five percent of live births have a Mendelian genetic disorder—the long tail of thousands of rare or orphan diseases. "We're not going to be spending $2 million on 5 percent of births!" Church says. He estimates that the total cost, including opportunity losses and caregiver costs, is a catastrophic $1 trillion worldwide per year, not to mention the collective pain and suffering.

To appease the sticker shock over its Zolgensma pricing, Novartis executives came up with a lottery. In December 2019, the company announced it would offer one hundred free doses of Zolgensma annually—four names drawn every fortnight—for patients outside the United States. Winners must undergo testing for AAV9 antibodies before AveXis sends the magic dose to the relevant hospital. "Imagine parents putting a child in a draw every two weeks to see if their life can be saved," sighed Lucy Frost, mother of an SMA child. "I think it could have been done much better."[36]

The molecular arsenal to combat cancer and tackle genetic diseases is expanding far beyond gene therapy. Cell therapy, RNA interference, and phage therapy all show promise. Former President Jimmy Carter is the poster boy for CAR-T therapy. A pair of FDA approved drugs, Kymriah and Yescarta, have produced near-miraculous results in some patients, although the effects are short-lived in many others.

Recently, an ancient therapy dating back more than a century has returned to the headlines. Bacteria, as we have seen, evolved their CRISPR defense systems to nullify phage invasions. But in diseases caused by antibiotic-resistant bacteria, we can weaponize these phages to become a new form of precision medicine.

In 1915, physician Frederick Twort discovered a bacteria-killing extract that he said contained an "ultra-microscopic virus."[37] Around the same time, Félix d'Hérelle at the Pasteur Institute was studying dysentery among French cavalry when he noted the bacteria in one of his cultures "dissolved away like sugar in water." He suspected an invisible microbe, "a virus parasitic on bacteria"[38] and coined the phrase bacteriophage (from the Greek *phagin*, "to eat"). D'Hérelle later performed the first successful phage therapy experiment, treating dysentery patients with phage isolated from another patient's stool samples. But the reaction of most of his peers ranged from indifference to scorn. D'Hérelle forged a collaboration with George Eliava in Tbilisi, the capital of the Republic of Georgia, who had independently discovered phages that kill cholera samples. Founded in

1923, the Eliava Institute in Tbilisi became the last refuge for decades for phage therapists.[39]

In September 2017, fifteen-year-old cystic fibrosis patient Isabelle Holdaway underwent a lung transplant at the Great Ormond Street Hospital in London. Although successful, pockets of antibiotic-resistant bacteria seized her liver and her surgical wound. Isabelle's health deteriorated as bacterial nodules broke through her skin. Her doctor, Helen Spencer, feared the worst. Isabelle's mother suggested a Hail Mary pass—phage therapy. The medical team contacted Graham Hatfull at the University of Pittsburgh, who had amassed a trove of 15,000 phage strains, housed in a pair of six-foot-tall freezers.[40] Hatfull identified a trio of phages from his subzero stockpile, named Muddy (isolated from a rotting eggplant),* BPs (from a storm drain), and ZoeJ (in a soil sample).[41] In June 2018, Isabelle received her first infusion of about one billion phages. Within six weeks, her liver infection had cleared up and the skin lesions were under control. A similarly miraculous outcome occurred in San Diego, when a phage cocktail saved the life of Tom Strathdee, who suffered a serious multidrug-resistant bacterial infection.

Gene therapy is on course to become a mainstream part of 21st-century medicine. Novartis bought facilities to manufacture the large quantities of engineered virus to deliver Zolgensma and other therapies—patients with a neuromuscular disorder require ten times more vector than needed for a localized disease in the eye. Sarepta Therapeutics licensed the rights to two other muscular dystrophy gene therapy programs developed by Mendell's team. "Our goal is to make Columbus the center of the universe for gene therapy," said Sarepta CEO Ed Kaye.[42] Meanwhile, Amicus Therapeutics, led by CEO John Crowley (portrayed by Brendan Fraser in the Harrison Ford film *Extraordinary Measures*), licensed ten programs for lysosomal storage disorders. Spark Therapeutics is one

* Scientists like to show off their warped sense of humor when it comes to naming genes in certain species (notably fruit flies) or, it turns out, newly discovered viruses. Examples (courtesy of Amanda Warr) include CaptnMurica, IceWarrior, Heffalump, PuppyEggo, BeeBee8, and Megatron. Rule #1 is literally: "Do not name your phage after Nicolas Cage."

of several companies developing a gene therapy for hemophilia. Fulvio Mavilio, an executive with Audentes Therapeutics, reported initial success of gene therapy for boys with the incurable disease X-linked myotubular myopathy (XLMTM). Children previously unable to sit up, let alone walk, can now take their first steps unaided, and speak after being taken off a ventilator.

Terry Flotte, the dean of the University of Massachusetts Medical School, is leading a trial for Tay-Sachs disease, which is most common in Ashkenazi Jews. Investigators used gene therapy to supply the missing enzyme in two children, injecting the virus into the brain either directly or via the spinal fluid. And in New York, Ron Crystal, another gene therapy veteran, is launching an ambitious trial to treat Alzheimer's patients, building on trailblazing work by Allen Roses two decades ago. Roses discovered an association between a rare version of the apolipoprotein E gene (*APOE4*) and the risk of Alzheimer's disease. Crystal's strategy is to deliver a different form of ApoE, which in principle will mop up the harmful variant. "If you're a mouse, we can cure you of your amyloid plaques," Crystal tells me. In San Francisco, one of David Schaffer's new-and-improved AAV vectors using direct evolution has been licensed by Adverum Biotechnologies for use in treating the wet form of age-related macular degeneration.

The renaissance of gene therapy was best illustrated in a 2019 cover story on Jim Wilson in *Chemical & Engineering News* on the twentieth anniversary of Jesse Gelsinger's death. The cover headline said it all: "The redemption of James Wilson."[43] His new operation, featuring a staff of two hundred working in multiple buildings at Penn, was more a production line than an academic lab. "Ten years ago, no one would touch Jim with a ten-foot pole. Now everyone is happy to work with Jim and gives him lots of money," said one biotech CEO.[44] Indeed, Wilson was finally able to commercialize the production of AAV vectors in a company, RegenXbio, which went public on the sixteenth anniversary of Gelsinger's death. The image of a sharply dressed Wilson and the company executives laughing as they were showered in confetti at the Nasdaq exchange dismayed Paul Gelsinger. "It really was all about the money," he said.

Despite this remarkable turnaround, gene augmentation is not a perfect therapy. The faulty gene still lurks in the patient's cells. More importantly, despite their excellent overall safety profile, Wilson and others have sounded the alarm about safety concerns using AAV vectors at higher doses.[45] In 2018, Wilson resigned as a scientific advisor to Solid Biosciences, because of concerns about toxicity linked to high AAV dosage. In June 2020, six months after its $3 billion acquisition by Japan's Astellas Pharma, Audentes disclosed the deaths of two young boys with XLMTM receiving the highest dose of the AAV8 vector. The children died from sepsis, and while pre-existing liver conditions might have been a factor, the FDA halted the trial.[46] The XLMTM tragedies are a humbling reminder that nature still has a say in what we can and can't do. Nicole Paulk, a gene therapy expert at UCSF, says we have to design viruses better so such extreme doses aren't necessary. (The boys who died in the trial received around four quadrillion—a million billion—viruses each.) "As scientists and clinicians," Paulk says, "we owe it to these boys to make sure this doesn't happen again."[47]

So gene therapy's renaissance is not yet complete. But that has not stopped the technology marching forward. What if we could build on the promise of the gene-editing technologies highlighted earlier and actually go into the cell to correct the corrupt code? What if we take our molecular scissors and repair some of the more than 75,000 mutations, deletions, and rearrangements that give rise to genetic diseases? What if, in the words of Chris Martin in fact, we could "fix you"—genetically speaking?

CHAPTER 12
FIX YOU

In April 2016, Sean Parker, the billionaire co-founder of Napster portrayed by Justin Timberlake in the film *The Social Network*, hosted a star-studded party at his $55 million Los Angeles home, which borders the Playboy Mansion. Among the celebrities in attendance were Tom Hanks, Peter Jackson, Sean Penn, "Mother of Dragons" Emilia Clarke, and Katy Perry. Musical performances included John Legend, the Red Hot Chili Peppers, and Lady Gaga slaying a rendition of "La Vie en Rose."[1] In the audience, Bradley Cooper was blown away, "like in that old Maxell cassette commercial," he said. That performance clinched her starring role in Cooper's remake of *A Star is Born*.

Parker was celebrating the launch of the Parker Institute for Cancer Immunotherapy (PICI), to which he personally pledged $250 million. The guest list also included medical talent from cancer centers in LA, San Francisco, New York, Philadelphia, and Houston. Two years after the launch party, PICI supported Carl June's team at the University of Pennsylvania Abramson Cancer Center in treating the first cancer patients in a CRISPR trial.[2] A Phase I trial is all about safety, but June is well aware that Chinese doctors are pushing ahead much faster. "We are at a dangerous point in losing our lead in biomedicine," June told the *Wall Street Journal*.[3]

The SINATRA trial* is an extension of June's pioneering work on CAR-T cells, arming the patient's own T cells to hunt tumor cells. Cancer patients are alive today because of these first-generation immunotherapies, but there is room for improvement. June's team performed three kinds of CRISPR gene edits: insertion of a gene into the patients' T cells that codes for a protein engineered to detect cancer cells while simultaneously removing the gene that interferes with this process. The third edit removes a gene that marks the T cells as immune cells and thus prevents the cancer cells from disabling them. Once edited, the manipulated cells are re-administered to the patient.

June's team published the initial results on three very ill patients who had endured multiple rounds of chemotherapy and bone marrow transplantation in February 2020.[4] The CRISPR therapy appeared safe—no toxicity, no cytokine storms, no neurological toxicity. June was relieved that there had been no adverse immune responses given the bacterial origin of Cas9. And while there was a low level of chromosomal rearrangements, he said that was similar to what astronauts who have been in space a few months endure.[5]

The journal *Science* treated the report as another major milestone. The cover headline read: HUMAN CRISPR.

Genome editing offers two enticing benefits for patients: first, it strikes at the root cause of a disease by correcting the code, repairing the faulty DNA sequence. Conventional drugs typically treat symptoms, not the root cause of a disease. Traditional gene augmentation therapy (as discussed in the previous two chapters) might compensate for, but does not repair, the underlying mutation. Second, the fix should in principle last forever, a one-and-done repair rather than chronic disease management. Of course, the CRISPR machinery still has to be delivered to the right tissues, safely,

* PICI likes to name its clinical trials after famous musicians. Along with Sinatra, it is also organizing trials named after Prince, Gustav Mahler, Cole Porter, Mozart ("Amadeus"), and country music star Tim McGraw, a longtime cancer research advocate.

without triggering the patient's immune system or causing any collateral DNA damage. None of those issues has been completely resolved, but the precision and safety of CRISPR is improving all the time, as can be seen in some of the early results in clinical trials.

A decade ago, the prospect of a universal cure for sickle-cell disease (SCD) didn't look hopeful. But now genome-editing strategies are showing promise, adding to various approaches—allogeneic stem cell transplant, gene therapy, derepressing γ-globin—discussed in the previous chapter, at least for patients in first-world countries.

In April 2019, Sangamo announced results for their first patient treated with an *ex vivo* gene-edited cell therapy, in which ZFNs disrupt the *BCL11A* enhancer in the patient's hematopoietic stem cells to boost γ-globin production. "I could not have imagined HbF this high in my wildest dreams," tweeted Sangamo alumnus Fyodor Urnov. "Thirty-one percent HbF is SPECTACULAR." Meanwhile, studies by Merlin Crossley's group in Sydney are developing an approach he calls "organic gene therapy," using CRISPR-Cas9 to recapitulate specific HPFH mutations to boost HbF levels.[6]

Stanford's Matthew Porteus has been on a quest to use genome editing in the clinic since the early 2000s.[7] Using CRISPR-Cas9 to break the DNA before fixing the sequence, Porteus says, is like fixing the headlight on someone's car by first busting the headlight with a hammer before repairing it. The goal is to ensure that about 20 percent of the stem cells repopulating the bone marrow are genetically fixed. That means harvesting some 500 million stem cells from any given patient. In a specialized facility, those cells are mixed with Cas9 and the AAV vector. The modified cells will be transplanted back into patients, after they have recovered from chemotherapy to make room for the modified stem cells. The biggest hurdle is to improve the efficiency of delivering the Cas nuclease into the appropriate cells.

"We'll cure people who have sickle-cell disease, not because they have a genetic defect but because they're human beings and deserve all the rights, responsibilities, and value we should confer on any human being," Porteus says.[8] He calls genome editing "an anti-eugenics program." The eugenics movement in the 20th century was designed to improve the gene pool by sterilizing or eliminating people who had genetic defects. A consequence will be that the frequency of the SCD variant will increase in the population, because now "we'll be taking people who normally die in childhood and allowing them to live to adulthood, have families." But it's a fair exchange, he says. "We should embrace this consequence."

One of the patients in Porteus's care is teenager David Sanchez, who charms with his eloquence and humor. "My blood just doesn't like me very much, I guess,"[9] he shrugs as he prepares for his monthly three-hour appointment with the apheresis machine, which replaces his warped erythrocytes with a fresh batch of healthy biconcave blood cells. His nurse likens David's visit to booking an oil change for a car. David endures his share of debilitating pain episodes but he exhibits no self-pity, even after enduring brain surgery. "I'm not just going to not play basketball. You can't *not* play basketball," he says in the film *Human Nature*.

Asked what he thinks about the possibility of fixing his disease using CRISPR, potentially eradicating the scourge of sickle cell altogether, Sanchez says, "Hmmn . . . that sounds cool." Then he pauses. "There's a lot of things that I learned having sickle cell," including patience and keeping a positive outlook. "I don't think I'd be me if I didn't have sickle cell."

A new biotech industry is rising fast in the belief that genome editing will provide the ultimate new weapon to deliver precision medicine. The industry is centered in Kendall Square, where the first three publicly traded genome editing companies—Editas Medicine, CRISPR Therapeutics, and Intellia Therapeutics—jostle for space and research talent with dozens of innovative biotech and gene therapy firms.

In July 2019, CRISPR Therapeutics (working with Vertex Pharmaceuticals) became the first company to launch a CRISPR-based clinical trial for a genetic disease in the United States. A thirty-four-year-old mother of four from Mississippi, Victoria Gray, was the first of dozens of volunteers in the trial. Diagnosed with SCD as a baby, her faith has sustained her during tough times. "I always knew that something had to come along and that God had something important in store for me," she told National Public Radio's Rob Stein.[10] At a clinical center in Nashville, Tennessee, doctors removed her bone marrow cells and administered chemotherapy. On July 2, 2019, hematologist Haydar Frangoul injected some two billion of Gray's gene-edited "supercells" back into her body.

After weeks in hospital, Gray was finally able to return home, leaving hospital wearing a blood-red T-shirt with I AM IMPORTANT emblazoned on the front and proudly sporting an invisible genetic modification. "I'm a GMO. Isn't that what they call it?" she said.[11] A month later, she returned to Nashville to receive her preliminary results, this time modeling a black sweatshirt that said simply, WARRIOR. Frangoul could not disguise his excitement.[12] Almost half of Gray's hemoglobin consisted of the fetal form, a spectacular result. That was still the case nine months after her therapy, with 80 percent of her bone marrow cells displaying the desired edit. Another patient with beta thalassemia in Germany has not needed a blood transfusion in fifteen months.[13] But it is still too soon to call it a cure.

CRISPR Therapeutics was the brainchild of Emmanuelle Charpentier. "I always had in mind that one day it would be nice if my research could lead to anti-infective strategies," she told me. "But the right application for me was human gene therapy."[14] As she was setting up her new lab in Berlin, Charpentier called her old friend, Rodger Novak, an executive at Sanofi who had been a postdoc with her at Rockefeller in the mid-'90s. "What do you think about CRISPR?" she asked. Novak didn't know what she was talking about. He suggested Charpentier talk to an investor friend, Shaun Foy, to appraise the technology. "Maybe you think I'm crazy?" Charpentier asked Foy. A month later, Foy called Novak with a simple suggestion: "You need to leave your job!" he said.

Charpentier partnered with Novak and Foy to create Inception Genomics in November 2013, which later became CRISPR Therapeutics. They initially selected Basel, Switzerland, as their pharma-friendly headquarters before relocating to Kendall Square for closer access to investors and talent. (They also founded another company, ERS Genomics—the name assembled from the initials of the three cofounders—to handle Charpentier's patent rights.) Charpentier had initially approached Doudna and Zhang about potentially teaming up to form a CRISPR company. Doudna demurred, agreeing instead to join forces with George Church and Zhang. Two other big names in Boston joined the band: Keith Joung, a pathologist at Mass General Hospital who had deep experience working on zinc fingers, and David Liu, a Harvard chemistry professor and Zhang's friend at the Broad Institute. The company's initial name was Gengine, but it was sensibly retired in favor of something with a bit more, um, gravitas: Editas Medicine.

The all-star quintet met in Boston for a company photoshoot but by the time Editas was announced, something appeared to have gone awry. One reporter noted that Doudna was notably absent from the official photos, a wisp of auburn hair in one photo the only evidence she'd been there at all. The reason soon became apparent: The Broad Institute had scooped Doudna and Charpentier by expediting Zhang's patent application. In May 2014 Zhang was awarded the big CRISPR-Cas9 patent. Doudna quickly cut ties with Editas. She later offered family reasons and the burden of travel, but the patent defeat, even if only the first round of a marathon contest, stung.

Editas was backed by three giant venture capital firms working together for the first time—Polaris, Flagship and Third Rock. Bill Gates joined the second round of investors who put up more than $100 million. Gates was fascinated by the potential of genome editing not only to treat genetic diseases but also to combat infectious diseases such as malaria and improve food production in developing nations.[15] As CEO Editas appointed Katrine Bosley, who had spent twenty-five years in biotech, most recently as chief executive of a cancer drug company, Avila Therapeutics, which is where I first met her.[16] When Bosley* first heard about genome editing, it sounded like science fiction. But during a few months as an entrepreneur-in-residence at the Broad Institute, she got to know Zhang and learn about CRISPR's potential. She was impressed by the technology's ease of use and wide applicability—something in common with the novel chemistry her team developed at Avila.**

---

* Bosley's father, Richard Bosley, designed and built the iconic Bosley GT MK I sports car in the 1950s.

** Avila Therapeutics developed drugs that bind covalently to their target, which is unconventional in drug discovery.

The lead program at Editas is for another form of Leber's congenital amaurosis (LCA type 10). The attraction of LCA10 as a target was for many of the same reasons that Bennett identified: because the eye is both small and accessible as a delivery target. Moreover, the desired gene edit—a small deletion—has a lower degree of difficulty. LCA10 is caused by a single-letter mutation in a gene called *CEP290*, first characterized in a consanguineous French-Canadian family. *CEP290* is expressed in the photoreceptors behind the retina; in LCA10 patients, the loss of the protein leads to degeneration of the outer segment of the photoreceptors. The LCA10 mutation, which resides in an intron that is normally spliced out of the gene transcript, results in an extra chunk of sequence being inserted into the RNA message. This cryptic exon includes a premature stop codon: so all of these RNA messages fail to produce a functional protein.

The Editas plan is to use CRISPR-Cas9 to deliver a pair of precise cuts flanking the mutation to restore normal gene splicing, thus producing the normal protein and rebuilding he photoreceptors. Like Spark Therapeutics, Editas uses an AAV vector to deliver the CRISPR machinery, dubbed EDIT-101. In July 2019, together with Allergan, Editas announced enrollment of the first patients in the Brilliance clinical trial.[17] The first surgery was performed in an hour-long procedure at the Casey Eye Institute at the Oregon Health & Science University in Portland in early 2020. It was called a new era in medicine—the first time CRISPR had been injected directly into a human patient, as opposed to the *ex vivo* approach employed for Victoria Gray.[18]

In early 2019, Bosley surprisingly stepped down as CEO. During her career, she had experienced the thrill of seeing drugs impact patients' lives, but she was foregoing that opportunity at Editas. (She did admit in one interview that five years at the helm seemed like a thousand.) She signed off on Twitter saying: "I'm proud of everything we achieved and built together, and I'll always be cheering them on—now as Editas Emeritas, as we say."

In late 2013, Nessan Bermingham, a rambunctious venture capitalist with Atlas Venture, conceived a new gene editing company. Bermingham grew up on an army base in County Kildare. After earning his PhD in London at St Mary's Hospital Medical School—something we have in common—he did postdoctoral research at Baylor College of Medicine before moving into finance. During an Atlas retreat in Miami, Bermingham struck up a conversation at the salad bar with John Leonard, a physician who had just retired after a successful career in big pharma.

Leonard moved to industry early in his career after his wife was diagnosed with multiple sclerosis. In the early 1990s, he ran the antiviral program at Abbott Laboratories during the AIDS crisis. "Our mission was absolute, and from that I learned how empowering belief can be," he said.[19] His team developed ritonavir, one of the first HIV protease inhibitors, and later Humira. After retiring as head of R&D at AbbVie in 2013, he dabbled in a couple of more low-key entrepreneurial ventures, including a craft cider company in Michigan. The trip to Miami was just a networking opportunity, or so he thought. A few months later, Bermingham flew to Chicago to take Leonard to lunch and lay out his ideas for Intellia Therapeutics. (The company's name came from the Greek "entelia," a state of pristine excellence.) "I couldn't think of a more promising and exciting technology than CRISPR," he said. Bermingham was confident he could secure Doudna's intellectual property (IP), which was good enough for Leonard, who joined as chief medical officer.[20]

In May 2014, Bermingham officially founded Intellia as CEO in partnership with Doudna's first company Caribou,* which holds Doudna's IP (and owns a stake of Intellia). Intellia's cofounders include Barrangou, Sontheimer, and Marraffini. Caribou is led by Rachel Haurwitz, who was the first student in Doudna's lab to work on CRISPR, but determined to launch a biotech company even before she finished her thesis. The other cofounders were Jínek and James Berger (now at Johns Hopkins).

* Caribou is an abbreviation for CRISPR-associated ribonucleic acid; the company logo sports a pair of antlers in the shape of a single-guide RNA.

Intellia emerged from stealth six months later, with support from Novartis before another round raised $70 million. In May 2015, officially divorced from Editas, Doudna joined the cofounder group along with stem cell biologist Derrick Rossi. A year later, Intellia went public, outperforming Editas by reaping some $110 million in the largest IPO for a Boston biotech that year. Haurwitz and Barrangou joined the team celebration as Bermingham ran the opening bell on the Nasdaq exchange.

Bermingham is a fierce competitive athlete, reveling in ultramarathons, boxing, and mountain biking. During an after-dinner speech at a medical conference in Washington, DC, in 2017, Bermingham drew parallels between extreme sports and precision medicine, presenting his vision to develop a one-and-done treatment paradigm where "freedom from genetic disorders is no longer an inherited privilege." His motivation came from the patients he hoped one day to cure:

> When I climb a hill in the final stages of a marathon, I push through because I know that my lungs will never have to fight harder for oxygen than those patients living with alpha-1 lung disease. When I step into the boxing ring, I'm fearless because I know that I'll not experience the pain children with sickle-cell disease fear every day. When continuing the race demands I choose to suffer, I remind myself it is by my choice and I will never have to face a preventative double mastectomy and living under the sword of breast cancer.[21]

Bermingham needed every piece of inspiration six month later competing in his first Roving Race in Patagonia. Athletes run twenty-five miles for four consecutive days, followed by a fifty-mile overnight leg and a trivial six miles on the final day, while carrying their own gear.[22] He placed in the top half out of more than three hundred competitors (dozens of whom didn't finish) in a time of thirty-nine hours—only nineteen hours behind the winner. Barrangou, no slouch as an entrepreneur, can only doff his hat. "Ness scares me! I can run with those guys, but this guy's in a different league," he laughed.

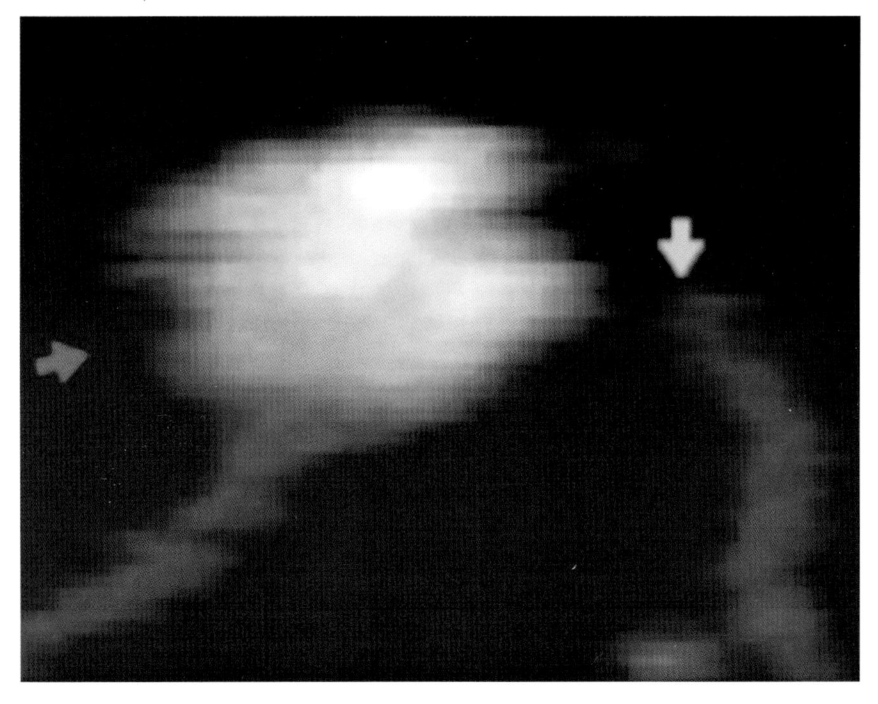

ABOVE: Caught in the act: Cas9 severs a strand of DNA, captured by Japanese researchers using high-speed atomic force microscopy. *Courtesy of Hiroshi Nishimasu.* BELOW: The Cas9 caress: Schematic shows the Cas9 protein (teal) holding the guide RNA (blue) scans for its target DNA (red) by first recognizing a PAM site (yellow). *Courtesy of HHMI BioInteractive.*

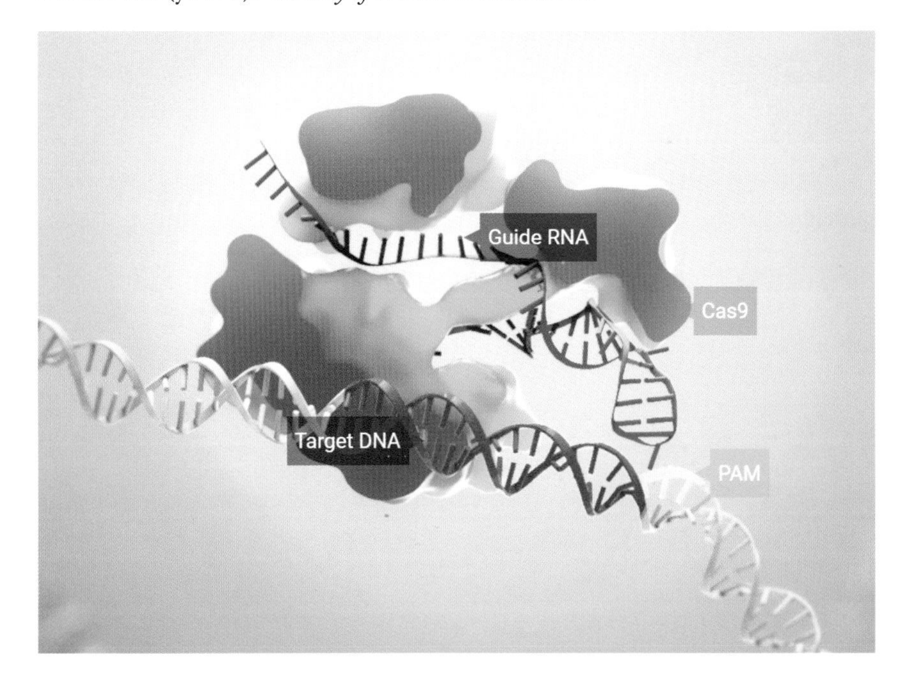

ABOVE: Francisco Mojica and the author at the salterns of Santa Pola, Spain, in 2019. BELOW: Microbiologist Luciano Marraffini, who collaborated with Feng Zhang in 2012, in his lab at The Rockefeller University, New York (September 2019).

ABOVE: Jill Banfield, UC Berkeley professor who, among other things, introduced her colleague Jennifer Doudna to CRISPR. *Courtesy of Derek Reich/ Human Nature.* BELOW: Plant biologist Caixia Gao in a field of CRISPR-edited wheat (to reduce susceptibility to powdery mildew) in Beijing, China. *Courtesy of Stefen Chow.*

ABOVE: Fyodor Urnov, the expert raconteur of genome editing, now at the Innovative Genomics Institute. *Courtesy of Derek Reich/Human Nature.* BELOW: Dana Carroll, biochemist who pioneered use of zinc finger nucleases.

ABOVE: Larger than life: George Church next to an ex-woolly mammoth in a hotel lobby in Yakutsk, the coldest city on earth in Eastern Siberia. *Courtesy of Eriona Hysolli.* BELOW: Church in his office at Harvard Medical School.

Jennifer Doudna (UC Berkeley and HHMI), cradling a model of Cas9, the molecular scissors of genome editing, opens the World Science Festival in New York in May 2019.

Emmanuelle Charpentier, founding director of the Max Planck Unit for the Science of Pathogens in Berlin, on a flying visit to New York (September 2018).

ABOVE: DC united: Doudna-Charpentier team photo on the steps of Stanley Hall, UC Berkeley, in 2011. (L-to-R): Charpentier, Doudna, Martin Jínek, Krystof Chyliński, and Ines Fonfara. *Courtesy of M. Jínek.* BELOW: Lithuania's Virginijus Šikšnys receives the 2018 Kavli Prize in nanoscience from King Harald of Norway, alongside Doudna and Charpentier (September 2018). *Courtesy of Fredrik Hagen/NTB scanpix.*

ABOVE: Feng Zhang holds CRISPR in his hands, with *60 Minutes*' correspondent Bill Whitaker. *Courtesy of the Broad Institute of MIT and Harvard.* BELOW: Winners of Canada's Gairdner Awards in 2016. (L-to-R): Anthony Fauci, Zhang, Charpentier, Rodolophe Barrangou, Doudna, and Philippe Horvath.

ABOVE: Base editing and more: David Liu in his office at the Broad Institute, under the watchful eye of Tony Stark. *Courtesy of Juliana Sohn, the Broad Institute of MIT and Harvard.* BELOW: First authors: Alexis Komor and Nicole Gaudelli, former postdocs in the Liu lab, in Victoria, British Columbia, February 2019.

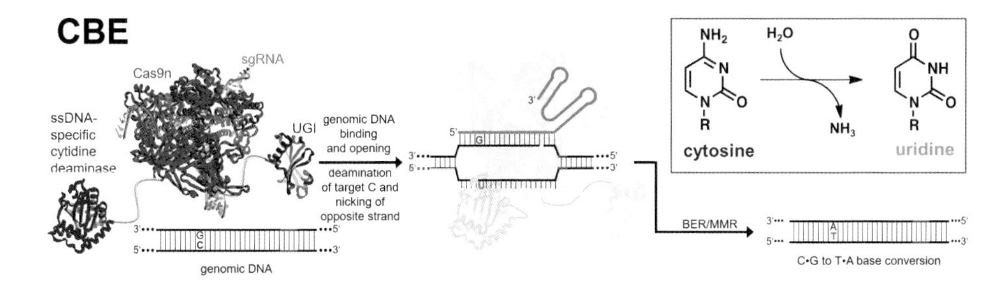

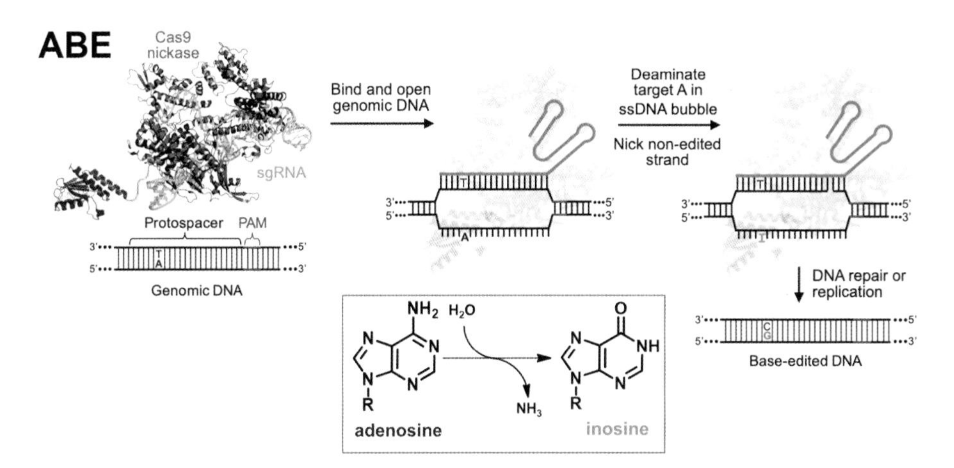

Base adjustment: the first base editors developed in the Liu lab perform chemistry directly on DNA (see chapter 22). The cytosine base editor (CBE) is a three-component molecular machine that converts cytosine to guanine (via uridine). The adenine base editor (ABE) catalyzes the transition of adenine to thymine (via inosine).

ABOVE: Pope Francis urges caution in applying CRISPR during an address at the Vatican in April 2018. *Courtesy of the Cura Foundation.* BELOW: Senator Elizabeth Warren quizzes former Editas CEO Katrine Bosley and Stanford University's Matthew Porteus at a U.S. Senate committee hearing (November 2017).

ABOVE: Lap-Chee Tsui, vice-chancellor of the University of Hong Kong, greets the author prior to the start of Hong Kong summit on genome editing, November 2018. BELOW: Feng Zhang discusses the potential of CRISPR to treat a wide spectrum of genetic diseases at the Hong Kong conference.

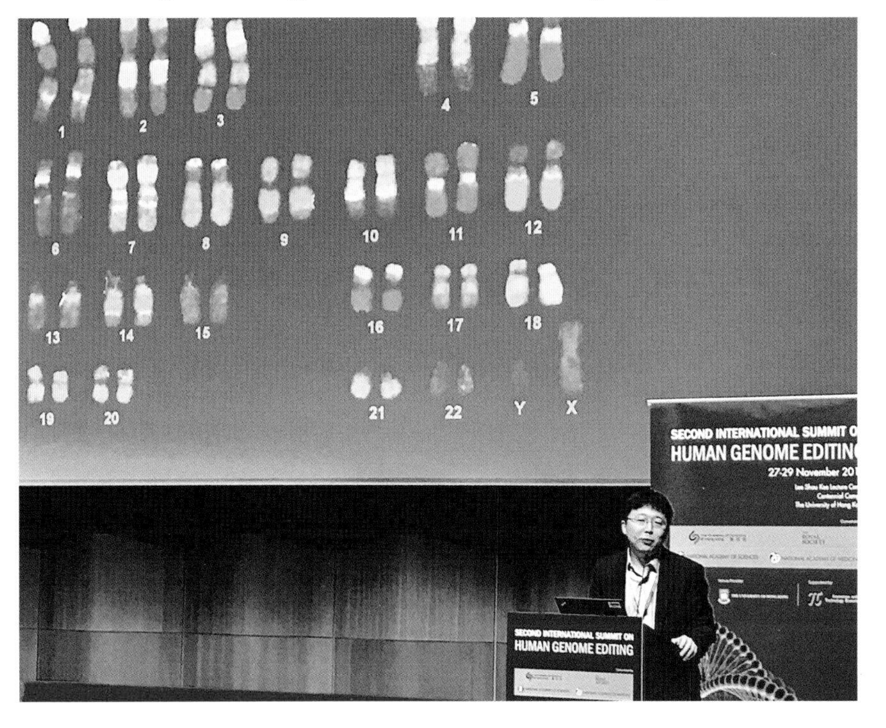

ABOVE: He Jiankui details the results of editing the DNA of Lulu and Nana, the CRISPR twins, at the Hong Kong summit, November 2018. *Courtesy of William Kearney/U.S. National Academy of Sciences.* BELOW: CCR5 variants: on the left, the structure of the normal CCR5 receptor and the Δ32 variant; on the right, the variants in Lulu and Nana (abnormal amino-acid sequences in red). *Courtesy of Sean Ryder/The CRISPR Journal.*

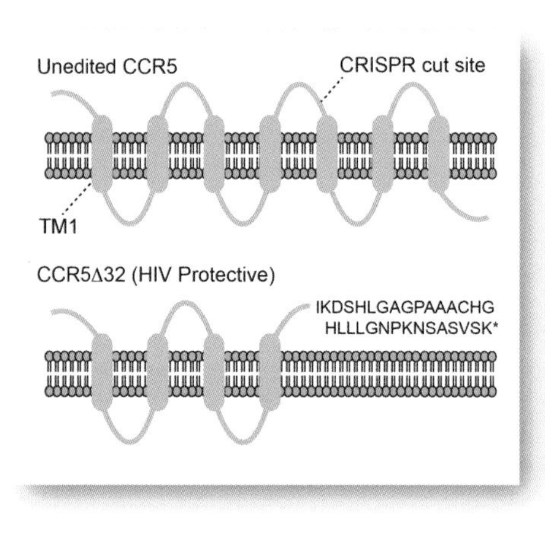

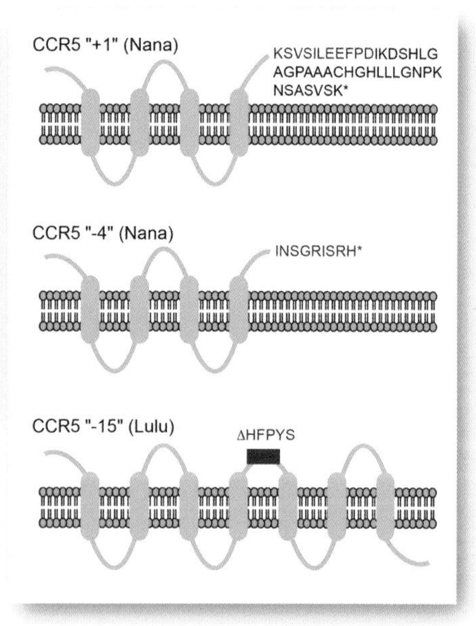

ABOVE: Meet the press: More than 200 photographers await the arrival of He Jiankui at the Hong Kong summit. *Courtesy of William Kearney/U.S. National Academy of Sciences.* On the right, Robin Lovell-Badge in the middle of a media scrum. BELOW: Home alone: He Jiankui spotted on a balcony in Shenzhen while under house arrest in December 2018. *Courtesy of Elsie Chen.*

ABOVE: David Sanchez, who has sickle-cell disease, sees hope for a cure in CRISPR. *Courtesy of Derek Reich/Human Nature.* BELOW: Dame Kay Davies (University of Oxford), co-chair of an international commission on heritable genome editing, with the author (no relation).

Perhaps exhausted, Bermingham resigned* as Intellia's CEO a short time later, passing the torch to Leonard. Barrangou thinks he's the ideal CEO: "John has done it before. It's not about being first in the clinic, it's about being best in the clinic."[23]

Intellia's biggest bet is on a liver disease called transthyretin amyloidosis (ATTR), which affects about 50,000 patients worldwide. (In 2016, Regeneron paid Intellia $75 million for rights in ten therapeutic areas, including ATTR.) This disease is adult-onset, typically fatal, caused by toxic amyloid protein deposits leading to heart failure and neuropathy. Patients typically survive just a few years after diagnosis. The plan is to use CRISPR-Cas9 to deactivate the *TTR* gene in the liver using a lipid nanoparticle for delivery, which unlike some viruses does not trigger an immune response. The strategy avoids leaving Cas9 hanging around too long, minimizing the risk of any off-target mistakes. If results in nonhuman primates hold up, this could represent a cure for patients with TTR.

Eric Olson, a professor at the University of Texas, may not be an expert in genome editing, but he knows muscle.[24] Olson started his own lab in the early 1980s at MD Anderson Cancer Center, and taught himself molecular biology. His first two grant applications were trashed by reviewers but it was third time lucky. Olson receives letters and emails almost daily from parents searching for hope, if not a cure, for muscular dystrophy. Olson's company, Exonics, is on course to make a difference.

Duchenne muscular dystrophy (DMD) is the most common and severe form of inherited muscular dystrophy. It is caused by one of thousands of different mutations in the largest gene in the human genome. As an X-linked disease, DMD affects mostly boys—about 300,000 around the world. The corresponding protein, dystrophin, sits under the membrane of muscle cells and acts like a giant shock absorber; without it, the membranes

---

* Bermingham has since launched a new company called Triplet Therapeutics, a start-up targeting diseases like Huntington's disease, which are caused by the expansion of triplet repeat DNA sequences.

start to leak, resulting in muscular weakness. Replacing dystrophin is a formidable task, but attempts to find a workaround, such as supplying a "mini" dystrophin gene or switching on a dormant related protein called utrophin, has met with little success.

The dystrophin gene consists of seventy-nine coding sequences, or exons, spread over 2.6 million letters on the X chromosome. Many of the 4,000 catalogued DMD mutations congregate in the middle of the gene between exons 45–50. These frequently result in a "frameshift" mutation, shifting exon 51 out of frame, which in turn compromises production of dystrophin. Interestingly, the middle portion of dystrophin—the spring, if you will—is less critical than the ends of the giant protein. We know this because patients with a milder form of DMD, Becker muscular dystrophy, have shortened springs but otherwise a partially functional protein, and have a longer life expectancy than DMD patients.

Olson's plan essentially is to use CRISPR to convert some of the DMD mutations in the central exons to the milder Becker form of muscular dystrophy. By using CRISPR-Cas9 to engineer a cut in exon 51 to excise one particular mutation, the resulting protein should be close to fully functional. Just as Jean Bennett found a canine model to test her gene therapy for LCA, Olson's team turned to another canine model. The beagle colony at the Royal Veterinary College near London looks and sounds as happy and healthy as their normal cousins, with tricolor coats and a trademark howl. But the affected dogs noticeably drag their hind legs.*

Leading the Exonics team is an expat from Moldova, Leonela Amoasii. The initial results in 2018 put a smile on everyone's face.[25] Amoasii treated one-month-old puppies and then compared results with controls two months later. The muscle fibers of the treated dogs express newly restored dystrophin protein, as much as 80 percent of normal. But the most compelling evidence is to watch the treated dogs run, jump, and play as happily as their wild-type cousins. "We're really excited," says

---

* The dystrophin mutation was first observed in a Cavalier King Charles Spaniel, then bred into a line of beagles, which provide a better physiological match to humans.

Amoasii. So too is Vertex Pharmaceuticals, which acquired Exonics in June 2019 in a deal potentially worth up to $1 billion—not bad for a company founded less than three years earlier.

For any company hoping to treat a patient with CRISPR genome editing, safety is perhaps the biggest concern. The technology relies on a bacterial enzyme that can slice and dice DNA. Several safety issues have surfaced that have given investigators—and occasionally investors—pause. The biggest concern is that Cas9, while roaming the genome to find the correct target gene, could accidentally latch onto a closely related sequence, perhaps with just a single mismatched base, and cleave in the wrong place. This molecular mistaken identity could be harmless, seamlessly patched up by the DNA repair network. Or it could be disastrous, deactivating a critical gene or potentially switching on a cancer-causing gene. That could mean the end not only of a promising clinical trial but worse, a devastating setback for the entire field.

These concerns are not unique to CRISPR: any programmable genome editor will have a slight chance of fixating on the wrong sequence. But the stakes are higher with CRISPR because of its widespread use and stratospheric expectations. Some off-target fears have been overblown. A study from a Stanford group documenting multiple off-targets in 2018 raised alarms, even though the data were collected on just three lab mice. Industry scientists and academic groups swiftly rebutted any notion that CRISPR-Cas9 was unsafe.[26] Subsequent studies suggested that if unintended edits do occur, they are no more frequent than those that inevitably result from background radiation and the errors that naturally accumulate during DNA replication and cell division.

Scientists have devised many ways to improve on-target specificity, from the design of the guide RNA to the choice and design of Cas9 nucleases.[27] Concerns were also raised about on-target effects: Allan Bradley, a highly respected mouse geneticist and former director of the Sanger Institute, prompted another mini CRISPR crisis when his group showed that CRISPR-Cas9 can cause larger deletions than expected at the target site.[28] Barrangou summed up the sentiment: "Keep calm and CRISPR on!"

Another safety issue is anticipating what happens when a bacterial protein is injected into patients. Porteus and colleagues found that many individuals carry antibodies to Cas9, suggesting they have been previously exposed to bacteria that express the protein.[29] This is not surprising; after all, the immune system is designed to detect foreign proteins. A variety of methods—including selecting Cas9 enzymes from different bacteria or modifying the surface of the protein to make it less immunogenic—should minimize the risk of an undesirable immune response.[30]

Yet another scare flared up in 2018, when two highly publicized reports suggested that genome editing with CRISPR might increase the risk of genomic instability and cancer predisposition by selecting against the function of p53, the most frequently mutated tumor suppressor gene. Among many commentaries mulling over the significance of these findings was David Lane, who dubbed p53 "the Guardian of the Genome." Land and Teresa Ho said that this episode was a "cautionary rather than apocalyptic tale," just like "every therapeutic endeavor of the past and future."[31] The case for using genome editing as part of a life-saving T-cell transplant in a 75-year-old cancer patient might be very different to editing cells to correct a Mendelian gene defect in a baby.

These issues were mostly taken in stride by the CRISPR companies. By July 2020, the big three biotechs had a combined market cap of $10 billion. And yet, the question of who actually owned the rights to the invention of the CRISPR revolution was still unresolved.

CHAPTER 13
PATENT PENDING

"A few years ago with my colleague Emmanuelle Charpentier, I invented a new technology for editing genomes. It's called CRISPR-Cas9."[1] Doudna raised a few eyebrows with that offhand remark during a TED talk in London in 2015, which made light of a billion of years of evolution, not to mention the competing efforts of a few other investigators. But it is fairly ingrained in the popular culture. In November 2019, Alex Trebeck read a question on *Jeopardy*:

> JENNIFER DOUDNA & EMMANUELLE CHARPENTIER ARE CO-INVENTORS OF THE REVOLUTIONARY TOOL CRISPR TO EDIT THESE IN THE BODY*

Lest we forget, bacteria clearly invented CRISPR many hundreds of millions of years ago. But the legal question of who invented CRISPR genome editing technology is very real—and really important. At stake are commercial rights to a technology potentially worth billions of dollars. Law professor Jacob Sherkow calls it "the most monumental biotech patent dispute in decades."[2] At the heart of the dispute is the Broad Institute, the base of Feng Zhang, versus the University of California (UC), home of Doudna at Berkeley, in collaboration with Charpentier and her host institution.

* The correct answer is "genes."

So who "invented" the revolutionary CRISPR gene-editing technology? The transatlantic team that devised the single-guide RNA and showed that Cas9 could be programmed to cut DNA on demand in June 2012? Or the group that first demonstrated editing of human DNA using CRISPR-Cas9 seven months later? What about other less-celebrated contenders, such as Virgis Šikšnys, who filed a patent application in March 2012? In his small office at the New York Law School in Tribeca, Sherkow enthusiastically guides me through the landmarks of the saga.

One might assume that, because the Doudna-Charpentier team published the first account of CRISPR-Cas gene targeting, the invention should belong to them. Had the CRISPR saga taken place twelve months later, that might be the end of the story. In April 2013, the criterion for awarding American patents changed to a "first to file" basis. Before then, however, applications were reviewed on a "first to invent" basis. Eight years after the first key publication, we still don't have a definitive answer. Doudna and Charpentier, together with Jínek and Chyliński, filed their provisional US patent application in May 2012, shortly before submitting their manuscript to *Science*.[3] The three patent co-owners were UC, the University of Vienna, and in an interesting twist, Charpentier herself. In Sweden, where Charpentier was based, academics enjoy "Professor's privilege," which grants them full rights to their own inventions. (Thus, Charpentier is both a co-inventor of her technology and a co-owner of it.) The triumvirate is known as CVC.

This was not the first patent application to mention CRISPR: in April 2004, the Danisco group of Barrangou and Horvath described a method for sequencing CRISPR regions in a dairy sample to fingerprint variants of *Lactobacillus acidophilus*. "The origins of the CRISPR patent estate, far from being forged in combat, were gently cultured in distant tuns," observes Sherkow.[4] Another interesting application was one from Erik Sontheimer and Luciano Marraffini. In his office at the University of Massachusetts Medical School, Sontheimer rifles through his computer files searching for what he calls "the best grant I'd ever written," which mirrored his patent application.

In January 2009, Sontheimer submitted a five-year, $1.8 million grant application to the NIH, entitled "RNA-Directed DNA Targeting in Eukaryotic Cells." "CRISPR interference could provide unique capabilities, if it can be ported to eukaryotic cells," Sontheimer wrote. "It could be easily programmed (and, when desired, reprogrammed) by the introduction of small RNA molecules." The goal was to manipulate the structure and activity of genomes of higher organisms using RNA that recognizes its target DNA using the well understood rules of Watson-Crick base pairing. The method "would have transformative and disruptive potential in biotechnology and medicine."[5] There were parallels to RNA interference, the technology codiscovered by the Nobel laureate next door, Craig Mello. But "CRISPR interference" was exciting because of the sequence specificity conferred by the twenty-four- to forty-eight-base spacers. Sontheimer and Marraffini's discovery of "a natural, highly specific, RNA-directed, DNA-targeting machinery in bacteria" suggested a route toward a "reprogrammable genome targeting system in eukaryotic cells."

Sontheimer's plan was to port CRISPR into eukaryotic cells and test its ability to target specific genes. He would start small, working in yeast, then progress to *Drosophila*, and eventually mammalian cells in the final two years of the funding by 2012–13. RNA-directed genome targeting "has the potential to transform biomedical research, biotechnology, genetic medicine, and stem cell therapeutics," he wrote. In retrospect, it was too much too soon. The proposal "hit a brick wall" and was rejected, and his companion patent application was abandoned. "The vision and idea were out there, but we hadn't reduced it to practice," said Sontheimer wistfully.[6]

Six months after CVC filed their patent application ('772) just prior to publication of their *Science* paper, the Broad Institute followed the same playbook, as Zhang filed an application covering the use of CRISPR editing in eukaryotic cells in December 2012. But in a nifty legal maneuver, the Broad lawyers paid the princely sum of $70—a round of cocktails at the local Meadhall gastropub—for an expedited "fast track" review, thereby leaping ahead of UC's application. In January 2014, the United States Patent and Trademark Office (PTO) provisionally denied the application, citing the "prior art" of the CVC application. Zhang quickly filed a forty-page "personal declaration" rebuttal. It paid off: three months later, in April, the PTO awarded patent number 8,697,359 to the Broad. The decision sent shockwaves across the biotech communities. Doudna resigned as an advisor to Editas, the all-star company she had cofounded with Zhang and others. CVC filed for an interference proceeding, seeking a PTO ruling to establish who truly invented the technology.

The Broad responded by filing a "priority statement" with the PTO in May 2016. Exhibit A was an internal Broad document—a "Confidential Memorandum of Invention"—that Zhang had signed in February 2011. In this document, Zhang declared he had the initial idea for CRISPR gene editing on February 4, 2011—shortly after hearing Gilmore's lecture—and first documented those ideas four days later.[7] Zhang wrote:

> The key concept of this invention is based on the CRISPR [repeats] found in many microbial organisms. Enzymes associated with the CRISPR complex use short RNA sequences to recognize specific target sites on the host genome and performs site-specific cleavage. The key novel feature of this invention is that it does not rely on the design of site-specific DNA binding proteins (i.e., zinc finger or TAL effector) and can be easily targeted to multiple sites through the use of multiple sequence-specific CRISPR spacer elements.

The Broad mounted an effective campaign to sway public opinion. "CRISPR itself cannot be patented," explained communications director Lee McGuire. "Cas9 is a naturally occurring protein and part of a naturally-occurring bacterial process, but this process, on its own, does not work in mammalian cells. What [Zhang] has patented are engineered components and compositions specifically altered from their naturally-occurring form to be useful in methods for editing the genomes of living mammalian cells."[8] There were shortcomings in both the CVC application as well the earlier Šikšnys application. These applications, the Broad argued, merely showed that "purified Cas9 protein and a certain purified RNA could cut a short piece of DNA in a solution in a test tube." And then the Broad dropped the hammer. Both applications "contained no disclosure of work in cells, no genomes, and no editing."[9]

No cells, no genomes, no editing. It was a brutal rebuttal to the CVC appeal. But would it work in court?

Showtime came on a rainy Tuesday morning in December 2016. Legal scholars including Sherkow and Robert Cook-Deegan, attorneys and journalists formed an orderly queue more than an hour before the doors opened at the PTO's headquarters in Alexandria, Virginia. They were there to witness history in the making: such courtroom dramas would likely become a thing of the past under the "first to file" rule.[10] The Patent Trial and Appeal Board (PTAB) was unprepared for the crush of spectators who rapidly filled the courtroom and two overflow rooms. "It was probably the most well-attended interference proceeding in USPTO history," said Sherkow.

The purpose of the hearing was not to establish who invented CRISPR genome editing. Rather, the three judges on the PTAB were trying to determine "what the *what* is," as Sherkow put it. If the PTAB concluded that there were indeed two separate inventions with two distinct timelines, then the question of who came first was redundant. But if the dispute was over the same invention, then all bets were off. There is a reason, says Sherkow, that for example Bell Laboratories required all its engineers to sign and date every page of their notebooks before they left work each day to ensure there was documentary evidence in the event of an interference proceeding.

The panel of three judges, led by Judge Deborah Katz, who has a doctorate in molecular biology, fired their sharpest questions at the CVC lawyer, who insisted that Doudna's discovery, restricted to DNA in a test tube, could be extrapolated to all organisms including humans.[11] Among hundreds of research papers, emails, PDFs, patents, and other documents, UC submitted statements from a pair of expert witnesses. One was from Nobel laureate Carol Greider, a Berkeley grad herself, who discovered telomeres (the DNA aglets that protect the ends of chromosomes from fraying). The other was from gene-editing pioneer Dana Carroll, who was compensated $500/hour to compile a brief totaling nearly two hundred pages.

Much of the deliberation centered on the contemporaneous commentaries of Carroll and other experts on the initial prospects of translating CRISPR gene targeting to human cells. Recall that in his September 2012 commentary, Carroll mulled over the prospects for CRISPR genome editing in human cells. In conclusion he wrote: "Whether the CRISPR system will provide the next-next generation of targetable cleavage reagents remains to be seen, but it is clearly well worth a try. Stay tuned."[12]

Table: Key Dates in the CRISPR-Cas9 Patent Dispute

YEAR	DATE	EVENT
2012	March 20	Šikšnys files patent ('739)
	May 25	CVC files patent application ('772)
	December 6	MilliporeSigma files patent
	December 12	Broad files first patent ('527)
2013	March 15	CVC files patent application ('859)
2014	April 15	Broad awarded '527 patent
	July 7	Rockefeller files CRISPR patent
2015	April 13	CVC files "suggestion of interference"
2016	January 11	PTO declares an interference
	December 6	PTAB interference oral argument
2017	February 15	PTAB interference decision favors Broad
	March 23	EPO grants CVC patent
	April 12	CVC appeals Interference in federal court
2018	January 18	EPO revokes Broad patent in Europe ('468)
	April 30	Appeal oral argument
	June 19	PTO awards CVC its first patent ('772)
	Sept 10	Court upholds PTAB appeal decision
2019	February 8	PTO awards CVC '859 patent
	June 24	PTO declares interference
2020	January 16	EPO dismisses Broad appeal of '468
	February 7	EPO upholds CVC
	May 18	PTAB hearing

* CVC, University of California/University of Vienna/Charpentier; EPO, European Patent Office; PTAB, Patent Trials and Appeals Board; PTO, U.S. Patent and Trademark Office.

Incredibly, the PTAB judges tried to unpack just what Carroll meant by "Stay tuned." Was Carroll skeptical that CRISPR would work in human cells, or suggesting it was only a matter of time? The judges dissected another line from Carroll's article where he said: "there is no guarantee that Cas9 will work effectively on a chromatin target." The judges summed it up: "We fail to see how 'no guarantee' indicates an expectation of success"—especially for someone having ordinary skill in the art. They concluded sternly that Carroll "did not have a reasonable expectation that the system would work." And if it had been a formality, then one would have to question why the editors of *Science* wanted to publish the Zhang and Church reports six months after they published the Doudna-Charpentier masterpiece.

Among the hundreds of documents and exhibits made public in the Broad-CVC saga, was a bizarre email that Doudna received in February 2015. It was sent by Lin Shuailiang, a Chinese student who had spent nine months at the Broad from October 2011 through June 2012 and was a coauthor on Zhang's *Science* article. Lin raised allegations about the provenance of the CRISPR work in Zhang's lab. Lin called the PTO decision "a joke" and claimed, "Feng is not only unfair to me but also the science history." He continued:

> I began the CRISPR project solely the first day I came to Feng Zhang's lab in 2011 as a side project for a visiting student. At that time, the whole lab except me focused on TALEN projects. I was working on the CRISPR project until I went back to China [in] 2012 June for my PhD . . . After seeing your [*Science*] paper, Feng Zhang and Le Cong quickly jumped to the project without letting me know. My lab notebooks, emails and other files . . . recorded every step of the lab's failure process . . . We did not work it out before seeing your paper, it's really a pity. But I think we should be responsible for the truth. That's science.[13]

The Broad dismissed Lin's assertions, noting that his visa was about to expire and he was out of a job at the Broad. Lin ended up joining a San Francisco nanomedicine biotech company called Ligandal. On the company's website, he was described as "the very first scientist in the world to work with CRISPR."

On February 15, 2017, the PTAB ruled decisively in favor of Zhang and the Broad Institute, finding no "interference-in-fact"—the Broad's patent did not overlap significantly with the CVC application.[14] Zhang's invention of CRISPR genome editing was "not obvious" because "one of ordinary skill in the art would *not* have reasonably expected a CRISPR-Cas9 system to be successful in a eukaryotic environment." As the rival patent applications did not overlap, the Broad's patents could stand. "Even though it seemed like both parties were there at the same time and they seemed to be working in relatively the same area, their patent applications, at the very least, did not interfere with one another," Sherkow summarized.[15] Editas's stock surged 30 percent, pushing its market cap above $1 billion.

Three months later, I was in Vilnius, meeting Šikšnys for the first time. On his desk was a newly arrived FedEx envelope. Before I left, he agreed to open it. Inside was an official copy of his newly approved U.S. patent, filed back in March 2012: "RNA-directed DNA cleavage by the Cas9-crRNA complex." DowDuPont (now Corteva) wasted no time in licensing the rights to Šikšnys' IP.

CVC might have lost the Battle of Alexandria, but it was still determined to win the war. Doudna put her own spin on the verdict: "They have a patent on green tennis balls. We will have a patent on all tennis balls."[16] The players regrouped and dug in for another nail-biting set.

Two months later, CVC filed an appeal in federal court, confident it would establish definitively that the Doudna-Charpentier partnership "was the first to engineer CRISPR-Cas9 for use in all types of environments" including human cells.[17] It was an uphill battle—the federal court's task was not to retry the case but merely to assess whether the PTAB had committed a legal error. At the appeal hearing in April 2018, UC's lawyer Donald Verrilli, a former U.S. solicitor general, argued that if Doudna and colleagues hadn't demonstrated the feasibility of genome editing in higher organisms, the multiple publications in early 2013 would have used different techniques. "This was like the California Gold Rush," Verrilli said. The competing researchers were all trying to demonstrate genome editing first. "If they thought making changes was necessary, they would have."[18]

But the three judges didn't bite. "You start with the common technology," declared Judge Kimberly Moore. She highlighted statements by Doudna and others expressing frustration about getting CRISPR to work in human cells. For example, in a profile for a Berkeley College of Chemistry magazine, Doudna openly admitted: "Our 2012 paper was a big success, but there was a problem. We weren't sure if CRISPR-Cas9 would work in eukaryotes."[19] In the end, the flood of reports in early 2013 showed that it was fairly easy to get genome editing to work in human cells based on the UC methods. But the key question according to Moore wasn't "Did it turn out to be easy to do?" It was: "What was the perception of the skilled artisan at the time?"

On September 10, 2018, the Court ruled in favor of the PTAB's earlier judgment. Sherkow said that while legally correct, he wasn't sure the decision meshed with the way biomedical research is actually practiced. "The PTAB's decision and Šikšnys's patent sit as the Scylla and Charybdis for UC's continued prosecution of its application," Sherkow wrote.[20] Some reporters thought it was game over: NPR declared "East Coast scientists" the victors.[21]

But not so fast: in another surprising twist in February 2019, the PTO assigned the CVC '859 patent—in limbo for six years—to a new examiner, who promptly allowed it with minimal comment.[22] By the end of 2019, CVC had been awarded some twenty CRISPR-related patents, gleefully punctuating each award with a press release. But questions lingered: was the key CVC patent overly broad (no pun intended)? And did it meet the enablement clause? Did it provide sufficiently detailed instructions to allow someone to edit a genome?

In June 2019, the PTO declared its own interference, hoping to resolve the dispute. In one corner, a dozen patents owned by the Broad (the senior party); in the other, ten awarded to CVC.[23] As the junior party, CVC still has the uphill task. The core of the case, Sherkow explains, is who can claim priority for the single-guide RNA?

Many have wondered why UC and the Broad have spent so many millions of dollars on this drag-out patent fight? The bills are actually paid by the biotech companies; Editas has paid tens of millions of dollars on legal fees on behalf of the Broad Institute. The thought that the Nobel committee is unlikely to award the inevitable CRISPR prize to the loser in the patent drama "could be one factor in the Broad's litigation strategy," says Hank Greely.[24] The Broad insists it has been open to a settlement for years, but UC hasn't shown any appetite for compromise. Indeed, CVC has filed motions alleging that Zhang did not include tracrRNA until after Doudna's paper was published, also questioning authorship discrepancies in some of Broad's patents in the United States and Europe.[25]

In the run-up to the latest PTAB hearing in May 2020, the Broad filed an expert brief by Chad Mirkin, a chemistry professor at Northwestern University. Mirkin, who runs a massive lab of seventy people, charged his standard consulting rate of an eye-watering $1,600/hour in support of the Broad's claims. He argued that the hypothetical "person of ordinary skill in the art" would not have successfully achieved genome editing in mammalian cells based on the CVC. During a deposition in early 2020, UC lawyers laid into Mirkin's CRISPR credentials. He was forced to admit that his lab had not actually performed a CRISPR experiment until 2016.

Whether this legal sparring will have any material bearing on the case is unclear. At the PTAB encore hearing in May 2020, the judges did not tip their hand. Whoever wins the next round, the other side is still likely to come out swinging.

In Europe, CVC has fared better, thanks in part to a curious event that happened early on in the legal drama. The original Broad application in 2012 listed four co-inventors: Zhang plus Le Cong, Ann Ran, and Marraffini, all coauthors on the 2013 *Science* paper. But in the later '359

patent application, Zhang was flying solo.* In July 2014, the Rockefeller University filed its own patent, essentially a carbon copy of the Broad's application, only with Marraffini's name restored.[26] It was an unorthodox gambit that violated PTO rules because Zhang's name had been included in this later filing without his notification or permission.

The dispute went to binding arbitration behind closed doors. On January 15, 2018, the Broad announced that it had prevailed. This meant that Marraffini, the microbiologist who had collaborated with Zhang throughout 2012, would not be listed as a co-inventor on the Broad's major patents regarding eukaryotic editing.[27] The Broad's press release stating that the two sides had "settled their disagreement" gave the appearance of an amicable resolution. But when I even broached the subject with Marraffini in his office months later, he shot me the universal my-lips-are-sealed gesture.

But the Broad's celebrations lasted barely seventy-two hours, as it was dealt a major setback in Europe. In March 2017, the European Patent Office (EPO) had awarded CVC a wide-ranging patent across all organisms—microbes, plants, and animals.[28] Just a few days after the Broad-Rockefeller arbitration results, EPO announced that it was revoking the Broad patents ironically because of the Marraffini dispute. As the list of inventors of the Broad applications had changed from the earliest application in December 2012, EPO deemed the earliest application to be invalid. With that off the table, the Broad's applications did not predate CVC. The EPO upheld that decision in January 2020 and reaffirmed the CVC patent three weeks later.[29]

The dispute has driven a wedge between the two biggest names and ex-industry colleagues in CRISPR in the United States. It's not a subject Doudna cares to discuss, saying only it is "very disappointing for me, as a scientist and as a person."[30] Lurking in the wings is another patent application, originally filed on December 6, 2012, by MilliporeSigma researchers, who claim they were the first to perform CRISPR editing in eukaryotes. Ironically, the PTO judged that claim to be obvious in light of the CVC

* The U.S. PTO is usually quite forgiving about such adjustments; less so in Europe.

work, in contrast to the history of the Broad application. MilliporeSigma has petitioned the PTO to declare another interference with CVC.[31]

Hopefully this saga won't be repeated as more CRISPR technologies are developed. In more recent iterations like base editing and prime editing (see chapter 22), the patent situation looks much cleaner. And the first to file rule is less contentious than first to invent. As patent attorney Joe Stanganelli says: "If there is any universally good news to take away from the CRISPR patent saga, it is this: It is unlikely to happen again."[32]

What hinges on these patent disputes? In the coming years, the CRISPR biotech companies will likely have several gene-editing therapies on the market, all using garden-variety Cas9. "There will be a lot of money changing hands to make sure those drugs get into patients' veins without infringement," says Sherkow, who is now a law professor at the University of Illinois. What about 2040? "We'll have a million CRISPR therapeutics on the market, every nuclease, RNA editors, a complete riot of options. Will we look back and say, 'What a godawful waste of money'? Absolutely. It's a joke."[33]

Joke or not, the patent wars hold an "inside baseball" fascination that will at some point result in some momentous decisions about who will be paying licensing fees to whom. But in 2018, a new danger flared up. Some scientists and observers considered it nothing short of an existential threat to the fledgling CRISPR industry as it took its first nervous steps into the clinic.

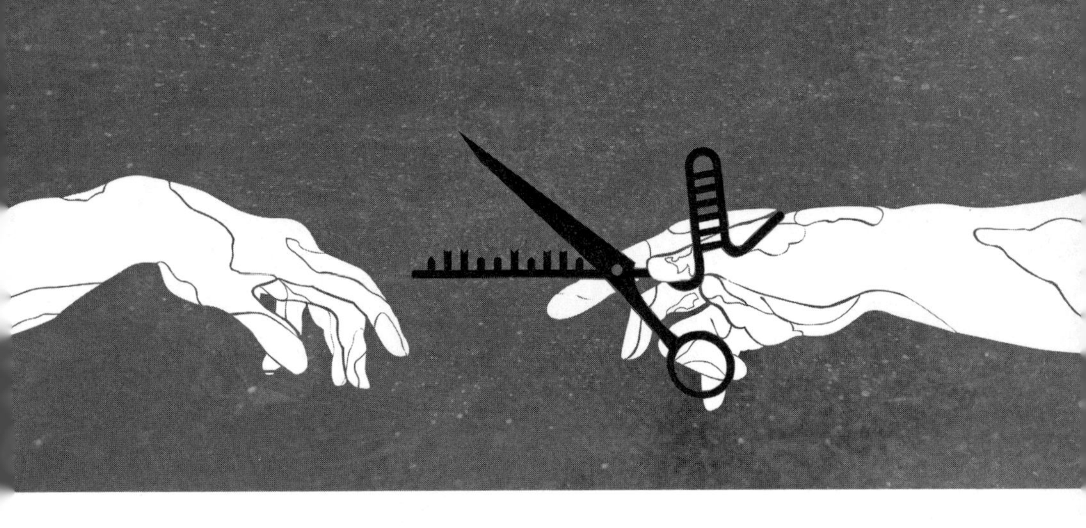

PART III

"What more powerful form of study of mankind could there be than to read our own instruction book?"

—Francis Collins

"Humankind ascended to the top so quickly that the ecosystem was not given time to adjust. Moreover, humans themselves failed to adjust."

—Yuval Noah Harari

Rita: Don't you worry about cholesterol, lung cancer, love handles?
Phil: I don't worry about anything anymore.
Rita: What makes you so special? Everybody worries about something.
Phil: That's exactly what makes me so special. I don't even have to floss.

—*Groundhog Day*

CHAPTER 14
#CRISPRBABIES

In April 2018, more than 15,000 scientists and physicians arrived in Chicago for the American Association of Cancer Research annual convention. In a hotel lobby, Marilynn Marchione, the chief medical correspondent for the Associated Press (AP), met an acquaintance, Ryan Ferrell. The year prior, Ferrell, working for a public relations company that represented Sangamo, had arranged for Marchione to have exclusive access to the launch of the first *in vivo* gene editing clinical trial in the United States.[1] Marchione's story shadowed Brian Madeux, an Arizona man with Hunter syndrome, who became the first patient to receive the groundbreaking therapy.

Ferrell's intention was obviously to orchestrate some headline news for Sangamo. But it was also a contingency plan: to build a relationship with a trusted medical reporter, making sure she knew the lead investigators in the event that there should be a serious adverse event in the trial. In Chicago, Ferrell asked Marchione if she was interested in working on another exclusive gene-editing story, this time nothing to do with Sangamo. This was a big one, he said, although he kept some cards close to his chest. In particular, he didn't mention his dinner two months earlier with a Chinese scientist he'd worked with before, named He Jiankui, who was moving into clinical genome editing. Nor did he mention a bombshell email he'd just received from the same scientist, whom he called JK.* That email revealed that JK had done something no-one else had dare attempt. And now a woman was pregnant, carrying a genetically edited fetus.

Ferrell had received an offer to shape the public revelation of a medical milestone that would put the Sangamo story in the shade: the first genome editing of a human embryo resulting in a pregnancy and twin births. Ferrell was already contemplating leaving his Chicago-based PR agency. In July, he handed in his notice and a few weeks later, flew to Shenzhen, having accepted a short-term contract from JK to advise his group.**

Ferrell knew he could trust Marchione to tell the story of the CRISPR babies accurately and responsibly, even in the midst of what was sure to be an international media frenzy. In the second week of October, after a major Chinese national holiday, four China-based AP reporters visited JK's lab at the Southern University of Science and Technology (SUSTech) in Shenzhen. While they interviewed JK, the photographer captured one of JK's colleagues, embryologist Qin Jinzhou, injecting a human embryo with a CRISPR construct targeting a gene called *PCSK9*. JK told the reporters about the successful pregnancy, although the babies had not yet been born. On camera, he said defiantly: "The world has moved onto the stage for embryo genetic editing. There will be someone, somewhere who is doing this. If it's not me, it will be someone else."[2]

Marchione began drafting her exclusive story, keeping everything confidential, the timing of its release still to be determined. For her, the most profound aspect of JK's experiment was "the enormity of the leap for humankind." For the first time, a scientist had dared to rewrite the recipe

---

* He Jiankui is widely known as "JK," so for convenience I have used this abbreviation (it is not intended as a term of familiarity). It is Chinese convention to write surname before given name.

** Pulitzer Prizes have been won for this type of medical reporting. In 2011, a pair of reporters at the *Milwaukee Journal Sentinel* shared a Pulitzer for exclusive reporting on the life-saving genetic diagnosis of a boy named Nicholas Volker, who suffered a mystery autoimmune disease. By sequencing Volker's genome, Howard Jacob and Liz Worthey identified a rare mutation. A bone marrow transplant almost certainly saved Volker's life, mercifully ending a protracted diagnostic odyssey.

of life, changing the script for this and future generations."[3] Even after she got word of the twins' birth, Marchione still needed confirmation of the births and evidence that JK had actually done what he was claiming. Although unlikely, scientists have been known to commit fraud. What if this was just one giant outrageous hoax?

JK planned to submit his research article detailing the editing and birth of the world's first genetically edited babies to *Nature*. The first page of his manuscript listed ten coauthors, including JK's American PhD mentor, Michael Deem, a professor at Rice University in Houston, Texas. JK was hoping that *Nature* would act with "Shenzhen speed" and rapidly review and publish the article. Shortly before the Hong Kong genome ethics summit in late November 2018, Ferrell sent Marchione a copy of JK's draft manuscript. Lulu and Nana were test tube babies, conceived in a dish, their genetic makeup customized by a human hand using a fledgling gene-editing technology built around a billion-year-old bacterial enzyme. If true, she was holding literally the scoop of the century.

Marchione had no way of independently confirming the twin births—the identities of the children and their family, as well as their location, were a fiercely kept secret. But she had in mind to verify the results JK described in the manuscript. Both Marchione and her editors at AP insisted on doing things by the book. "We planned to publish only when—and if—we were satisfied that there was reasonable evidence that [JK's] claim was legitimate, or if he made it public in some fashion," she told me.[4] With JK scheduled to speak at the summit, he would never have a better chance of commanding the world's attention. How could JK resist the opportunity to pull the equivalent of a Steve Jobs mic drop: "And one more thing . . ."

Marchione sent extracts of the manuscript in confidence to a trio of medical and scientific experts—George Church, University of Pennsylvania cardiologist Kiran Musunuru, and Scripps Institute cardiologist Eric Topol. Musunuru said his heart sank as he opened the file. He called it "a soul-destroying moment."[5] This wasn't a hoax. If anything, it was worse.

In the end, Marchione's hand was forced. Shortly after touching down in Hong Kong, she got wind of a breaking story in *MIT Technology Review*. The reporter didn't know about the births, but he knew that JK was on the verge of making history by overseeing a pregnancy using gene-edited embryos. By the time I arrived in Hong Kong on the eve of the summit, the world had changed. I spent the taxi ride to my hotel trying to keep calm as I saw a trending hashtag—#CRISPRbabies.

Antonio Regalado, a tenacious science reporter at *MIT Technology Review*, is tall and lean, with hollow cheeks and eyes befitting a journalist not averse to pulling an all-nighter in service of a scoop. With a physics degree from Yale, a degree from the New York University School of Journalism, and nine years at the *Wall Street Journal*, he has impeccable credentials. But he admits to drawing much of his inspiration from the supermarket tabloids. "This is my sensibility," he says. "Are my stories worthy of *Tech Review* and *Weekly World News*?" Ideally, he wants his readers to come away from a story unsure if what they've just read is true. "But it is true! That's the sweet spot for me."[6] Stanford's Hank Greely quips: "He wants to be Woodward and Bernstein combined."[7]

In pursuing a massive story such as editing human embryos, Regalado followed the advice Walter Gretzky once gave his precociously talented son, Wayne: "Skate to where the puck is going to be." With eight of the first ten published reports of genome editing on human embryos emanating from labs in China, a country that seemingly lacked the legal restrictions or oversight of many countries in the West,[8] Regalado was in no doubt where the first attempts to produce a gene-edited baby were going to take place. The opportunity to investigate his hunch came in October 2018, one month before the Hong Kong conference.

Regalado got the chance to tour China with a pair of documentary filmmakers, director Cody Sheehy and Samira Kiani, an Iranian expat physician-scientist at the University of Pittsburgh. Along with George Church and biohacker Josiah Zayner, Regalado was going to be a central character in their documentary, *The Human Game.*[9] Providing ground support was Sheehy's brother-in-law, Nicholas Shadid, a China-based consultant who spoke Mandarin. Shadid had scheduled interviews with scientists in Shanghai and Guangzhou, a city of 13 million people in southern China, including a visit to Sun Yat-sen University to interview one Huang Junjiu.[10]

In 2015, Huang published the first attempt to perform CRISPR genome editing on human embryos. He was motivated by the public health threat of thalassemia in his home city. His experiments were little more than a proof of principle: they had been conducted on malformed embryos rejected by an IVF clinic—"tripronuclear zygotes" in the trade—with no intention of implantation. The results were underwhelming. For all the supposed ease of use of CRISPR, Huang's team reported only partial success in editing the target beta globin gene. Making matters worse, there was also evidence of random off-target edits in the genome.

Nevertheless, this was a world first so Huang could be forgiven for trying his luck by submitting his group's report to *Nature* and *Science.* Both journals rejected the manuscript in quick succession. With rumors of the Chinese study spreading, *Nature* published an op-ed written by Ed Lanphier, CEO of Sangamo, and colleagues titled "Don't Edit the Human Germline." Lanphier, Fyodor Urnov, and colleagues called for a ban on editing human embryos, partly for moral and ethical reasons but also because of the negative impact it could have on the future of somatic gene editing.[11] Huang eventually published his report in a China-focused journal called *Protein & Cell.*[12] It is unclear how rigorous the peer review process was: the time from manuscript submission to acceptance was just forty-eight hours.

Huang had kept a low profile since his fifteen minutes of fame in 2015, so his willingness to be interviewed was a minor coup for Regalado. As the filmmakers settled into their Guangzhou Airbnb, Kiani received a surprise email from Ferrell. Earlier in the year, Kiani had asked Ferrell for access to film one of Sangamo's gene therapy patients. In his email, Ferrell began by apologizing for not being able to help her film project, but all was not lost. He continued:

> I've taken a post in a Chinese lab working on the safety of CRISPR gene editing at the time of embryo fertilization. My

> goal is to push the lab to engage the world more directly in science and ethics given the controversial nature of the work and the potential to disrupt others working on somatic therapies. Would you be open to reconnecting . . .?[13]

Kiani couldn't believe her luck. Ferrell's new Chinese lab was in Shenzhen, just a sixty-minute train ride from Guangzhou. "We are in China! Why don't we meet?" she replied. Ferrell agreed to get together the next day, adding that he would be traveling with a Chinese colleague. Regalado did what any enterprising reporter would do: he googled the name of Ferrell's companion, who had been copied on the email.

It was He Jiankui.

The meeting took place in the afternoon at the Westin Hotel, a short walk from the major train station in Guangzhou. Shadid led the group to a dimly lit corner of the lobby, telling the hotel staff that they were guests awaiting a very important Chinese scholar. Ferrell allowed Regalado to join, providing their conversation was off the record.

JK was initially reserved, wary of the intense American journalist scribbling notes. His English was passable, having spent four years in the United States. After a while, Regalado asked Shadid to switch to Mandarin. Shadid asked if JK was interested in taking part in the film, opening up his life "to show the world that he is a person first and a scientist second, with dreams and a moral point of view." Shadid wanted to show a leading Chinese scientist as being similar to his Western counterpart, rather than the stereotypical boogeyman, "the personality-free Chinese lab rat with no individuality or moral agency."

JK and Ferrell insisted they did not want to be portrayed as "evil scientists doing something immoral." They wanted to change the perception of how China's work in gene editing is perceived in the West and give it legitimacy. Kiani and Shadid suggested a return visit in the New Year, when they could film JK with his family. He seemed to like this idea. "He had a crazy twinkle in his eye," Regalado recalled. He talked about a public opinion survey his team had conducted showing the Chinese public was largely in favor of gene editing. Ferrell then handed Kiani a sheet of paper.

It was a list of five ethical principles that JK felt should shape the future of genome editing in human embryos. They hoped the guidelines would be published just before the Hong Kong summit. They were:

1. Empathy for patients—for some families, early gene surgery may be the only way to cure disease.
2. Only for serious disease, never for vanity.
3. No one can control a child's life. A gene-edited child retains the same rights as a "normal" child.
4. Genes do not define us. DNA does not predispose us.
5. Everyone deserves freedom from genetic disease, regardless of wealth.

As the discussion turned to possible examples, JK said that editing the *CCR5* gene to in effect immunize babies from getting HIV would fit his ethical criteria. HIV was a huge public health problem in China, particularly in western China. He opened some slides on his laptop, which showed preclinical gene-editing experiments on a few hundred human embryos. It was debatable, however, whether targeting *CCR5* was tantamount to curing or merely preventing a disease. Privately, Regalado thought it was an extremely weak choice.

While JK believed editing this gene offered an important health benefit, the primary objective had to be safety. JK's expertise was in genome sequencing, which lent him some credibility in screening for errant edits. The other nagging issue was mosaicism, a phenomenon that occurs when not every cell in the developing embryo carries a particular gene variant or edit. JK acknowledged the concern but he didn't think it was a showstopper. He was more irked by one of Regalado's magazine stories, which featured a full-page illustration of a baby reduced to a series of billiard balls or atoms. JK said the image, which looked like the baby was being blown apart, was disgusting.

"Is this being tested in humans?" Kiani asked him. "We have no human trials," JK replied, while hinting that his plans would depend on the kind of response he receive at the summit in a few weeks' time. But later, JK conceded there was a gene-edited monkey fetus in a womb somewhere.

So how close was he to producing a human baby? "Almost there," he said. Regalado shivered. *Almost there . . .*

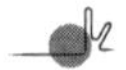

After JK and Ferrell left the hotel to return to Shenzhen, Sheehy turned on his camera while his friends exhaled and debriefed. "What just happened?" Kiani asked, her head spinning. "I think we just found out about this secret embryo editing project that we were looking for," Regalado said incredulously. "It was a hell of a meeting." Here, halfway around the world, he had stumbled upon the "promised land," meeting the ambitious young scientist running one of the largest human embryo editing programs in China. Moreover, he had signaled his intentions to move from monkeys to humans. But when?

The Huang embryo editing report in 2015 triggered a serious worldwide reaction led by Doudna, culminating in a major international ethics summit in Washington, DC. "The unthinkable has become conceivable," David Baltimore had cautioned opening the meeting. How would we as a society choose to use this capability—if at all? That put China back on its heels a bit. "They went from leading the science to being the caboose on the ethical conversation," Regalado said. He picked up the sheet of paper listing JK's five ethical guidelines for human germline editing. "This document is a list of moral principles *after* they did the experiment of editing embryos!" he said. "This is the necessary groundwork for producing a child with this technology." JK was trying to prevent an ethical backlash by retrospectively framing his "principles" to stay one step ahead.

Regalado felt it was only a matter of time before we were talking about a genome edited fetus in a womb somewhere. "The monkey is the tryout for the real thing. We might be having that conversation in the near future." Shadid agreed. Chinese government regulations meant there was only ad hoc enforcement, he said. "The Government can step in when they want and stop whatever's going on or they could try the wait-and-see approach."

Shadid said JK could be the human face of science and medicine in China, "the guy behind the first gene edited baby." They could film JK in his own words, the man "behind editing the first baby while holding his baby daughter." It was a tantalizing prospect. "I've never seen a more affable, expressive scientist in China," said Shadid, not that the bar was particularly high. *CCR5* offered a safe target and HIV a prevalent disease. To win the Chinese Central Committee's approval, JK was going to need bioethicists, government officials, and hospital administrators to approve the clinical trial. "They need a reason why this is necessary. It all adds up."

The next day, Regalado interviewed Huang as scheduled and duly wrote up a story for his magazine, but this wasn't where the puck was going—he was desperate to find out what was going on in Shenzhen. Back in the United States, Regalado dug into JK's story, hoping it would be the centerpiece in his article that would be the curtain-raiser before the Hong Kong summit. With one week to go, Regalado checked in with Kiani: Ferrell had told her that JK was planning on recruiting women in the New Year for trials targeting two genes—*CCR5* and *PCSK9*.

On the Sunday morning of Thanksgiving weekend, November 25, two days before the Hong Kong conference started, Regalado finally hit the mother lode. Searching "He Jiankui" and "CCR5," Google finally served up an entry in the Chinese Clinical Trial Registry—helpfully listed in Mandarin and English—that was backdated to November 8. The study leader was He Jiankui and the title said it all: "Evaluation of the safety and efficacy of gene editing with human embryo CCR5 gene."

The trial inception date was listed as March 2017. The applicant was Qin Jinzhou, JK's embryologist. The study rationale cited the Berlin patient as creating "a new medical model for HIV elimination." JK would recruit HIV-positive patients with infertility, seek informed consent and ethical approval from the hospital. It went on: "Through the CCR5 gene editing of the human embryo in a comprehensive test system, we set to obtain healthy children to avoid HIV providing new insights for the future elimination of major genetic diseases in early human embryos."[14]

And there was more. There were two documents linked to the registration. One was an ethics statement (in Mandarin) dated March 7, 2017, submitted to the Harmonicare Shenzhen Women and Children's Hospital, entitled simply "CCR5 gene editing." Seizing on the just-published National Academies of Sciences (NAS) report on genome editing, which "for the first time approved the ethics application for a major disease in an embryonic editing study," JK outlined his plans to target *CCR5* in embryos at risk of HIV. The use of CRISPR editing "will bring a new dawn for the treatment of untold numbers of serious genetic diseases." And then:

> We ardently expect that . . . [the project] will occupy the commanding elevation of the entire field of gene-editing technologies. Like the point of an awl sticking out through a bag, the project will stand out in the increasingly intense international competition of gene editing technologies. This innovative research will be more significant than the IVF technique which won the 2010 Nobel Prize and bring about the dawn of the cure for untold severe genetic diseases.[15]

But the real smoking gun, Regalado found, was in of all things an Excel spreadsheet. It was a table in Chinese but there was a familiar scientific term in the title: "cfDNA." Now Regalado got goosebumps. cfDNA stands for circulating free DNA, the raw material for noninvasive prenatal testing (NIPT), the new standard procedure for performing genetic analysis on pregnant women. The table reported DNA analysis in maternal blood at three timepoints: 12, 19, and 24 weeks. NIPT was developed independently by Dennis Lo in Hong Kong and Stanford's Stephen Quake—JK's postdoc supervisor. JK's expertise was in precisely this type of next-gen sequencing analysis that would be used to perform NIPT. Regalado concluded that JK was monitoring a pregnancy in a genome-editing trial. That could only mean one thing.

Regalado called Shadid in China, unconcerned it was the middle of the night in Shanghai. "Is this what I think it is?" he asked. Shadid said yes. Over the next few hours, Shadid worked on translating the ethics document. Next Regalado called JK and asked if any gene-edited babies had been born. JK declined to answer and referred him to Ferrell. Shadid alerted Kiani that Regalado was planning to publish his explosive story "as soon as people are awake in China." Kiani called Ferrell, who had been asleep. Ferrell scanned his phone to find multiple messages from Regalado. When they finally spoke, Ferrell wouldn't confirm anything but urged Regalado to delay his story while offering more access to JK.

But Regalado wasn't interested. For about four hours, he'd been more or less the only person in the world who knew what JK had done and was ready to publish it. As he reviewed the story one last time, he thought: *This is either the best story I've ever written, or it's the last!* But here it was—*Weekly World News* meets *Tech Review*.

China was waking up. Regalado clicked publish.

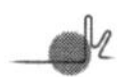

The term "exclusive" is hopelessly overused in the popular media, where every story and sound bite is tagged as breaking news. But in case anyone was in any doubt, Regalado prefaced his bombshell story with the word in all caps—and included it in the URL for good measure:

> EXCLUSIVE: Chinese scientists are creating CRISPR babies[16]

At 7:15 P.M. (Eastern time) on Sunday, Regalado tweeted: "BREAKING: Chinese scientists making CRISPR babies." The trial registration news was a bombshell, and while Regalado couldn't say unequivocally that gene-edited babies had been born, there was an excellent chance that one or more had been conceived."

Regalado's scoop left Marchione and the AP with an unforeseen dilemma. She had been stealthily preparing her own exclusive, waiting to release her story at the right time, either when JK presented on stage in Hong Kong, or when his *Nature* manuscript was finally published. After consulting her editors, she posted her own story, just a few hours after Regalado. The AP story contained two crucial exclusive details: the births of Lulu and Nana, genetically edited twins.[17]

Few people reading either story knew who JK was, but suddenly we could see and hear him in his own words. JK uploaded five short videos onto YouTube—a site ironically banned in China—in which he revealed the names of the twins and defended his decision to carry germline editing to term. Beaming like a proud parent, JK announced the birth of two beautiful Chinese girls, Lulu and Nana."[18] The twins were safe and healthy, the gene surgery had worked with no off-target effects. "As a father of two girls, I can't think of a gift more beautiful and wholesome for society than giving another couple the chance to start a family," JK said.

Gene surgery should only be used to alleviate disease, JK insisted; any desire to deploy the technology cosmetically or to try to enhance a child's IQ should be banned, as it is "not what a loving parent does." He closed with the understatement of the year: "I understand my work will be controversial, but I believe families need this technology, and I'm willing to take the criticism for them."

One of the few people who was not shocked by the sensational headlines was Doudna. Since 2015, she had galvanized the scientific community and led the ethical debate about CRISPR and its potential misuse on human embryos. She thought human germline editing was almost inevitable although she had resisted calls for a moratorium, fearing it could drive the research underground. In fact, she all but predicted the CRISPR baby saga. During an informal interview in 2017, she revisited her efforts to encourage ethical debate: "I didn't want to see someone giving birth to a 'CRISPR baby' so that they could be famous and then having that lead to all kinds of health problems for that child that would then cause a backlash," she said. "You could imagine this scenario."[19]

Such a scenario could prompt fierce backlash from the public and legislators—not unlike the clampdown on federal stem cell funding in the wake of the birth of Dolly the sheep in 1998 and overblown fears of human cloning. Other commentators saw reason to be worried. In a TEDx talk in 2015, UCLA stem cell biologist Paul Knoepfler reacted to the first Chinese report of editing in human embryos. "I think someone is going to go that extra step and continue the GM human embryo work and maybe make designer babies," he predicted.[20]

Doudna's biggest concern wasn't so much the safety of CRISPR but the negative publicity that would accompany any rogue effort at germline editing, not to mention the government scrutiny that might shut down this new technology before it could help patients. In a discussion with NPR's

Joe Palca at Berkeley, Doudna discussed launching the debate around CRISPR and germline editing at her Napa retreat in 2015. You can hold ethical debates, Palca said, but how are you going to stop somebody who is not here? "Apart from the lab police running around to stop them?" "I don't even think we have the lab police," Doudna replied to much laughter from the audience.[21]

But there was no laughing now. On the day after Thanksgiving—appropriately Black Friday—Doudna received a terse email from her acquaintance, He Jiankui, that left her feeling physically sick. The subject line read simply: "Babies Born."[22]

She headed to the airport, while confiding the news with a very close circle of colleagues including Barrangou, who was on a family vacation in London. She was preparing for a Hong Kong showdown.

CHAPTER 15
THE BOY FROM XINHUA

The most common picture conveyed by the media of the unheralded scientist at the center of the #CRISPRbabies firestorm was that of a modern-day Victor Frankenstein, an unscrupulous rogue operator secretly experimenting on human embryos in his underground lair surrounded by flasks of colorful bubbling liquids and galvanic sparks of electricity. In other narratives, He Jiankui was a 21st century Sorcerer's Apprentice, naïve, ambitious, reckless, and hopelessly out of his depth. But as more details emerged about JK's confidants and movements leading up to 2018, we glimpsed a more complex, nuanced narrative. As one Chinese commentator suggested, "the story that emerges is more Elizabeth Holmes"—the disgraced founder of Theranos—"than Dr. Frankenstein."[1]

So who is He Jiankui and where did he come from?

He Jiankui was born in 1984, of all years, in Xinhua, a small village in central Hunan province in southern China where the average income was just $2 a week. His parents were both farmers, working in the fields bordering the Yangtze river. In the summers, JK would have to pick leeches off his legs while helping his parents. He shone academically and attended the best high school in the region.[2] Poor and with no political connections, his only way out was to excel in the *gaokao*, China's punishing college entrance exam held over two consecutive days each June. He earned a coveted spot at the University of Science and Technology of China (USTC) in Hefei, the "Caltech of China." With some 30 percent of graduates continuing their education abroad, USTC earned another nickname: "United States Training Center."

JK enrolled at USTC in the summer of 2002, one year after the publication of the first draft of the human genome sequence. He wasn't able to enroll in his first-choice program, but after a freshman year in the Department of Precision Machinery and Precision Instrumentation, he transfered to modern physics. His father bragged to the Chinese media that his son always came top of his class (although this may not be entirely true). Whenever JK came home, "everyone wanted to meet him and hear his tips for good grades."[3]

In 2006, scholarship in hand, JK joined the flock of migrating USTC graduates heading to America, winning a PhD slot in the bioengineering program at Rice University in Houston. JK's interests were shifting to the life sciences, believing that the golden age of physics had passed. He joined the lab of Michael Deem, who had studied at the real Caltech. Deem's ambition was to define an example of Newton's famous F=ma equation for biology. His diverse research interests—genetic engineering, mathematics, physics, and astronomy—suited JK. Together they coauthored five papers published in various physics and biology journals on an eclectic range of subjects, including the hierarchical evolution of animal body plans; the impact of globalization and recessions on the world trade network; and a computational analysis of emerging flu strains to guide vaccine development, another of Deem's principal interests.

It was during his PhD that JK first got his first taste of CRISPR—Deem and JK published a mathematical model of CRISPR arrays in 2010 in a respected physics journal. The paper, which had nothing to do with gene editing, featured lots of advanced mathematics that went over my head, including the Lebowitz-Gillespie algorithm, Latin hypercube sampling, and the Shannon entropy.[4] Three months later, there was a double celebration. JK was awarded his PhD for his thesis, "Spontaneous Emergence of Hierarchy in Biological Systems," completed in very quick time. And he got married to Yan Zeng in the Rice University Chapel. A local Chinese-language newspaper, the *Southern Chinese Daily News*, covered the nuptials under this headline: "With outstanding morals, excellent academics, and infinite potential, a union for life complete with both good looks and scholarly talent."

Soon thereafter, JK and his new bride moved to the Bay Area, where JK took up a postdoctoral position at Stanford University with leading biophysicist Stephen Quake, the entrepreneurial cofounder of a string of biotech companies including publicly traded Fluidigm, Verinata Health, a non-invasive prenatal testing (NIPT) platform, and a next-generation sequencing (NGS) company called Helicos.[5] One colleague said Quake was "out to hunt death down and punch him in the face."[6] After developing the technology for a new NGS instrument, the Heliscope, Quake volunteered to be one of the first people to have his genome fully sequenced for an estimated $50,000.[7] He posed for the cover of a biotech magazine in front of the state-of-the-art instrument,[8] while a paper on the Quake's genome, aka "Patient Zero," appeared in the *Lancet*.[9] His colleague Euan Ashley told Quake that he carried a gene that predisposed him to cardiomyopathy, but Quake wasn't comfortable following all the well-intentioned medical advice. "We still haven't found the compliance gene," joked another colleague, Atul Butte.[10]

NGS, Quake said, was a cross between a microscope and a washing machine. Helicos was based on a proof-of-principle report that Quake published in 2003, describing the sequencing of "a ridiculously small" string of bases—four to be exact—along a single molecule of DNA. The article was spotted by a biotech entrepreneur named Stanley Lapidus, who flew out to San Francisco to persuade Quake to build a company around this embryonic single-molecule sequencing technology.

Five years later, Lapidus' company had built a prototype "DNA microscope" and set course to take on the market leader, Illumina. But the Heliscope was a monstrous slab of hardware—some customers had to reinforce their lab floors to accommodate the instrument—with an equally hefty $1 million price tag to match. Worse, there were chemistry glitches that compromised accuracy. Helicos went public in 2007 but that too was a miscalculation. "We were trying to IPO with no revenue. We got killed!" recalls Steve Lombardi, who succeeded Lapidus as CEO.[11] By the time JK arrived at Stanford, Helicos was delisted from Nasdaq and shedding staff—a rare defeat for Quake.

Another Quake innovation would have a direct bearing on JK's future path. In 2008, graduate student Christina Fan led an NGS study to survey the cell-free (cf) DNA in maternal blood to measure the relative contribution of each chromosome. By literally sampling millions of free-floating DNA fragments, she could diagnose cases of trisomy 21 (Down syndrome) and other chromosomal disorders or aneuploidies.[12] This breakthrough—similar results were reported by Dennis Lo and colleagues in Hong Kong—gave birth to the technology of NIPT. Quake's spin-out, Verinata Health, was later acquired by Illumina.

JK spent twelve months in the Quake lab, working mostly as a computational biologist, publishing a paper on the sequence analysis of genes in the immune system.[13] But he didn't have to wait long before he landed a golden opportunity to return to his homeland. In late 2011, JK met Zhu Qingshi, the former president of USTC, during a recruiting trip. Zhu had been named founding president of an ambitious new private university in Shenzhen modeled on Stanford—the Southern University of Science and Technology, or SUSTech.

JK returned home on the wings of the Peacock Plan, a multimillion-dollar program to attract outstanding talent to Shenzhen. (The name comes from an old Chinese proverb, "The peacocks fly to the southeast.") JK announced the "He Jiankui and Michael Deem Joint Laboratory" on his blog before he'd even left Stanford, along with the SUSTech Gene Sequencing Center, which would feature "a world-class next-generation DNA sequencing platform."

The location of SUSTech in Shenzhen, neighboring Hong Kong, was significant. Just twenty-five years earlier, Shenzhen had been little more than a fishing village. In 1980, the Chinese government declared Shenzhen an economic prosperity zone. The subsequent expansion has been astonishing: today, Shenzhen is a sprawling city of 20 million people, dominated by high-tech, manufacturing, and China's largest genomics company, BGI (formerly the Beijing Genomics Institute).

BGI was officially born in September 1999—or more precisely, at "9 seconds past 9 minutes past the 9th hour of the 9th day of the 9th month of the 99th year of that century," as the founding chairman Yang Huanming told me several years ago.[14] We met following a party at Boston's famous tourist spot—the Cheers pub—to mark the opening of BGI's operations in the United States. The cultural entertainment included a sand painter, artfully sculpting pictures in sand on an overhead projector, charting the course of history from the big bang to the birth of BGI. The institute had grown from a bit player in the international Human Genome Project supplying about 1 percent of the total genome sequence, to one of the world's genomics powerhouses. BGI deciphered the rice genome in 2002 and six years later published the genome sequence of the first Asian person, an anonymous individual named "Y.H." ("If someone says it's me, it's a rumor," Yang smiled.)

Under the leadership of Yang, president and cofounder Wang Jian, and a talented young bioinformatician, Wang Jun, in 2011 BGI bought a fleet of 130 Illumina NGS instruments to establish the world's largest genome sequencing center in Hong Kong.[15] BGI compiled a "genomic zoo" of plant genomes, including rice, cucumber and soybean, and animals, including the honey bee, silkworm, and of course, China's beloved giant panda.[16] The secret to BGI's success—apart from quality, speed, and price—was scale. "We must be the biggest," Yang said.

In 2013, BGI became the first Chinese organization to buy an American public company when it acquired Complete Genomics, a San Francisco–based company with its own NGS technology, which now features in a new spin-off. This company, MGI, has released its own benchtop sequencer that has quickly attained a domestic market share approaching 50 percent, and in 2020 announced it had reached the "$100 genome."[17]

In addition to pushing into the clinical market, BGI embarked on a Cognitive Genomics project, in collaboration with Michigan State physicist Stephen Hsu, to investigate the genetics of cognition and intelligence. The controversial project has since been shelved, but it epitomizes Wang Jian's attitude to the genetics market. In the West, "you have a certain way," Wang said. "You feel you are advanced and you are the best. Blah, blah, blah."

Wang has no interest in following the fussy rules and regulations decreed by American institutes and think tanks. "You need somebody to change it, to blow it up," Wang said. "For the last five hundred years, you've been leading the way with innovation. We are no longer interested in following."[18]

That spirit and resolve was infectious. Whether studying the genetic basis of intelligence or running experimental clinical trials or editing human embryos, Chinese scientists were determined to push the envelope, with little concern for the cultural, ethical, or regulatory strictures that might govern such endeavors in the West.

With BGI dominating the genomics landscape in China, Shenzhen was a dynamic location for an ambitious young researcher looking to make his mark both in academia and business. JK was lavished with millions of dollars' worth of government funding—$5 million in 2016 alone.[19] "Shenzhen has transformed itself from labor-intensive industry to high tech," JK told my former colleagues at *Bio-IT World* in 2015. "The government has ambitions. They're trying to switch from 'Made in China' to 'Invented in China.'"[20]

Part of that reinvention was to retool the dormant Helicos sequencing technology following the company's demise in 2012. While a Boston company called SeqLL bought up Helicos's hardware in a fire sale, JK licensed Quake's intellectual property to build a new company. Following in BGI's footsteps, he aimed to release a domestic DNA sequencer deployed in Chinese hospitals addressing domestic health problems, servicing the largest population in the world.

To assist with the company's launch, JK contacted HDMZ, an American PR company that had produced a magazine article on sequencing technology for *Nature*. The company's director of science communications, Ryan Ferrell, flew to Shenzhen in 2015. JK's preferred name for the company was Public Genomics, which signaled its aim to serve the Chinese people. But the leadership team voted for Direct Genomics, or Hanhai in Chinese. "We're a new generation of entrepreneurs," JK said. "We've had great discussions with the Chinese FDA . . . They really hope our Chinese brand could be used in hospitals."[21]

JK invited Deem, his PhD mentor, to become a scientific advisor. He redesigned the Helicos platform so that the target enrichment for a simple diagnostic test could occur on the same flow cell that performed the DNA sequencing. With improved optics, lenses, and cameras, his company shrank the sequencer from the size of an industrial fridge to a portable dehumidifier, and named it the GenoCare analyzer.

In December 2015, JK posted a photo on WeChat, the ubiquitous Chinese social media platform, taken next to prototype instruments alongside Deem and Quake, everyone wearing lab coats and disposable booties. He also persuaded Helicos's former chief science officer, Bill Efcavitch, to join the advisory board. JK and Deem traveled across China surveying the exploding clinical market. The GenoCare still suffered some limitations, but in specific applications such as searching for cancer gene mutations or hepatitis, JK was confident the technology would suffice. Others were skeptical: "The company is just dodging GenoCare's faults as a sequencer by picking clinical targets where its flaws are irrelevant," *Bio-IT World* concluded.[22]

Later that year Quake called Lombardi, who had moved to Connecticut and launched a consulting business after Helicos folded, to see if the former CEO would like to serve as JK's business mentor. For the next two years, Lombardi helped JK manage the financial and strategic aspects of growing his company. He introduced him to some leading Wall Street analysts covering genomics. "He bought my rolodex," Lombardi told me.[23] In July 2017, JK proudly posed for photographers next to the prototype GenoCare at the instrument's official debut in Shenzhen. The launch of the world's first "third-generation sequencer" featured congratulatory speeches by various government dignitaries and senior Chinese scientists. By 2018, the company had raised nearly $35 million and claimed to be manufacturing 1,000 instruments a year to meet burgeoning demand.

All very impressive, as was JK's involvement in several other business ventures, including a liquid biopsy company called Vienomics. But apparently it was not enough. Like Icarus, the son of a master craftsman, the boy from Xinhua was setting a course perilously close to the sun.

A few months after the furor surrounding the first Chinese embryo editing study in 2015, the British government granted its official seal of approval to a very specific form of germline editing called mitochondrial replacement therapy (MRT). Sometimes called "three-parent IVF," MRT was pioneered by Douglas Turnbull at the University of Newcastle. His inspiration was meeting Sharon Bernardi, a mother from nearby Sunderland, who had buried seven children in three different cemeteries because of a mitochondrial disease called Leigh syndrome. Her fourth son Edward lived until twenty-one, but was often writhing in pain. "I don't want my son to have just died for nothing," Bernardi said. "We're trying not to pass it on to children and make it better for future families."

Mitochondria are the capsule-like power-stations inside our cells that create adenosine triphosphate, ATP, the molecule that provides energy for myriad cellular processes. Mitochondria contain a tiny circular piece of DNA, almost insignificant compared to the size of the nuclear genome, encoding just a few dozen genes—less than 0.2 percent of the human genome. But just like the 20,000 or more genes on our twenty-three pairs of nuclear chromosomes, these mitochondrial genes can suffer damaging mutations. These mutations result in a range of debilitating, sometimes hard-to-diagnose genetic diseases, including Leigh's disease, that affect 1 in 5,000 births. Importantly, mitochondria are only inherited through the maternal line. That means if the mother has a mitochondrial disorder, she will inevitably pass it onto her children, boy or girl.

Designed to help at risk pregnancies, MRT is a modified version of IVF. Before fertilization, the nucleus of an egg from the affected mother is swapped into an enucleated egg from a donor that has healthy mitochondria. Following IVF, the resulting embryo can claim three parents—the complete set of twenty-three chromosomes provided by both father and mother plus the smidgeon of mitochondrial DNA from the woman who donated the hollowed oocyte. (If you want to get mathematical about it, it's not so much "three-parent" as "2.001 parent.")

In October 2015, the British parliament became the first country to legalize MRT with the passage of the Human Fertilization and Embryology Act. With almost fifteen years of research and preparation in the balance, Turnbull was understandably stressing out in the public gallery when the House of Commons voted on the bill. As the vote totals were announced showing comfortable passage, cheers of delight burst out from Turnbull's guests, patients, and family members, followed by tears of relief. The bill was ratified in the House of Lords, where Viscount Matt Ridley voiced his support. "Britain has been the first with most biological breakthroughs. In every case we look back and see we did more good than bad as a result," he said. Failure to act to prevent suffering would be on their consciences, adding: "There is nothing slippery about this slope."[24]

But there were still regulatory hurdles to overcome before the first clinical experiments could take place. In April 2016, the world's first "three-parent" baby boy was born to Jordanian parents in Mexico City. The mother had lost two children and suffered four miscarriages because of Leigh's disease. John Zhang, a Chinese-born, British-educated fertility expert based in New York, performed the spindle nuclear transfer procedure that produced five embryos. Only one developed normally, which Zhang implanted. He defended his decision to do the procedure in Mexico, saying "to save lives is the ethical thing to do."[25]

Through his spin-off company, Darwin Life, Zhang attempted to offer a similar service for older women to transfer the nucleus into the egg cells of a younger woman. The FDA quickly stepped in, citing a 2015 Congressional amendment that forbids the agency from considering applications in which a human embryo is deliberately created or modified to include a heritable genetic modification. Zhang's company complied but he still believes that germline editing will be useful in the future. Any technology that will eventually benefit humankind should be allowed, he told the *Washington Post*. "Look at history: People were against antibiotics, general anesthesia, vaccines."[26] Zhang has partnered with a Ukrainian physician, Valery Zukin, as reports of MRT babies in countries like Ukraine and Greece continue.[27] [28]

Turnbull and his colleagues weren't overly upset by Zhang's scoop. "The translation of mitochondrial donation to a clinical procedure is not a race but a goal to be achieved with caution to ensure both safety and reproducibility," said Alison Murdoch.[29] In 2018, the British Government's Human Fertilization & Embryology Authority (HFEA) granted permission for the first MRT procedures to take place for two women. Within twelve months, a dozen more applications were approved. We await news of the first pregnancy.

Meanwhile, using philanthropic funds, Dieter Egli at Columbia University has used MRT to create "three-parent" embryos for several women with mitochondrial disorders. Those embryos remain frozen in legal limbo until US regulations change,[30] either through legislation or possibly litigation. Whether MRT is a natural extension of IVF or a crossing of the Rubicon is open to debate. But the families affected insist they're not trying to fashion designer babies or playing God. They just want a healthy, biologically related child.

"Extend your arm. Expose a vein. Make a fist. And it's 50 yuan."

In the early 1990s, China's rural Henan province was the epicenter of a devastating black-market blood drive. "Blood heads" set up stations across the province to buy blood from hapless peasants, from which plasma could be separated and sold abroad. For the farmers, the "plasma economy" was a godsend worth the princely sum of fifty to seventy yuan, enough to buy kilos of rice, save for a portable television, pay school bills, or even build a small house. Tragically and predictably, the use of unsterilized needles and the mixing of blood samples before reinfusing to volunteers led to an HIV epidemic, a "nameless fever," to which the local authorities turned a blind eye. It took a local doctor, Shuping Wang,* to finally get the attention of authorities in Beijing, which has provided free HIV drugs since 2003.

In 2016, JK visited Wenlou—the "AIDS village" four hundred miles south of Beijing.[31] Hundreds of villagers—almost one in two blood

* Shuping Wang, a naturalized American citizen, died in 2019 after suffering a heart attack while hiking outside Salt Lake City. Her whistleblower experience inspired a play, *The King of Hell's Palace*, which premiered in London. Wang publicly called out the pressure placed on her family in China from authorities to call off the production. Twenty-five years later, China again showed its disdain to the medical whistleblowers of the coronavirus pandemic.

donors—contracted the disease. By the end of 2015, more than two hundred residents had died. Similar tragedies played out in dozens of villages across the province. Aside from the gravestones, JK saw few signs of a national scandal. There are no longer emaciated patients walking in the streets. Many of the orphaned children have left or been sent far away to start a new life, sometimes under a false identity to escape discrimination. In the elderly villagers' faces, he might have seen flashbacks to his own family and childhood.

Much like China's initial response to the coronavirus outbreak twenty-five years later, the government's response to the AIDS crisis was denial or inept. But more recently, the cause of HIV awareness received a high-profile champion—China's First Lady, Peng Liyuan. In 2017, she received the UNAIDS award for her work fighting HIV discrimination. Surely success in developing a bold new strategy to combat HIV would have support in the highest echelon of the Chinese government.

With Chinese scientists launching the first gene-editing experiments on human embryos, JK saw an opportunity to make an even bigger name for himself. His biggest inspiration was an Englishman almost sixty years his senior—Nobel laureate Robert "Bob" Edwards, the father of IVF. After Edwards died in 2013 at eighty-eight, *Nature* wrote: "Several scientists have made discoveries that have saved millions of lives. Robert Edwards helped to create them."[32]

Edwards met Patrick Steptoe, an obstetrician who worked at Oldham General Hospital, near Manchester, at a conference in 1968 in London. The same year, using oocytes recovered from a woman at Edgware General Hospital* in north London, Edwards performed IVF for the first time. Edwards and Steptoe published their initial results in *Nature* in 1969, prefaced by one of those classic understatements for which Brits (and *Nature*) are famous:

> Human oocytes have been matured and fertilized by spermatozoa *in vitro*. There may be certain clinical and scientific uses for human eggs fertilized by this procedure.[33]

---

* Also my birthplace, it should be noted.

The study didn't mention "test tube babies." But *Nature*'s savvy editor, John Maddox, played up the study in a piece for the *Times* that appeared on Valentine's Day. Subsequent newspaper headlines fretted about a "human time bomb," a "test tube baby factory," and "the end of human beings as a wild breeding race." Third World nations could improve their influence and wealth by breeding "a race of intellectual giants." All this was, I suspect, much to Maddox's journalistic amusement. It ensured that Edwards and Steptoe's future work was carried out under the intense glare of the media, not to mention the church and the government.

In 1971, Edwards reported his first success in culturing human blastocysts. "There should be no criticism in giving these [infertile] couples their own children: comments about overpopulation seem to be highly unjust to such an underprivileged minority," he wrote.[34] Despite having their funding proposal rejected, Edwards and Steptoe persevered. Their first attempt at re-implanting an IVF embryo in 1976 resulted in an ectopic pregnancy. But on July 25, 1978, Steptoe delivered Louise Brown, weighing in at 5 pounds, 12 ounces, by cesarean section. Edwards and Steptoe suggested Louise's middle name be "Joy," for what she would bring to so many couples. Among the worldwide well-wishers was the Queen's gynecologist.[35] Only two more IVF babies were born in the first couple of years after Brown's birth, the first in the United States not until 1983. IVF wasn't legalized in the UK until 1990. But from such small beginnings, the technology blossomed. In the first forty years since Louise Brown's birth, an estimated eight million IVF babies were born.[36]

Edwards was finally awarded the Nobel Prize in physiology or medicine in 2010. The honor came too late for Steptoe, who was deceased, and almost too late for Edwards who, suffering from dementia, was unable to travel to Stockholm. His Nobel Prize was roundly criticized by the Vatican but his friends said the honor delighted Edwards—and many grateful fans. One comment left on the Nobel website said: "Dr. Edwards, thank you for my life."[37]

JK listed Edwards among his medical role models, along with the time it had taken for them to be duly recognized, in a PowerPoint presentation that he showed to his lab:[38]

Christiaan Barnard: Domestic, 3 weeks, International, 1 Year
Robert Edwards: 7 years
Edward Jenner: 1 year.

Drawing encouragement from these examples of scientific trailblazers, JK imagined his own future reflected in their image: initially controversial perhaps, but ultimately celebrated as having been in the vanguard. It was all about having the courage to take a controversial first step—breaking the glass—and thereby pushing science and humanity forward.

CHAPTER 16
BREAKING THE GLASS

In March 2016, Michael Deem visited Shenzhen to speak at a small symposium on "Biodynamical Systems." In the official group photo, Deem stands tall in the center of the group. His smiling protegé, He Jiankui, is easily recognizable. Deem's lecture was on CRISPR, the same topic on which JK and Deem had once published together. How far CRISPR had progressed since then: it was now a universal genome editing tool, spawning several biotech companies hoping to cure cancer and genetic disorders.

Six months later, SUSTech announced that Deem, listed as a member of the university's Department of Biology, and He Jiankui had won a $5.7 million grant as part of Shenzhen's Peacock Plan for further development of their sequencing system. The thrill of building a company like Direct Genomics with the prize of cracking China's clinical market would have satisfied the ambitions of most biotech entrepreneurs. But although he did not have any hands-on experience using CRISPR, JK was contemplating an even more daring venture.

JK first discussed his idea to edit human embryos privately with Quake, his former Stanford supervisor, in 2016. The response wasn't what he was expecting. "That's a terrible idea. Why would you do that?" Quake told him.[1] JK reluctantly agreed with Quake's advice that, at a minimum, he should obtain the appropriate ethical approval and patients' informed consent. JK promised to do so, emailing later: "I will take your suggestion that we will get a local ethic approve [sic] before we move on to the first genetic edited human baby. Please keep it in [sic] confidential." Quake thus became the first of a small but not insignificant "circle of trust"—scientists and ethicists in whom JK confided and sought advice. Their reactions varied upon learning of JK's intentions, but no one broke a confidence or blew the whistle.

That summer, JK traveled to New York to attend the biggest conference on CRISPR and genome editing, held at Cold Spring Harbor Laboratory on Long Island. During one of the breaks, JK introduced himself to Doudna, one of the meeting co-organizers, and posed for a selfie with her sitting in the front row of the Grace auditorium, which he posted on WeChat.[2] Doudna took such requests in stride: she is about as popular as Taylor Swift, graciously allowing photographs from admiring students and scientists at every event.

In late 2016, JK reached out to Mark DeWitt, a lesser known scientist at Doudna's Innovative Genomics Institute (IGI), who had recently published a paper on genetically editing cells for sickle-cell disease.[3] After meeting for lunch, DeWitt accepted JK's surprise invitation to visit China to give a lecture at SUSTech, but not before reiterating that Berkeley had no interest in supporting experiments on human embryos. "I'm a little reluctant to speak about it again," DeWitt told me by phone. "It's important but it's not what I want to be known for."[4]

In January 2017, JK was honored to be invited to attend a small workshop at IGI organized by Doudna and Stanford bioethicist William Hurlbut on the topic of public education and engagement around human genome editing. JK was the youngest scientist at the meeting, and other than delivering a short talk presenting preliminary data on editing in human embryos, focusing on the safety of CRISPR technology, kept a low profile. On his blog, he posted a summary of his talk, which highlighted five safety issues, including off-target effects and mosaicism, that would require further study. He signed off his report by stating that the behavior of anyone performing germline editing would be "extremely irresponsible," unless and until these important safety issues were resolved. He also noted something else of interest: the Valentine's Day publication of the National Academy of Sciences report on germline editing, which (as we will see later) in JK's view gave a "yellow light" for human germline editing.[5]

The meeting began with a bang: a public lecture by George Church, provocatively titled "Future, Human, Nature: Reading, Writing, Revolution."[6] Church spoke openly about germline therapy, speculating that mitochondrial germline therapy would move into treatments for infertility. Compared to our ancestors, "we are already augmented," Church said, mostly through physics and chemistry rather than genetics. He noted humanity's ability to see, to hear, to fly, to reach altitudes, to plumb the ocean's depths, further than our ancestors could have dreamed. But geneticists would get their chance: near the end of his talk, Church displayed a list of "protective alleles"—variants in certain genes that offer beneficial traits with no known or serious downsides. Among those genes was *CCR5*. "Wow, that was wonderful," said Doudna, Church's host for the evening, probably echoing JK's sentiments in the audience.

During the meeting, JK also met William Hurlbut and his son Ben, a science historian at Arizona State University. Looking back, Ben Hurlbut concluded that JK's motivations were familiar and mundane,

> driven by the high-octane milieu of contemporary biotechnology, both in the United States and in China. He internalized ideas that led him to believe that his experiment would elevate his status in the international scientific community, advance his country in the race for scientific and technological dominance, and drive scientific progress forward against the headwinds of ethical conservatism and public fear.[7]

Two years later, JK shared with Hurlbut that one particular statement at the IGI conference really affected him. "Many major breakthroughs are driven by one or a couple of scientists . . . cowboy science," JK said. "You need a person to break the glass."[8]

JK was following progress in the arena of CRISPR gene editing. China was racing ahead in offering CRISPR-based somatic therapies to dozens of cancer patients "unhampered by rules," although it was unclear how successful these early efforts were.[9]

But JK was interested in going much further, to "break the glass." Engineering a permanent, heritable change to human DNA as a means to treat or prevent a disease not only in a patient but also in their children forevermore. In the three years following the groundbreaking human embryo experiments performed in Guangzhou in 2015,[10] nine more human embryo editing studies were published. All but two came from research groups in China.[11, 12, 13, 14, 15, 16, 17] In March 2017, a group at Guangzhou Medical School reported editing a pair of disease genes in human embryos. Although none of the studies had been designed to actually implant an edited embryo, there was a feeling that it was only a matter of time. When *Nature Biotechnology* asked leaders in the genomics field whether germline editing was inevitable, including Doudna, Dana Carroll, Hank Greely, Robin Lovell-Badge, and Craig Venter, they concurred it was.[18]

In contrast to the Chinese reports on gene editing in human embryos, two other studies—one in Europe, the other in the United States—were given top billing in *Nature*. One was led by Kathy Niakan at the Francis Crick Institute in central London, the $1 billion crown jewel of British molecular biology research. The institute was officially opened by the Queen in 2016, who unveiled a portrait of Crick that had been commissioned by Jim Watson. Niakan is a widely respected developmental biologist who studies the fundamental mechanisms by which embryos develop in the critical first few days and weeks after fertilization. Understanding how specific genes orchestrate cell growth and development could help identify risk factors for miscarriages and other complications of pregnancy, and improve the success rate of IVF.

In 2016, the UK HFEA granted Niakan limited permission to use CRISPR-Cas9 to engineer DNA alterations in human embryos.[19] She used spare embryos created during IVF and donated for research, growing them for no more than fourteen days. "I promise you she has no intention of the embryos ever being put back into a woman for development," insisted her colleague, Lovell-Badge. In September 2017, using CRISPR, Niakan's team reported in *Nature* a key role for a gene called *OCT4* in human embryonic development.[20]

A bigger splash came just a month earlier. ONE GIANT STEP FOR DESIGNER BABIES was the exclusive front-page headline in a British tabloid. The scoop, from science journalist Steve Connor, also ran in *MIT Technology Review* under a more sober title: "First human embryos edited in U.S."[21] Connor had heard the news several weeks before *Nature* was due to publish a report from a leading American IVF center. The journal editors were "incandescent that their thunder had been stolen," which is music to any science reporter's ears.*

The study's lead author was Shoukhrat Mitalipov of the privately funded Oregon Health & Science University in Portland, so was not subject to the federal funding prohibition on human embryo research. Originally from Kazakhstan, Mitalipov trained in Moscow before emigrating to the United States in the mid-1990s. An expert in stem cell and embryo research, his group reported the first cloned monkeys and the first individualized stem cells (by genetically reprogramming a patient's skin cells). His work in mitochondrial replacement therapy earned him recognition as one of *Nature*'s top ten personalities of 2013.

So why did *Nature*, despite the many Chinese human embryo precedents, elect to publish Mitalipov's paper? It surely wasn't just because the controversial field of human embryo editing had been carried out on American soil.[22] Mitalipov's work stood out for two reasons: first, the team's results were technically impressive. Mitalipov's group had created dozens of embryos by fertilizing donated eggs with sperm from a man with hypertrophic cardiomyopathy. (This disease can cause sudden, fatal heart attacks, notably in young athletes such as former Boston Celtics star Reggie Lewis.) The Oregon team used CRISPR to repair the faulty copy of the *MYBPC3* gene, which had a four-base deletion. Of the fifty-eight treated embryos, forty-two were apparently repaired with two healthy copies of *MYBPC3*, without any serious off-target effects.

---

* Sadly it was Connor's last major scoop; he died of prostate cancer a few months later, aged sixty-two.

The other novel finding was the manner in which the mutant gene had been corrected. Mitalipov had expected that if gene editing occurred, the CRISPR machinery would cut the *MYBPC3* gene and paste the guide sequence that was co-injected into the embryos. But instead, the repaired sequence appeared to emanate from the corresponding gene on the partner chromosome. This suggested a novel DNA repair mechanism—a gene replacement rather than an edit. Consequently, Mitalipov didn't use the word "editing" in the title of his report but "correction." "Everyone always talks about gene editing," Mitalipov said. "I don't like the word *editing*. We didn't edit or modify anything. All we did was un-modify a mutant gene using the existing wild-type maternal gene."[23] Others weren't so sure. Embryologist Tony Perry argued that the genomes of egg and sperm were too far apart for this mechanism to work. "The genomes are separated by what are, in cellular terms, intergalactic distances," he said.[24]

But most observers judged the work to be an important step forward. "It feels a bit like a 'one small step for (hu)mans, one giant leap for (hu)mankind' moment," said Doudna.[25] And she predicted correctly that Mitalipov's results would be "encouraging to those who hope to use human embryo editing for either research or eventual clinical purposes."* Asked what was the biggest challenge in his research, Mitalipov said it was nothing technical but rather the regulatory approval required from three different committees vetting the research, which could be a forerunner to clinical trials in the future. He matter-of-factly said that dominantly inherited disorders would be prime targets. This would include cancer predisposition genes such as *BRCA1* and *BRCA2*, which when mutated increase the odds of breast and/or ovarian cancer.

Mitalipov did not revel in the spotlight, appearing uncomfortable during a live interview on the Charlie Rose television show. But in his "giant leap for (hu)mankind," Mitalipov felt his research offered hope for millions of people suffering from or carrying serious genetic diseases who worried about transmitting those genes to their children. "We now have a chance

* If Mitalipov's observations on the DNA repair mechanism hold up, then investigators will have more work in cases where both copies of the gene are mutated. They will have to figure out a way to coax the embryo to accept the externally provided repair patch.

to prevent disease at the earliest possible stage of life," he said.[26] But given the backlash that had greeted the first human embryo tinkering in 2015, who would dare take the next step?

In Shenzhen, JK read the headlines and hype around Mitalipov's study with incredulity bordering on disgust. Two years earlier, American scientists had rushed to condemn China's initial exploration of human embryo editing, which had used nonviable embryos, but here they were basking in the media spotlight with an article in *Nature*, no less, using viable embryos, without anything like the same criticism. JK charged this was a double standard, nothing less than "scientific racism against the Chinese."[27] Feeling a sense of patriotic duty, JK was emboldened to push on.

In mid-2017, JK started talking to Lombardi about the idea of launching a gene-editing company. He wanted to hire a CEO for Direct Genomics so he could free up time to focus on the science of gene editing. He asked Lombardi for help to attract funding so that his "NewCo" would be an international company. "What do you want to do?" Lombardi asked. "We're going to gene edit humans," JK said. Lombardi was almost speechless. "You realize there's a moratorium, right?"[28]

JK didn't share Lombardi's concern. The 2017 National Academies report on germline editing had surprised many observers in suggesting that germline editing could be permitted under certain conditions. JK sent Lombardi a rudimentary business plan, focusing on addressing the HIV health crisis in Southeast Asia. According to the plan, China was setting up a district that would do medical tourism around gene editing. "My stuff will be part of this," JK told Lombardi. For the next twelve months, Lombardi scheduled exploratory meetings, with the intent of letting JK discuss his commercial aspirations with potential investors. Each time, JK canceled his trip at short notice.

One reason was that JK was starting to enroll couples in his clinical trial. Working with a Beijing-based organization that helps Chinese people with HIV, JK recruited eight couples where the father was HIV-positive. One couple subsequently dropped out. On a Saturday in June 2017, JK met two suitable couples interested in volunteering for his trial. For just under an hour, JK walked the couples through the informed consent process, presenting a document that described the risks of the procedures. The study was billed as an "AIDS vaccine development project." JK was the project leader, while funding was provided by SUSTech. The project would knock out the *CCR5* gene "to help these CCR5 gene-editing [sic] babies to obtain the genotype of the Northern European to naturally immunize against HIV-1 virus." Each couple would receive about $40,000 for medical costs for the IVF procedure, genome editing, and hospital visits.[29]

While the document outlined the risks of genome editing, JK explained his team was not responsible for any off-target effects, which were "beyond the consequence of the existing medical science and technology." And only JK's team had the right to make any public announcement about the trial. Sitting at the end of the boardroom table watching JK invite these couples to make history were two notable observers.[30] One was Michael Deem. The other, casually dressed in a blue T-shirt as if he'd just dropped in after a 5K run, was Yu Jun, a very prominent Chinese scientist, one of the four cofounders of BGI. Yu had sold his share of the organization, and was based in Beijing where he is on the faculty of the Chinese Academy of Sciences (CAS)—China's equivalent of the NIH.[31] Moreover, he is the chief editor of the official genomics journal of the Genetics Society of China. It was a glaring illustration of the high-profile friendships that JK had forged during his short time back in China.

Two months later, JK returned to the Cold Spring Harbor Laboratory, this time as an invited speaker. He stood at the podium where countless legends of molecular biology have presented, under the gaze of a large portrait of Jim Watson.* JK talked about the improving efficiency of gene editing, measuring off-target mutations by genome sequencing in monkey and human embryos. There was no indication that JK was about to rush into clinical germline editing, at least not until his penultimate slide—a

* The Watson portrait was removed 18 months later after Watson uttered racist comments in a PBS documentary. Cold Spring Harbor Laboratory subsequently cut all ties with its former president and chancellor, removing his name from the graduate school.

New York Times magazine headline marking the death of Jesse Gelsinger. Human germline editing could occur "in the near future," JK said, but "we should do this slow and with caution. A single case of failure may kill the entire field, just like this case—'the biotech death of Jesse Gelsinger' for gene therapy."[32]

JK's final slide—the traditional acknowledgements to collaborators and funding agencies—thanked Deem, DeWitt, and William Hurlbut. He answered a few technical questions from the audience, but no one packed into the auditorium seriously thought the young Chinese scientist was about to break from the international consensus on germline editing. Many (like me) were just hoping the evening session would wrap up on time so they could head down to the bar for a well-earned beer.

Three months later, during an advisory board meeting for Direct Genomics, JK welcomed his latest prize recruit, Nobel laureate Craig Mello, to Shenzhen. JK asked Mello about the possibility of using CRISPR to prevent HIV transmission in human embryos, but Mello didn't think JK was really serious. Meanwhile, one of JK's PhD students, Feifei Cheng, sent a series of emails to Penn cardiologist Kiran Musunuru, for advice about using CRISPR to target another gene, *PCSK9*: individuals with rare mutations in *PCSK9* exhibit dramatically lower cholesterol levels. Musunuru replied to her first email but ignored the others. He forgot about the exchange completely until twelve months later, when Marchione emailed him JK's draft paper.[33]

Back in China, JK's profile and reputation were growing fast. The Central Government selected JK for its top multidisciplinary science program, Qianren Jihua (the Thousand Talents Plan), self-described as one of the world's most prestigious science awards. CCTV-13, the biggest news channel on Chinese state television, aired a fawning four-minute video showcasing JK based on a Chinese idiom, "The waves behind drive on those before." JK is shown joking with colleagues, and proudly displaying an encyclopedia of books he had published containing a complete human genome sequence, dozens of volumes stacked in a double helix tower configuration. There were also scenes of him enthusiastically playing five-a-side soccer as his wife and baby look on.[34]

Early in 2018, JK was back in San Francisco. He told DeWitt over dinner that he had received approval from his institutional review board (IRB). He also met Stanford's Matthew Porteus, revealing in confidence that he was starting work on implanting genetically edited human embryos. A stunned Porteus berated JK, telling him it was a terrible idea. "I told him he was putting the entire field at risk through his reckless actions," Porteus recalled, and he needed to talk to more experts.[35] JK sat poker-faced; he had expected a more supportive reaction. In retrospect, Porteus says he wishes he'd raised the alarm more publicly. Between a trio of Stanford faculty, Deem, Mello, DeWitt, Ben Hurlbut, and presumably others, JK took a significant number of American scientists into what Ferrell called his "circle of trust."[36] He was popping up everywhere like Zelig, Regalado said.

JK visited William Hurlbut several times throughout 2017 and 2018, enjoying some lengthy and intense ethics discussions. JK wanted to understand the objections to heritable genome editing in the West. On one occasion, JK brought his thumb and forefinger together and asked Hurlbut if something that small could be as important as a baby? Hurlbut thought JK was "humble and well-meaning," trying to advance science and help others. But Hurlbut didn't suspect that JK was seriously contemplating clinical work. "He's an idealist," Hurlbut said. "He's an inexperienced, perhaps naïve, optimist. I kind of knew [He] was involved in something of significance. But it's unfortunate that it had to happen this way. Sad, really, because he seems like a guy with good intentions."[37]

Hurlbut's sympathetic assessment was not widely shared.

In February 2018, JK took unpaid leave from his university. He was a man on a mission. He wanted to make his mark on the world stage, and to help his fellow citizens, especially those at risk of HIV. And he wanted to deliver a victory for Chinese science. He later told Ben Hurlbut that he had no interest in waiting for an international societal consensus that in his view was never going to happen. "But once one or a couple of scientists make

[the] first kid, [it's] safe, healthy, then the entire society including science, ethics, law, will be accelerated . . . So, I break the glass."

Following IVF, JK's team implanted thirteen embryos into five of the seven women enrolled in the trial. Two women became pregnant. In April 2018, JK had some exciting news to share with a few of his American confidants. He sent virtually identical emails to Mello, Quake, and DeWitt, with an emphatic subject line: "Success!"

> Good News! The women [sic] is pregnant, the genome editing success! The embryo with CCR5 gene edited was transplanted to the women [sic] 10 days ago, and today the pregnancy is confirmed.[38]

Quake decided to share JK's email with a colleague but didn't receive any concrete advice. Mello was also put in an awkward position. "I'm glad for you," he replied, "but I'd rather not be kept in the loop on this. You are risking the health of the child you are editing . . . I just don't see why you are doing this. I wish your patient the best of luck for a healthy pregnancy."[39] Mello's disapproval was tempered by his respect for JK. "I know you mean well," he wrote. Meanwhile, DeWitt was in shock and didn't know what he should do.

In August 2018, JK made another quick trip to the United States. In Boston, JK finally got to meet Feng Zhang, seeking advice on methods to reduce off-target effects in mouse and human embryos. "He was having the same challenges as other researchers around lack of efficiency and lack of precision," Zhang told a reporter in Hong Kong.[40] "I told him that the technology is neither efficient nor precise enough for real-world application in embryos, including in human IVF applications." From there, JK traveled to New York, where he visited Chengzu Long, a Chinese CRISPR scientist at New York University. He made quite the impression. "It was hilarious," the fast-talking Long recalled. "He had a male personal assistant open the car door for him and carry his briefcase. Very Chinese!"[41]

JK also traveled to Pennsylvania to meet executives of Geisinger, a leading healthcare company. The two-hour meeting on August 16 was hosted by two distinguished geneticists, Hunt Willard and David Ledbetter. JK's team was interested in starting a project called "Gene Achieve" in China, based on Geisinger's MyCode precision medicine biobank, in which volunteers opt in to receive genetic and other medical results. Willard said JK discussed his CRISPR research without mentioning his embryo work. "It wasn't obvious to me how [or] why Geisinger would want to get involved in any collaborations with either Direct Genomics or the hospital system," Willard told me.[42]

Despite his reservations, Mello traveled to China to attend a Direct Genomics scientific advisory board meeting. On November 19, 2018, a smiling Mello, sitting next to JK, posed for a photograph with company staff in a boardroom. As usual, JK posted the photograph on WeChat. It was just ten days before his scheduled appearance at the Hong Kong summit. But JK was keeping a massive secret.

A few weeks earlier, in late October, a visibly nervous JK had taken a flight north from Shenzhen to an undisclosed destination. Ferrell noticed there was a distinct look of relief upon his return. He carried the news that, by emergency caesarean section, not one but two CRISPR babies had been born.

CHAPTER 17
A MACULATE CONCEPTION

On November 27, 2019, I walked from my hotel to the University of Hong Kong (HKU) campus for the opening of the Second International Summit on Human Genome Ethics. Campus is a bit of misnomer, as the university is squeezed into the precious real estate of Hong Kong Island. Entering the university, I hurried past the Pillar of Shame, an arresting statue that commemorates the Tiananmen Square protests.

As I entered the impressive Lee Shau Kee Lecture Centre, I spied the jovial face of Lap-Chee Tsui, the vice chancellor of the university and one of the summit organizers. Thirty years earlier, working at Toronto's Hospital for Sick Children, Tsui led the discovery of the cystic fibrosis (CF) gene, teaming up with Francis Collins. It was a tour de force of genetic detective work, a lifeline for CF patients, and a magical moment for Tsui, who is originally from Taiwan. As a graduate student in Bob Williamson's rival team in London hunting the CF gene, I'd enjoyed sharing a few beers with Tsui at various conferences. If anyone was going to scoop us, we thought, let it be him.

Tsui asked me what I thought of the magnificent venue as the audience began drifting in. He couldn't resist telling me that he'd ordered the relocation of a reservoir serving more than 100,000 people in order to create space for this jewel of HKU's centennial campus. Tsui's success in literally the parting of the waters was impressive. But as Acts of God go, it was completely overshadowed by the revelations that were scheduled to take place on the stage in front of us.

The decision to hold the summit in Hong Kong was a compromise after the Chinese Academy of Sciences (CAS), the original hosts, pulled out. Feng Zhang suspected that the switch in venues signaled that the Chinese government probably knew about the CRISPR babies. "For some reason, [CAS] couldn't find an auditorium to fit five hundred people in Beijing," Zhang said sarcastically.[1] However, this particular conspiracy theory doesn't hold water: Tsui had received an official invitation to host the summit in December 2017, following months of exploratory talks, well before the fateful pregnancy.[2]

Taking my seat in the section roped off for the media near the back of the hall, the only question on my mind was whether JK was going to show up. By now, everyone had heard the news of the CRISPR babies, foreshadowing a dark new chapter in the annals of medicine. But Doudna, Tsui, and the other conference organizers had two urgent concerns: Was JK telling the truth in those extraordinary YouTube videos or was this an implausibly elaborate hoax? And was he still going to appear to talk about the experiments leading to the twin births?

To answer those questions, Doudna, Alta Charo, Robin Lovell-Badge, and University of Sydney developmental biologist Patrick Tam arranged to meet JK for dinner on the eve of the conference at Le Méridien Cyberport hotel, where most of the speakers were staying. "He arrived almost defiant," Doudna recalled.[3] For the next hour or so, Doudna and colleagues peppered JK with questions: Why had he chosen HIV? How were the families recruited? What was the informed consent process? Why was there no public disclosure of any preclinical data? JK tried to answer and gave a preview of his presentation, dropping the news that a second woman was also newly pregnant with a gene-edited baby.

Around the table, JK's inquisitors were dismayed by his offhand demeanor—what Doudna described as a combination of hubris and naïveté[4]—his stubborn self-belief and refusal to concede any mistakes or ethical lapses. "I got the strong impression that he saw Robert Edwards as a kind of hero, a paradigm breaker, a disrupter, and that he wanted to model himself after that," Charo said.[5] But JK's dreams of emulating Edwards were fading fast. "JK was hoping to make a big splash with the news of the twin birth," Tam told me.[6] He asked whether he should present the results when he gave his scheduled talk on Wednesday morning? "We were all like, 'uh, yes'," said Doudna. Eventually JK's patience during the dinner interrogation wore thin. He left some cash on the table and walked out of the restaurant before checking out of the hotel.

As I observed the ceremonial photographs and formal opening speeches by, among others, Carrie Lam, the chief executive of Hong Kong, I was confused. There was no acknowledgement of the elephant in the room: no mention of He Jiankui or the scandal that had broken. The only reference to the swirling drama was an oblique comment from organizing committee chairman David Baltimore, who said that gene-editing tools would be used in the clinic in the future. "We may even hear about an attempt to apply human genome editing to human embryos," he said coyly. Then, as if on cue, he brought up *Brave New World*: "Although Huxley could not have conceived of genome engineering . . . we should take to heart the warning implicit in that book."

Later that morning, we heard the first direct rebuke from the podium of JK's actions. It came surprisingly from a fellow countryman: Qiu Renzong, the eighty-five-year-old senior statesman of China's bioethics community and a member of the Chinese Academy of Social Sciences in Beijing. Qiu wasted no time in condemning his countryman. "There is a convenient and practical method to prevent HIV infection," he said. To resort to germline editing was like "shooting a bird with a cannon." Qiu accused JK of violating China's Ministry of Health regulations on assisted reproduction and questioned whether he had obtained the correct IRB approval. His final salvo sent shivers down my spine:

"How could Dr. He and [his] team change the gene pool of the human species without considering the need to consult other parts of the human species?"[7]

It was a mild surprise then when George Daley, the Dean of Harvard Medical School, offered a more measured, even conciliatory, perspective. "Just because the first steps [in germline editing] are missteps doesn't mean we shouldn't step back, restart, and think about plausible and responsible pathway for clinical translation," Daley said. Also speaking that day were Feng Zhang and Doudna. Zhang was mobbed outside the hall by a scrum of reporters after his talk, a smaller huddle around Doudna. But for once, neither of the CRISPR pioneers were the star attraction. The main event was still to come.

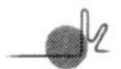

Twenty-four hours later, the atmosphere in the packed auditorium was electric. Almost three hundred journalists, photographers, and film crews had by this time descended on Hong Kong to capture science history. The photographers and camera crews were herded into a media pen extending along one entire wall of the auditorium—far more than I remember attending the reasonably historic announcement of the Human Genome Project in June 2000. The organizers shuffled the program to devote sixty minutes exclusively to JK before lunch. Shortly before JK was scheduled to appear, I skipped down the aisle steps past the banks of photographers to the microphone in front of the stage to ask a question. It was just a ruse so I could slip into one of the few empty seats in the front row, where I sat next to Hong Kong's most famous geneticist, Dennis Lo. I briefly wondered if security might come and have a word, but they had more pressing concerns.

As Robin Lovell-Badge from the Crick Institute prepared to introduce JK, I noticed the security guards stationed at both ends of the stage, a surreal indicator of the extraordinary event about to unfold. As a rookie editor at *Nature*, I had published Lovell-Badge's career-making discovery—the identity of the male sex-determining gene on the Y chromosome—in 1991. Sounding like the headmaster of an English boarding school, Lovell-Badge warned the audience to be polite and even show JK some respect. HKU has a strong tradition of free speech, he said. Any outbursts from the audience would result in the session being cut short. As Lovell-Badge invited JK to come to the stage, there was an awkward pause that lasted thirty seconds as if, even at the last minute, JK had changed his mind.

When a side door opened, hundreds of people craned their necks to get a first glimpse of the man of the hour. As JK climbed the stairs stage right, there was a hush in the hall and almost no applause—just the clatter of high-speed camera shutters and the stroboscopic flashes from the legions of photographers camped on the opposite side of the hall. JK walked briskly across the stage, dressed casually in an open-neck shirt, carrying a tan leather briefcase. He looked more like a commuter hurrying to catch the Star Ferry in the Hong Kong humidity than a scientist at the center of a massive international storm. At the podium, he shook Lovell-Badge's hand, and took in the scene. He'd imagined giving a lecture to a worldwide audience in front of Nobel laureates and scores of cameras, but not like this.

JK pulled his speech from his briefcase and for the next twenty minutes, delivered what was, on the surface, a fairly typical scientific lecture, accompanied by slides showing experimental data. The contents were anything but ordinary. He began with an apology—not for the manner in which he had conducted the study or obtained ethical approval or treated his patients, but that the story had become public. "First, I must apologize that this result leaked unexpectedly," JK said. It was an acknowledgement that the carefully orchestrated plans to trumpet his success with a major science publication had collapsed. He also thanked his university, which he said was unaware of his gene-editing studies. He didn't get much further before an irritated Lovell-Badge stood up and remonstrated with the photographers. The persistent clatter of camera shutters, coupled with the poor sound near the front of the stage, was making it difficult to hear. "Stop photographing!" Lovell-Badge ordered. "You must have a good photo by now!" The cameras went quiet.

After presenting results of preliminary gene-editing studies in mice and monkeys, JK went onto summarize his clinical work. The choice of gene target—*CCR5*—was perplexing. His goal had been to mimic the natural Δ32 deletion in *CCR5* that renders many people immune to HIV infection. Without fanfare, JK presented the genetic sequences of Lulu and Nana that had been engineered at conception by human hand. The peaks and valleys of Lulu and Nana's edited DNA sequence traces, like a multicolored polygraph, laid bare in graphic molecular detail the slipshod way the twins' *CCR5* genes had been modified. He did not show any photographs or reveal any further details about the twins or their health.

Had JK faithfully engineered the precise Δ32 deletion commonly seen in the human population, the public response might have been different, at least in some quarters. But while he had targeted the right spot in the *CCR5* gene, he had no control over the edits, which consequently produced novel sequence variants with uncertain effects. Instead of precisely excising the 32 letters of genetic code, it was as if the editor had shut his eyes and slashed at the page with a red pen, hoping to erase the right words.

Following his prepared remarks, JK sat on stage between Lovell-Badge and Matthew Porteus to answer questions for a further forty minutes. He appeared slightly nervous but stoic and unapologetic. The biggest revelation was when JK confirmed there was an additional early ("chemical") pregnancy. Baltimore led the criticisms. "I don't think it's been a transparent process," Baltimore said. "We only found out about it after it's happened and the children are born. I personally don't think it was medically necessary . . . I think there's been a failure of self-regulation by the scientific community because of a lack of transparency."

The first question from the audience was posed by Harvard's David Liu, respectful but visibly angry. "What was the unmet medical need?" Liu wanted to know. JK's protocol already included a sperm washing step that would essentially guarantee uninfected embryos. JK ducked the question, justifying his choice by stressing the public health menace of HIV affecting millions of patients in China. Indeed, many of his responses suggested that he still did not fathom the gravity of what he had done. For the final question, Lovell-Badge asked if JK would have performed the experiment on his own child? "If my baby might have the same situation, I would try it on them first," JK replied. It was a calculated answer. I wasn't sure I bought it, and nor did JK's acquaintance, Chengzu Long. "He doesn't give a shit about HIV. How dare he say that!"

A few moments later, to a smattering of applause, the Chinese scientist shook hands with his interrogators and exited through the same door he had entered sixty minutes earlier, evading any questions from the press. He was soon traveling back across the border to Shenzhen, where he would face an official university investigation while under house arrest. I wondered if we would ever see or hear from him again. In Hong Kong and around the world, the objurgation was just beginning.

The reaction to JK's appearance was furious, as if a mass atrocity had occurred. "Disgusting" and "abhorrent" were some of the milder terms used to describe a laundry list of ethical and technical concerns—excessive secrecy, minimal informed consent, negligible medical necessity, amateurish molecular editing, and suggestions of fraud. And yet, the initial Chinese reaction had been positive, even euphoric. On November 26, the *People's Daily*, the official mouthpiece of the Chinese Communist Party, had hailed JK's breakthrough as a moment of national pride: "A milestone accomplishment achieved by China in the area of gene-editing technologies."[8] But the story was swiftly deleted as the condemnation reached a crescendo. It was replaced with a story indicating that an official investigation into JK's actions was underway.[9] Exhibit A in the global uproar, perhaps surprisingly, was an open letter signed by 120 Chinese scientists, which said: "This is a huge blow to the global reputation and development of Chinese science, especially in the field of biomedical research. It is extremely unfair to most scholars in China who are diligent in research and innovation and adhere to the bottom line of scientists."[10]

Inside and outside the HKU auditorium, reporters and camera crews rushed to interview scientists for their reactions. Lovell-Badge was mobbed by a scrum of reporters and camera crews outside the auditorium. Not surprisingly, Doudna was in constant demand, giving a string of interviews in the harsh glare of the television lights, acting as the conscience of the scientific community. Unaccustomed to criticizing a fellow scientist, let alone on camera, she told Bloomberg she was "a bit horrified, honestly. And disgusted, really disappointed, that the international guidelines that many people worked so hard to establish were ignored." JK wanted to be the first to achieve an historic milestone but "that's an inappropriate reason to do something as momentous as this without appropriate oversight."[11]

Charpentier was not in Hong Kong but sent me a brief statement. "We are still at a very early stage of understanding the full implications of gene editing in human cells, and it would be irresponsible to apply the technology in the human germline," she said, while insisting that basic research on human embryos was still justified.[12] But one thing was clear. JK had left an indelible mark on science in the 21st century. As Urnov told *Al Jazeera*: "We live in an age where there is such a thing called Microsoft Word for human DNA. That cat is out of the bag, the genie is out of the bottle."[13]

Criticism rained down from all quarters. David Liu called the saga "an appalling example of what not to do with a new technology that has incredible potential to benefit society." Feng Zhang called for a moratorium on clinical embryo editing "until we have come up with a thoughtful set of safety requirements."[14] Kathy Niakan, the only researcher in Europe to have conducted genome editing on human embryos, said "it is impossible to overstate how irresponsible, unethical, and dangerous this is at the moment." She added, "There is a real danger that the actions of one rogue scientist could undermine public trust in science and set back responsible research."[15]

NIH Director Francis Collins accused JK of trampling on ethical norms, while reiterating that NIH did not support the use of gene-editing technologies in human embryos. "The project was largely carried out in secret, the medical necessity for inactivation of CCR5 in these infants is utterly unconvincing, the informed consent process appears highly questionable, and the possibility of damaging off-target effects has not been satisfactorily explored," he charged.[16] A letter to the *Lancet* signed by senior members of the Chinese Academy of Medical Sciences argued that JK had violated multiple government regulations on the ethics of human stem cell and sperm banks dating back to 2003.[17]

On social media and the blogosphere, reaction was even more intense. UCLA stem cell biologist Paul Knoepfler decried the use of the term "genome editing" to describe JK's work. "He did not *gene edit* these babies . . . what He did really was to *mutate* those twin girls, particularly since he was changing a normal wild type gene to a mutant form. We should call it what it is."[18] British broadcaster and author Adam Rutherford called

JK's actions "morally poisonous." Oxford University philosopher Julian Savulescu said the CRISPR babies were being used as "genetic guinea pigs. This is genetic Russian roulette."[19] You get the picture.

The former FDA Commissioner Scott Gottlieb warned of future government intervention. "The scientific community failed to convincingly assert, in this case, that certain conduct must simply be judged as over the line." And veteran chemistry blogger Derek Lowe called JK's actions criminal. "This experiment should not have been done yet," Lowe wrote. "We're going to alter the human genome, of that I have no doubt. But there was no reason to alter it now, like this, under these conditions. He Jiankui has just made life more complicated for everyone working in the field, and for what?"[20]

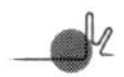

A few hours before JK's public appearance, I received an email from Ryan Ferrell suggesting we meet after JK's talk at a hotel on the south side of Hong Kong Island. Following JK's exit, I suspected the afternoon sessions would be an enormous anticlimax, so told Ferrell I'd meet him that afternoon. It was a thirty-minute taxi to Ferrell's hotel, which puzzled me when there were so many hotels near the university. I entered his small hotel room to find, sitting on the bed, the documentary team, Samira Kiani and Cody Sheehy, who had filmed JK's speech and interview a few hours earlier. Cases of video equipment took up most of the floor space, so I sat on one. Ferrell apologized for the remote location but he was literally on the run. His visit to Hong Kong had turned into a scene from *The Bourne Identity*, as enterprising Japanese journalists had twice located Ferrell's hotel and forced him to take evasive measures.

Ferrell was clearly exhausted and emotional, at times distraught. He blamed himself for the situation that JK found himself in. Ferrell had moved to Shenzhen on a mission. Part of his motivation was his sister, who suffers from a genetic disorder. Ferrell had wanted to help showcase the groundbreaking medical work of gene-editing trailblazers like JK. It was Ferrell who agreed to let Regalado sit in on JK's meeting with the filmmakers a few weeks earlier. And he blamed himself for not being alert to the posting of JK's CCR5 trial notice on the Chinese clinical trial registry website, which turned out to be Regalado's smoking gun.

After an hour or so talking, we went to the hotel bar for a much-needed drink. Ferrell motioned to the *Nature* manuscript in his briefcase. Forty-eight hours earlier, he'd been confident that JK's report would soon be published. But JK's inability to provide any compelling rationale for his actions suggested that was unlikely. One option was for JK to post his manuscript on a preprint repository such as bioRxiv*, a popular venue for scientists to post drafts of their manuscripts. But given the growing scandal, bioRxiv's administrators were understandably reluctant about showcasing such a controversial article, especially given the ethical issues. Those concerns were summed up by the Mayo Clinic's Stephen Ekker: "Sorry preprint enthusiasts," Ekker tweeted, "but there needs to be an ethics check first. Stop giving [He] a megaphone, or others will do the same."[21]

I disagreed. I didn't see any point in sweeping JK's actions under the carpet. It was far too late for that.

The morning after.

Day three of the summit felt like a heavy hangover after a raucous all-night party. The auditorium was less than half full and most of the media had dispersed in the wake of JK himself. One Asia-based reporter for ABC News stopped me to ask if I'd seen Michael Deem. She looked upset when I told her Deem hadn't attended the conference and continued her search.

The one major item on the agenda was for the summit's organizing committee, including Doudna, Porteus, Daley, and Lovell-Badge, to deliver

* BioRxiv is a preprint repository for biologists where they can post drafts of their manuscripts prior to peer review and publication in a journal. The site, launched by Cold Spring Harbor Laboratory in 2012, is modeled on arXiv, the popular physics preprint server. Posting a preprint allows authors to receive a timestamp of submission, solicit advice from other scientists, and disseminate results many months ahead of official journal publication. bioRxiv volunteers apply a quick screen before posting a preprint, checking for obscenity, nonsense, and any ethical red flags. The repository and its sister site, medRxiv, have come into their own during the COVID-19 pandemic, posting thousands of early results.

a closing statement when almost everything that could be said had been already. With the full committee seated on stage, Baltimore read the statement, which had been hastily drafted the previous evening.[22] The committee judged JK's work to be "deeply disturbing" and "irresponsible," while calling for an independent assessment "to verify this claim and to ascertain whether the claimed DNA modifications have occurred." The glaring flaws in JK's study included "an inadequate medical indication, a poorly designed study protocol, a failure to meet ethical standards for protecting the welfare of research subjects, and a lack of transparency in the development, review, and conduct of the clinical procedures."

Given the current state of the technology, Baltimore said the risks were "too great to permit clinical trials of germline editing at this time." But not forever: germline editing could become acceptable in the future if there was strict independent oversight, a compelling medical need, long-term patient follow-up, and attention to societal effects. And echoing Daley's earlier remarks, there should be a translational pathway setting standards for preclinical evidence and editing accuracy, the proficiency of the investigator, and robust partnerships with patients and advocacy groups.

But the committee's call for an independent investigation rang hollow. Who would do the investigating? The committee lacked any jurisdiction to entice any cooperation from JK. When I asked Baltimore what actions he hoped to see from JK, he stressed the importance of examining the CRISPR babies, but that would be difficult given the absence of any public information about the twins' identities and whereabouts.

We soon learned that JK was in no position to cooperate with any sort of external inquiry. In the weeks following his return to Shenzhen, although in email contact through his gmail account with reporters[23] and journal editors, his precise whereabouts were unknown. Several media outlets reported that JK had disappeared, based on his hasty departure from Hong Kong and the unceremonious removal of his university web page, where he had invited people to write to the world's first gene-edited twins at a special email account: *DearLuluandNana@gmail.com.*

Speculation about JK's whereabouts was understandable, although suggestions he had been executed seemed fanciful. In 2018, several prominent figures in Chinese society had "disappeared," sometimes for months on end, including Meng Hongwei, the former president of Interpol. One of China's most famous actresses, Fan Bingbing, vanished for nine months before returning to public view in April 2019, offering a groveling apology for tax evasion.

Four weeks after his Hong Kong bombshell, *New York Times* reporter Elsie Chen went searching for JK on the SUSTech campus. Acting on a hunch, she visited the main guest house, typically used to house faculty and visitors. That's where she fortuitously spotted JK on the balcony of a fourth-floor apartment and was able to grab a couple of photographs.[24] He was guarded by a dozen unidentified men who barred reporters from getting too close. Inside the apartment was a woman, presumably his wife, and a young baby. The lobby was full of guests checking in for a conference, oblivious to the fact that China's Most Wanted was under house arrest in their midst.

Around this time, JK told Lombardi he was being sequestered in academic housing for his own protection. Lombardi offered some final advice: stop looking for commercial funding, at least until you can present two healthy babies: "Put your head down and do your best, most ethical science for the next five years. And when you think you can put your head up again, keep it down another year."[25]

On January 21, 2019, the world learned JK's fate. As reported by Xinhua, the Guangdong state news agency, the provincial investigation into JK's actions concluded that, beginning in June 2016, JK had conducted prohibited reproductive experiments in secret while seeking "personal fame and fortune." His behavior "seriously violates ethics and scientific research integrity, and seriously violates relevant state regulations, causing adverse effects at home and abroad," the report stated. The head of the investigation said that JK, his colleagues, and relevant institutions would be "dealt with seriously according to the law." SUSTech immediately jettisoned JK. The university posted a terse statement on its website: "Effective immediately, SUSTech will rescind the work contract with Dr. Jiankui He and terminate any of his teaching and research activities."[26]

I briefly wondered if JK could continue his work in industry. But six months later, in July 2019, Direct Genomics announced that JK had resigned and sold his stake in the company.[27] Another theory I heard from a member of the "circle of trust" was that JK's disappearance might be part of a Faustian bargain: don't incriminate any Chinese government officials in exchange for professional rehabilitation when the time is right.

Several months later at the World Science Festival in New York City, on a panel with Doudna, William Hurlbut indulged in some speculation about what might have happened had Regalado not sensationally revealed JK's activities. What if JK had been able to publish and orchestrate coverage of his work as planned? JK expected some criticism from Europe and the United States, but not in his own country. If JK had not felt the wrath of scientific society, "the outcome might have been different."[28] If the results had been published the way JK dreamed, accompanied by a syndicated AP exclusive, Hurlbut ruminated, "I think JK would not have been in so much trouble. And in China, maybe not any."

Hurlbut offered a glimpse into JK's emotions and contrition. In an email sent to Hurlbut shortly after his confinement in Shenzhen, JK belatedly showed some remorse for his actions rather than just the manner in which they had been made public. He felt badly and wished he had waited, acted more carefully, and selected a different gene target. But this hardly qualified as a public apology. More important is whether he will apologize to the twins, Lulu and Nana, and a third edited baby, believed to have been born around May 2019. We may never know.

Three months after losing his university job, and disappearing from public view, JK was named one of *Time* magazine's 100 Most Influential People of the Year. It is a rare honor, although "influential" doesn't necessarily mean laudable. The winners' biographical sketches are usually written by previous honorees. JK's entry was penned by none other than Doudna, who had received the honor in 2015. It was not pretty:

> He Jiankui showed the world how human embryo editing is relatively easy to do but incredibly difficult to do well. Going against the consensus in the scientific community that CRISPR-Cas9 technology is still too experimental and dangerous to use in human embryos, he applied it to forever change the genomes

> of twin girls to give them immunity to HIV. His reckless experimentation on the girls in China not only shattered scientific, medical, and ethical norms, it was also medically unnecessary . . . He's fateful decision to ignore the basic medical mantra of "do no harm" and risk the unintended consequences will likely be remembered as one of the most shocking misapplications of any scientific tool in our history.[29]

Many of the *Time* 100 attended a black-tie gala at Lincoln Center in New York. What JK wouldn't have given to walk on the red carpet with the *Game of Thrones* actress Emilia Clarke, or pose for photographs with Liverpool soccer legend Mohamed Salah, or dance to Taylor Swift, even discuss CRISPR with *Rampage* star Dwayne Johnson. It was not to be. There were no more celebrity selfies on WeChat. On that star-studded evening, the only JK in attendance was Jared Kushner.

CHAPTER 18
CROSSING THE GERMLINE

The CRISPR babies scandal triggered a torrent of outrage and op-eds castigating JK. "Sooner or later it was bound to happen," physician and author Eric Topol wrote in the *New York Times* hours after JK's last public appearance in Hong Kong. "The potential risks to the babies grossly outweighed any benefits to them, and to science," judged Hank Greely, a Stanford law professor.

But it was science writer Ed Yong who best framed the sheer scope and magnitude of JK's transgressions. Writing in the *Atlantic*, Yong compiled a list of the major medical and ethical criticisms leveled at JK's work and the community's potential culpability. Under normal "rogue scientist" circumstances, a Top Ten list should have sufficed to capture the violations. Without much difficulty, Yong came up with no fewer than fifteen "damning details" of the JK affair. I can't improve on that, so here's the full list:[1]

1. He didn't address an unmet medical need.
2. The actual editing wasn't executed well.
3. It's not clear what those new mutations will do.
4. There were problems with informed consent.
5. He operated under a cloak of secrecy . . .
6. . . . but organized a slick PR campaign.
7. A few people knew about He's intentions but failed to stop him.
8. He acted in contravention of global consensus.

9. He acted in contravention of his own stated ethical views.
10. He sought ethical advice and ignored it.
11. There is no way to tell whether He's work did any good.
12. He has doubled down.
13. Scientific academies have prevaricated.
14. A leading geneticist (George Church) came to He's defense.
15. This could easily happen again.

As the saying goes, there's a lot to unpack there. We had reached a turning point in human history. The fictional and hypothetical warnings over decades about the perils and immorality of human genetic engineering were suddenly manifest in the form of twin babies. By excavating the germline of a pair of human embryos, JK had crossed a red line of medical practice and ethics. Had he put his ideas up for public debate or previewed them before the scientific community, he would have been shot down before he could even pick up a pipette.

Let's take the first three items on Yong's list. Just as David Liu had asked JK directly in Hong Kong, what was the unmet medical need for this couple? Why did JK pick HIV and *CCR5*? Why was genome editing even necessary given that JK's team performed sperm washing prior to IVF to remove any risk of HIV? JK could not give Liu a straight answer, so instead he painted the scourge of HIV in broader terms. "I truly believe that not just for this case but for millions of HIV children, they need this protection because a vaccine is not available. I have personally experienced 'AIDS village' where 30 percent people were infected. They had to give their children to relatives." JK said that he was proud of his work; the twins' father had been depressed but had pledged to work hard and take care of his new family.

AIDS has claimed an estimated 35 million lives. There are an estimated 37 million people worldwide infected with HIV, including more than 800,000 in China. The advent of a cocktail of antiretroviral drugs is one of the great recent medical success stories. These drugs can suppress the titer of HIV to undetectable levels and effectively abolish the risk of HIV transmission through sex. An advertisement from antiviral drugmaker

Gilead, spotted in a Broadway theater playbill, said: "Dear HIV, We didn't give up. XOXO, Science."

Earlier discussions about which disorders might prove candidates for heritable gene surgery seldom considered HIV, which was stuck in the "definitely maybe" category. Inactivating *CCR5* to enhance protection against HIV isn't in the same category as correcting a disease-causing mutation. But JK had identified two major reasons for his choice: safety and "real world medical value."[2] He cited two decades of research on *CCR5* and related clinical trials.[3] He also considered his gene surgery as relatively simple, seeking to disrupt the *CCR5* gene to scuttle the HIV receptor rather than trying to perform precision DNA surgery, swapping one letter in a gene for another, as would be needed if attempting to treat patients with muscular dystrophy or sickle-cell disease.

Worldwide about 100 million people carry the *CCR5* Δ32 mutation, with the highest concentration among northern Europeans. All evidence suggests the deletion is safe. If JK had edited the *CCR5* gene to mimic the naturally occurring Δ32 mutation, the scientific community's concerns might have been muted. But the mutations JK engineered in Lulu and Nana did not produce the Δ32 deletion, raising serious questions about the impact of these man-made genetic alterations, never before seen in humans or tested in an animal model. JK targeted a spot in the *CCR5* gene where he could land the Cas9 nuclease, like pointing a cursor on a particular word. But he couldn't accurately control the extent of the edit itself. "These two lives are now an experiment, a matter of scientific curiosity, which is an outrageous way to relate to human lives," said Ben Hurlbut.[4]

Sean Ryder, an RNA biologist at the University of Massachusetts Medical School, vented his fury on Twitter during the summit and subsequently in the *CRISPR Journal*. "Doudna warned us this might happen," Ryder wrote. "I had trusted that my fellow scientists would ensure that human trials involving embryonic genome editing would be done transparently, ethically, and with strong moral purpose."[5] But no.

In Nana, both copies of *CCR5* harbor so-called frameshift mutations: one copy carries a small deletion of 4 bases (-4), whereas the other copy actually has an insertion of a single base (+1); both mutations disrupt the natural phase of the genetic code.* It is possible, even likely, that these mutations would abrogate the function of CCR5 similar to the Δ32 mutation. But we don't know for certain. These mutations are "never-before-seen variants of unknown significance," said Ryder. "They might inactivate CCR5 activity and block HIV uptake, but they might also incur new risks."

The situation in Lulu was even less clear-cut: one copy of the gene appeared to be untouched, while the other carried a fifteen-base deletion. The resulting loss of five amino acids in the middle of an otherwise intact protein might abolish the function of CCR5 but this has never been tested. Moreover, in both twins, the edits appeared to be mosaic, meaning that not every cell in the twins carried the gene edits. Mosaicism is a natural phenomenon in biology—think heterochromia, a condition that results in different colored eyes. But no doctor, certainly not JK, could say with conviction that he had first done no harm.

There were also serious questions about off-target effects, despite JK's denials, including some evidence that such an effect had occurred in Lulu, and others could be uncovered if more rigorous whole genome sequencing were performed. Ryder closed his *CRISPR Journal* commentary with an impassioned plea:

> It is my fervent hope that Lulu and Nana will lead long healthy lives. It is certainly possible, even plausible, that they will remain healthy based upon what is known about *CCR5*. However, there are enough uncertainties about the function of the mutations to raise clear scientific objections to the work, in addition to the moral and medical objections. I pray that Lulu and Nana are never exposed to HIV. I hope we never have a chance to find out if the experiment 'worked.' To me, the best outcome is that

* The genetic code is read out in triplets, so if a DNA sequence is mutated by the insertion or deletion of 1 or 2 bases (any number that is *not* divisible by 3), it will fall out of phase, changing the corresponding amino-acid sequence. If a mutation is a small deletion or insertion of 3 bases (or any multiple of 3), then the results could still be devastating, but the sequence beyond the mutation will still be intact. A good example is the most common mutation in the cystic fibrosis gene, called ΔF508, the deletion of three bases that encode the amino acid phenylalanine.

> the babies are unaffected by the procedures of the He lab. Worse outcomes are possible.[6]

Even the assumption that deactivating *CCR5* neatly thwarts HIV with no other health consequences started to crumble. The Δ32 mutation arose in northern Europe but as Lovell-Badge pointed out "there are almost no people with the Δ32 mutation living in China. Therefore, it is necessary to ask why?"[7] Although it is likely that the deletion hasn't had time to spread through Asia, another possibility is that the *CCR5* mutation might have some other impact on health. We know Δ32 increases the risk of West Nile and possibly influenza. And some have even questioned the dogma that inactivating CCR5 protects against HIV. Apparently there are rare strains of HIV, including one called X4, that bypass CCR5 by using a different portal into the cell.[8]

Months after the twins' birth, more disturbing scenarios hit the news. "China's CRISPR twins might have had their brains inadvertently enhanced," screeched a trademark Regalado headline.[9] Alcino Silva, a UCLA neurobiologist, said the CCR5 mutations might damage cognitive function in the twins—based on his work in mice. His team's latest data suggested that individuals lacking CCR5 recover more quickly from strokes, while carriers of a single inactive *CCR5* gene perform better in school.[10] Meanwhile, a pair of Berkeley researchers mining the UK Biobank—a charity- and government-funded database of genomic and medical information on more than 500,000 Brits—reported that individuals with two copies of the Δ32 mutation were 20 percent more likely to die before age seventy-six than people with one normal copy of *CCR5*.[11] Media reports pounced on the supposed threat to the CRISPR babies' life expectancy (ignoring that the twins' edits were not the Δ32 deletion). But some excellent sleuthing by Sean Harrison, an epidemiologist at the University of Bristol in the UK, highlighted a systematic error in the study, leading eventually to its retraction.[12]

The CCR5 story spotlights a critical issue: how can we be certain that editing *any* gene will not have some unforeseen, knock-on, collateral damage? Our 10 trillion cells are sacks containing vast interconnected networks of protein-protein, RNA, and other biomolecular interactions laden with intracellular hubs and vesicles, membranes and molecular machines. When I studied biochemistry in college, we naïvely thought the key metabolic pathways of a cell could be displayed neatly on a wall poster. Today, we know that each of the more than 20,000 proteins in a cell interacts with dozens if not hundreds of other proteins. Predicting what happens next when editing any one of these important cogs and spokes may be possible one day with advances in artificial intelligence, but not today.

But before we condemn all future efforts to perform germline editing, let's remember one important safety issue. A typical somatic gene-editing procedure in an adult might involve fixing the DNA of 100 million cells or more. "Each one of them is a different CRISPR event, any one of which could hit a tumor suppressor gene," cautions George Church, who co-founded a company to do clinical genome editing. The consequences of that event as a cell amplifies could be serious. "But when you put it in an egg, it is a single event, so the a priori probability of hitting a tumor suppressor exon is a billion times lower. How is that more risky?"[13]

Another egregious aspect of the CRISPR babies' saga was the suspect manner in which JK obtained informed consent from the volunteer couples and the intense secrecy in which he conducted the experiments. "This was not a case of science outpacing ethical guidance or the law," said Oxford University ethicist Dominic Wilkinson. "There were guidelines in place that warned against research of this sort. This appears to be a researcher who had no interest in attending to ethical guidelines relating to scientific research."[14] "The whole angle about germline editing per se is a red herring," tweeted Leonid Kruglyak, chair of human genetics at UCLA. "It's the blatant disregard of all the rules and conventions we have in place for how one should approach *any* proposed intervention."

"Unborn children obviously cannot consent to an embryo-altering procedure," said New York University bioethicist Art Caplan,[15] although fetuses aren't able to consent to anything much, including conception.

Contravening accepted ethical standards, JK personally led the informed consent discussion. "The simple fact that he was directly involved in trying to get consent from the patients is a huge problem. You should never do that," said Lovell-Badge. It is common for an independent third party to explain the risks to the patient or volunteer.

But the notion that JK somehow pressured hapless Chinese couples to take part in his quest for glory is diminished by the haunting words of one of the couples that participated in the trial. They framed their participation against the pervasive backdrop of HIV stigma and discrimination in Chinese society. In a letter shared with Benjamin Hurlbut, they wrote:

> We were never misled. It was a form of compromise. The object of the compromise was society, and one could even say with the whole world. As an AIDS sufferer and family member, we firmly and deeply know that it is possible to use a preventative drug to have a healthy child . . . That drug can cure disease, but it cannot cure prejudice . . .
>
> For everyone who is listening, please hear this. At a certain level our participation in the experiment was indeed forced, but we were not coerced by any person in particular. We were coerced by society.[16]

JK conducted his work in great secrecy. Indeed, it is remarkable that rumors did not leak out until Thanksgiving weekend 2018. Greely unequivocally condemned JK's secret experiment that "against a near universal scientific consensus, privileged his own ethical conclusions without giving anyone else a vote, or even a voice."[17] And yet, while JK shielded his true ambitions from the scientific community, we now know he confided in his "circle of trust." This included a trio of Stanford faculty—Quake, Porteus, and Hurlbut—as well as Hurlbut's son Benjamin, DeWitt, Deem, Mello, and consultant Steve Lombardi.

Quake initially shrugged off media inquiries in the wake of the JK scandal, but as news of an official Stanford University inquiry spread, he finally broke his silence sharing communications and emails with the *New*

York Times dating back to August 2016. Quake viewed JK as bright and ambitious—not uncommon among Stanford postdocs—but someone who might cut corners if needed. Quake insisted that he would have been "very aggressive about telling people" about JK's intentions if he had any hints of misconduct. As Quake learned more about the pregnancy, he confided in two prominent scientists, but neither suggested he "notify the mythical science police." Quake trusted that JK had the necessary ethical approvals and conceded he could have done more.[18] The next day—Quake's fiftieth birthday—Stanford officially cleared its three faculty members of any wrongdoing.[19]

Porteus wished he had done more in hindsight. The idea of breaking a confidence troubles him but if a patient plans to hurt themselves or others, a physician has an obligation to overrule the doctor-patient confidentiality and disclose. "Perhaps this fits that scenario where it's such an egregious overstepping of bounds that it's worth violating the unwritten culture," he reflected.[20]

DeWitt habitually ignored media requests, so I was pleasantly surprised he agreed to talk to me. He doesn't want his own career to be remembered for his interactions with JK. "It wasn't just me telling him 'no.' He ignored everyone," DeWitt said.[21] "It was an evil and narcissistic quest to make this happen one way or another. . . . He should've been treated more like a criminal. He did things that were completely and utterly unacceptable and deeply unethical." Now in Los Angeles, DeWitt is helping to treat sickle-cell patients using genome editing. In retrospect he wishes he had done more to report JK. "But it's not exactly clear who I talk to?" Who indeed? The Dean? The director of the World Health Organization or the NIH? Antonio Regalado? Nobody knew what to do or where to turn.

It took a freedom of information request filed by the Associated Press to uncover Mello's involvement with JK,[22] although JK's WeChat timeline left a clear trail. The Nobel laureate hastily resigned from JK's company Direct Genomics after the CRISPR babies' revelation. "Whistleblowing is never easy but we need a system where this does not happen again," said Joyce Harper of University College London.[23]

Michael Deem has kept a vanishingly low profile since the JK scandal while Rice University conducts a confidential investigation. Deem's lawyers, who specialize in white-collar crime, have sought to distance their client from any direct involvement in the gene-editing trial, despite Deem's own admission (and visual evidence) that he participated in the informed consent process. Incredibly, eighteen months after the twins' birth, my inquiries to Rice University are returned with an unobliging, unwavering note: "The investigation is ongoing."

JK's decision to implant genetically edited embryos was in flagrant defiance of every pronouncement and report on the ethics of germline editing. For almost four years, scientists and ethicists had grappled with the implications of CRISPR technology that were hurtling researchers toward the unthinkable. The warning signs were there before CRISPR burst on the scene. For example, in a 2009 story in the *New York Times* on the first gene therapy trial using genome editing, Porteus said: "In principle, there is no reason why a similar strategy could not be used to modify the human germ line," quickly adding that he didn't think society was ready for such a proposal.[24]

In January 2015, Doudna hosted a small retreat in Napa, California, where some fifteen invited experts, all Americans, discussed the potential misuses of CRISPR, including the prospect of engineering permanent, heritable fixes into human embryos. The guests included Asilomar veterans and Nobel laureates David Baltimore and Paul Berg, bioethicist Alta Charo, Carroll, Greely, Daley, and Church. Doudna's concerns were amplified by her newfound celebrity status, which brought her to the attention of desperate parents. Amidst the emails pleading for help to find a cure for their child's rare genetic disease were letters expressing unconditional love for the genetically disadvantaged. Doudna lost sleep worrying whether CRISPR could do more harm than good. "When I'm ninety," she told Michael Specter, "will I look back and be glad about what we have accomplished with this technology? Or will I wish I'd never discovered how it works?"[25] Or as Specter put it, "Not since J. Robert Oppenheimer realized that the atomic bomb he built to protect the world might actually destroy it have the scientists responsible for a discovery been so leery of using it."[26]

It is to Doudna's credit that rather than shirk her responsibility or focus on research, she set about starting the conversation. In their recap of the Napa meeting,[27] Baltimore, Doudna, and colleagues warned against any attempts to initiate human germline experiments to allow more debate among many stakeholders, including the public, and more research on the safety of CRISPR gene editing. But at least one meeting attendee wondered, "There may come a time when, ethically, we can't *not* do this."[28]

In seeking a sensible middle ground, a "prudent path" as they put it, Baltimore and Doudna inevitably drew flak from both flanks. Robert Pollack, a Columbia University geneticist, argued their recommendations were inadequate, opening the door to eugenics. "The best in the world will not remove the pain from those born into a world of germline modification but who had not been given a costly investment in their gametes," Pollack said. Only a complete ban on germline modification would prevent such a powerful force for individualized medicine from becoming "the beginning of the end of the simplest notion of each of us being 'endowed by our Creator with certain inalienable rights.'"[29] But Stanford's Henry Miller dismissed such abstract concerns as insensitive to the suffering of patients in need. Germline gene therapy should be used sparingly, he said, but "we don't need a moratorium. We do need to push the frontiers of medicine to rid families of monstrous genetic diseases."[30]

One of the recommendations from Doudna's retreat was the need to widen the debate. A major conference on the ethics of genome editing took place at the National Academy of Sciences, overlooking the National Mall in Washington, DC, that December. It was a rare joint appearance (excluding prize ceremonies) by the three biggest names in CRISPR circles—Doudna, Charpentier, and Zhang. "We are close to altering human heredity," said Baltimore. "The overriding question is when, if ever, we will want to use gene editing to change human inheritance."[31] Baltimore quoted Huxley's *Brave New World*, in which he imagined "a society built on the selection of people to fill particular roles in society with environmental manipulation to control the social mobility and behavior of the population."

Huxley "couldn't have conceived of gene editing," Baltimore continued, but we should still heed his warnings as we assess "this new and powerful means to control the nature of the human population."[32]

For three days, scientists, physicians, ethicists, and philosophers debated the dangers and potential applications of human germline editing. Although in the minority, some spoke out strongly in favor of germline editing, at least in principle. Manchester University philosopher John Harris quipped, "If sex had been invented, it would never have been permitted or licensed . . . it's far too dangerous!" But the most electrifying moment came not on stage but from the audience. Sarah Gray rose up and spoke tearfully about the birth of her son with anencephaly, who suffered seizures for a week before his death. "If you have the skills and the knowledge to eliminate these diseases, then freakin' do it!" she pleaded.

Summit organizers argued long into the night before issuing a closing statement agreeing on two main principles: first, issues of safety and efficacy must be resolved before contemplating germline editing. Second, there should be a broad societal consensus about each application. Germline editing mistakes can't be corrected with a product recall notice. The ensuing *New York Times* story by Nicholas Wade opted for the *M* word, the headline reading: "Scientists seek moratorium on edits to human genome." But summit organizers had deliberately refrained from issuing a direct call for a total ban on germline editing research. Rather, they felt it would be "irresponsible to proceed" with embryo editing until there was a broad societal agreement on the safety, value and wisdom of the technology. The *M* word debate would resurface a few years later.

The DC conference triggered a lengthy succession of ethics reports from scientific societies and ethical groups in Europe, Asia, and North America. More than sixty organizations issued reports and policy statements, some running two hundred pages or more. But the net result was more confusion than clarity.[33] More than half of these reports concluded that no clinical germline editing should be performed, concerned about safety, medical necessity, and informed consent. Among them was a short statement from President Obama's chief scientist, John Holdren, who declared "The Administration believes that altering the human germline for clinical purposes is a line that should not be crossed at this time."[34]

Amid the noise, two reports—one American, one British—stood out for their provenance and thoroughness. On Valentine's Day, February 2017, the National Academies of Science, Engineering and Medicine (NASEM) released a detailed report on human genome editing.[35] The bill creating the National Academy of Sciences was signed into law by President Lincoln in 1863, at the height of the Civil War. The blue-ribbon committee, cochaired by Rick Hynes, a British expat molecular biologist at MIT, and the bioethicist Alta Charo, worked for more than a year to draft the report.

Surprisingly, the NASEM report left ajar the slim possibility of future clinical germline editing applications to correct genetic diseases—although not for enhancement. Trials could occur for "the most compelling circumstances," subject to comprehensive oversight that would protect patients "and their descendants," with safeguards against "inappropriate expansion" to less compelling uses. The committee offered a top ten list of criteria to support any future use of clinical germline editing, including: no reasonable alternatives to preventing a serious disease; genes should be edited to known DNA variants associated with ordinary health; and appropriate monitoring of patients and privacy protections.

Despite the preconditions, the report struck a much different tone than the DC summit fifteen months earlier. Far from saying germline editing was irresponsible and needed broad societal consensus, the NASEM committee concluded there was nothing inherently wrong with using germline editing to treat disease or disability. In a wood-paneled room at NAS, Hynes justified the change in stance. "In the past, there has been a line drawn by many that says one should refrain editing heritable features. That was mostly because there was no way of considering how to do that at all . . . so nobody was arguing that it should be done."[36] This was a brand new day.

It was a significant course correction from "impermissible as long as risks have not been determined" to "permissible if risks are accurately determined."[37] The report even left open the possibility of genetic enhancement, but only after public discussion. "Caution is needed, but being cautious does not mean prohibition," said Charo.

One year later, the UK's Nuffield Council on Bioethics concluded that germline editing might be "morally permissible" if it respected the well-being of the edited individual and did not exacerbate discrimination. "So long as heritable genome editing interventions are consistent with the welfare of the future person and with social justice and solidarity, they do not contravene any categorical moral prohibition,"[38] the report said. Being British, the report politely suggested that talks on an international governance framework should happen "sooner rather than later." The editor of the *Lancet*, Richard Horton, welcomed the findings. "Given the current state of the world, where national solipsism is on the rise and scientific developments can sometimes take place unchecked, there is no time to waste: the recommendations should be put into practice immediately."[39]

For all of their prestige, authority, and credentials, ultimately the NASEM and Nuffield declarations came down to the results of small groups of experts with a strong Western bias trying to reach a consensus. As Sarah Chan, a bioethicist at the University of Edinburgh, put it, there is no shortage of edicts in genome editing. The National Academy listed seven principles, the Nuffield report two more. There were five dozen more reports stacked high in her office. "We don't need more 'principles'!" Chan said.[40] But for JK, the refusal to issue a blanket prohibition of germline editing from either of these two prestigious bodies was tantamount to lifting a velvet rope. Or as Ed Yong put it, "it's as if he took the absence of a red light as a green one."

In October 2018, JK submitted an essay on his own ethical guidelines—built around the five bullet points he had shown Regalado and Kiani—to the *CRISPR Journal*. I communicated with Ryan Ferrell in advance and invited submission, thinking the essay might offer an interesting ethical perspective on genome editing from within China, although I'd never heard of JK, the group leader. The authors didn't say anything about conducting clinical research or implanting human embryos, nor did they disclose any conflicts of interest. The absurdity of that became clear the moment the CRISPR babies story broke. Specter called JK's ethical guidelines "admirable . . . if only He had spent more time reading them over, he might have skipped this stunt."[41] In retrospect, the commentary appeared to be a sly effort to lay the ethical groundwork for the blockbuster report on the CRISPR babies' birth. A few weeks later, the journal's chief editor, Rodolphe Barrangou, ordered a retraction.[42]

Where do we go from He? Or here? Two new highly credentialed commissions were launched in the aftermath of the JK fiasco. The World Health Organization (WHO) established an expert advisory committee tasked with "Developing global standards for governance and oversight of human genome editing." The geographically diverse eighteen-member group kicked off in March 2019 in Geneva, Switzerland, under the leadership of Edwin Cameron, a South African supreme court justice, and former FDA commissioner, Margaret Hamburg. That first meeting stopped short of calling for a moratorium, but the committee did propose a global registry of germline editing research.

Another blue-ribbon committee was co-organized by NASEM and the Royal Society, chaired by Dame Kay Davies (no relation) and the Rockefeller University president Rick Lifton.[43] Both efforts received rare bipartisan support on Capitol Hill. Senator Dianne Feinstein tabled a resolution condemning JK's actions and supporting the work of the commissions. "It's vital that the United States lead the way in creating ethical standards for gene-editing research," said Senator Marco Rubio, who cosigned the bill.[44]

These reports will be important but unlikely to be the final word. We debate the pros and cons of human embryo editing without having any real comprehension of the research under discussion. In August 2018, I attended a conference at Cold Spring Harbor. (One of the sponsors creatively offered vials of mosquito repellant to help the throngs of scientists survive the outdoors.) During a debate on editing of human embryos, Dieter Egli, a Swiss developmental biologist at Columbia University who works with human IVF embryos,[45] asked for a show of hands: "Who has any direct experience of working with human embryos?" Not a single arm went up. It was a stark reminder of how little we know about the earliest moments of life, even as we debate how we should proceed.

Amid the chorus of condemnation pointed at JK, one or two figures struck a different note. "It seems like a bullying situation to me," commented George Church.[46] "The most serious thing I've heard is that [JK] didn't do the paperwork right." Church concluded the saga could go in one of two directions: a medical tragedy on a par with the death of Jesse Gelsinger or a landmark advance in medical technology similar to the birth of Louise Brown. "As long as these are healthy, normal kids, it's going to be fine for the family and for the field," he said.

Samira Kiani, who had met JK, agreed that a piling on JK was "not going to be useful in the long term." She said, "We're not eliminating future trials [by vilifying JK], we're just pushing them underground. Similar scientists will not come forward."[47] Kiani's concerns made sense but her fears were misplaced. In Russia, as we'll see, a bear was stirring. And in the Middle East, an early sign of genuine interest in replicating JK's endeavors. One week after his revelation, on December 5, 2018, JK received an email from a fertility and gynecology clinic in Dubai. The email, shared publicly by William Hurlbut at the World Science Festival, read:

> Dear He Jiankui,
>
> Congratulations on your recent achievement of the first gene editing baby delivered by your application! . . . Our embryologist is interested in partaking in a course regarding CRISPR gene editing for Embryology Lab Application. Does your facility offer this type of course? . . .

Hurlbut advised JK not to reply.

CHAPTER 19
GOING ROGUE

Today, *Nature* is part of a huge German publishing empire, Springer Nature. But in 1990, when I nervously began my first job as a science editor, *Nature* was a quintessentially British company, tucked away off famed Fleet Street in central London. I arrived bright and early on my first day wearing a sharp Italian double-breasted suit, expecting to walk into the sleek high-tech headquarters of the world's most prestigious science journal. My heart sank as I entered a Dickensian newsroom resembling a primary school classroom with desks crammed together, books and newspapers spilling onto the floor. The only other pinstripe suit belonged to chief editor John Maddox, his presence evident from the plume of cigarette smoke wafting from his corner office.

This was before the invention of email or the World Wide Web. Manuscripts were mailed in quadruplicate hard copies. Maddox would smoke a pack of Marlboros and drink a bottle (or two) of wine while meeting his weekly editorial deadline. Geneticist Magdalena Skipper was appointed editor in May 2018, becoming *Nature*'s first woman chief editor. Assuming the position, she had many issues on her plate—open access publication, data reproducibility, women and diversity in science. But nobody anticipated the problem that landed on her desk six months later.

It was logical that JK would select *Nature* as the prize venue for his CRISPR babies report. The journal had published the double helix of course, but also the seminal IVF papers of JK's idol Robert Edwards, as well as the first (and only) two human embryo editing studies conducted outside China. For many scientists, a *Nature* paper is the ultimate validation, the Oscar of scientific achievement. (Chinese scientists are also handsomely rewarded with a cash incentive for publishing in top-tier journals.)

JK uploaded his manuscript in November 2018, about a week before the Hong Kong summit. He emailed Craig Mello on November 22 to say that the paper had been sent out for peer review. Although *Nature* staunchly preserves the confidentiality of the review process, extracts of the manuscript were eventually published by *MIT Technology Review* and by Kiran Musunuru, one of the experts originally contacted by the AP to vet JK's claims.[1] JK naïvely assumed his manuscript would be published expeditiously to worldwide acclaim. Months later, Fyodor Urnov trashed the manuscript as nothing more than "sets of pieces of paper."[2]

Dissection of JK's unpublished manuscript—"Birth of twins after genome editing for HIV resistance"—reveals a naïveté and arrogance on the part of the authors, as if assuming that the *Nature* editors would fall over themselves to publish this medical milestone. There were ten coauthors, JK positioned at the end of the list, signifying his senior status. The penultimate author was Michael Deem—indicating a major role in driving the work—although his lawyers later insisted that Deem had withdrawn his authorship name from the list of authors.

From the article's opening summary, JK and his coauthors failed to accurately represent their work and reveal their true intentions. In reporting "the first birth from human gene editing," JK wrote that his team had used CRISPR to "reproduce a prevalent genetic variant of the CCR5 gene," when in fact the editing had failed categorically to reproduce the Δ32 deletion. JK asserted that his team had witnessed no off-target or cancer-causing mutations. And his revolutionary therapy would not only "control the HIV epidemic" but also "bring new hope to millions of families" desperate to avoid inheriting a genetic disease. Urnov could barely hide his disgust: "The profundity of the delusion and the hubris is overwhelming," he said.[3]

JK reiterated his goal of trying to combat the spread of HIV worldwide, without addressing the overwhelming success of HIV antiviral drugs or the impracticality of addressing the epidemic via editing one embryo at a time. Nor did he dwell on the fact that during the IVF protocol, the father's sperm were "washed thoroughly to remove infectious seminal fluid." This was standard protocol, ensuring that any babies carried to term would not contract HIV.

From four healthy embryos, JK's team analyzed the two that had been edited at the *CCR5* gene. They reported just one off-target edit, unlikely to affect the activity of any genes. But brushed aside was the Catch-22 of embryo editing: the only cells tested from the blastocyst would, by definition, play no further role in the development of the embryo. The remaining cells of the embryo were not analyzed, making it impossible to measure what off-target effects might exist in those cells.

In the manuscript, JK said the babies were born in November 2018, even though reporting says the babies were born in late October. Perhaps this was a deliberate feint designed to throw off any amateur detectives trying to identify the CRISPR babies. JK wrote that the twins would undergo periodic medical testing at least until they were eighteen, including tests to ensure they were immune to HIV infection. Such experiments should have been conducted before Lulu and Nana became human guinea pigs. "This egregious violation of elementary norms of ethics and of research borders on the criminal," Urnov fumed.

In his book *The CRISPR Generation*, Kiran Musunuru described the visceral emotion of being one of the first people outside JK's group to read the manuscript, which he received from Marchione on the Monday after Thanksgiving. His dismay grew until he opened a figure that showed the DNA sequences of *CCR5* obtained from the CRISPR-treated embryos. The first image (Lulu) should have shown two distinct traces—the normal sequence of *CCR5* on one chromosome, a fifteen-base deletion on the other. This figure showed three. Musunuru knew that meant the embryo was mosaic, before emitting a guttural scream. It was a similar story with Nana's embryo—three traces rather than the two edits JK had claimed.

The phrase that stuck in Musunuru's mind was "hack job"—not an unusual occurrence among inexperienced researchers learning to use

CRISPR, but unconscionable in a human embryo. Mosaicism happens naturally, the result of a random mutation occurring in one cell in the early embryo being passed down as that cell repeatedly divides. In principle, if applied early enough to a single-cell embryo, the CRISPR edits should be transmitted to all daughter cells. But Cas9 takes time to find the correct target sequence, so there is a possibility that the fertilized egg will have undergone cell division before the edits have taken place. If the CRISPR-Cas components are not shared evenly as the cell divides, genes will not be edited uniformly in the developing embryo.

Musunuru's despair turned to anger as he saw signs of mosaicism at the off-target location in Lulu's embryo, indicating that "both embryos were flawed." The only silver lining: the mosaicism showed that Lulu and Nana were not a hoax. No scientist trying to invent such a sensational story would present such a mediocre editing job. He summed up: "The knowledge that somebody had gone ahead and made the first gene-edited babies—for better or for worse, a historic event for humankind—would have been distressing enough, even if it had been done perfectly. The fact that it had been done with flawed embryos, in such a careless fashion, made it a hundred times worse."

Within a week, *Nature* closed the file as the scale of the scientific and ethical transgressions exploded around the world. *Nature* does not comment on the fate of specific articles but the guidelines were laid out by the editor of one of its sister journals:

> Any manuscript reporting genetic modification of human embryos or gametes would need to follow strict scientific and ethical guidelines . . . On the basis of the available information, He's research would not have met the editorial criteria adopted by *Nature* journals.[4]

JK decided to try again, resubmitting to the *Journal of the American Medical Association* (*JAMA*). In an extraordinary move, the *JAMA* editor sent the manuscript to eleven outside experts for review in early December, including George Church, before it too rejected the paper.

A year after being spotted on a guest house balcony, we finally learned JK's fate. On December 30, 2019, the Nanshan District People's Court of Shenzhen found JK and two colleagues guilty of "illegal medical practice." (There is no law in China that expressly prohibits editing human embryos.) JK was sentenced to three years in prison, a fine of 3 million yuan (about $430,000), and a ban from further research in "assisted reproductive technologies."* Qin Jinzhou (the first author on JK's manuscript) and Zhang Renli were handed fines and suspended sentences of eighteen and twenty-four months, respectively. "The three accused did not have the proper medical certification to practice medicine, and in craving fame and wealth, deliberately violated national regulations in scientific research and medical treatment," the court stated. "They have crossed the bottom line of ethics in scientific research and medical ethics."[5]

Reactions to JK's incarceration ran the gamut. "What JK did was criminal: he broke laws of ethics and medicine, endangered lives, and stained an entire field for no reason other than hubris," Urnov said. "What needs to be imprisoned, metaphorically speaking, is the entire enterprise of human embryo editing for reproductive purposes."[6] Doudna took a slightly more diplomatic tone. "As a scientist, one does not like to see scientists going to jail, but this was an unusual case," she said.[7] "Jail isn't the right punishment [for JK], but just because we can do something with science doesn't mean we should," tweeted Scott Gottlieb, the ex–FDA commissioner. William Hurlbut expressed mixed feelings. "Sad story—everyone lost in this (JK, his family, his colleagues, and his country), but the one gain is that the world is awakened to the seriousness of our advancing genetic technologies," he wrote.[8]

And then there was Josiah Zayner, the face of the biohacker community. Zayner sought to frame JK's transgression as a minor speed bump on

* "Illegal medical practice" has three grades in China: For an offense in which no one was harmed, the maximum sentence is three years (plus a fine). If a patient has been harmed, the maximum sentence is ten years, longer in cases with a loss of life. Whether the CRISPR babies were harmed in any way remains to be seen. JK's sentence seems more than a slap on the wrist. Perhaps the Chinese government is sending a message of the dangers of money poisoning science.

the road to human germline engineering, insisting the Chinese scientist would be remembered more than any scientist of his day. "As long as the children He Jiankui engineered haven't been harmed by the experiment, he is just a scientist who forged some documents to convince medical doctors to implant gene-edited embryos. The four-minute mile of human genetic engineering has been broken. It will happen again."[9]

So who is right? Zayner the biohacker, who believes that thousands of edited human embryos will be born during this century? Or Urnov, the creator of GMO humans, who insists that germline editing is a classic case of "a solution in search of a problem"?

From the instant his name hit the headlines, He Jiankui was branded a rogue scientist. Writing in the *New York Times*, Eric Topol said JK should be "castigated" and warned that there was no "foolproof way to rein in such rogue efforts."[10] In Hong Kong, Harvard Medical School dean George Daley said, "scientists who go rogue carry a deep cost to the scientific community." JK's university, hospital, and the Chinese authorities swiftly denied any knowledge of, or responsibility for, his actions. The picture of JK as a rogue agent—secretly seeking fame and fortune, lusting obsessively like a young Jim Watson for a Nobel Prize for himself and glory for his proud nation—spanned the globe.

It's a convenient narrative, but is this the full story? I have my doubts.

First, JK was not some anonymous researcher emerging from some shadowy secret lair but a highly recruited talent lavished with state and national funding. His commercial success and hungry ambitions in genome sequencing were widely feted and highly publicized. He was featured on state television on a program called "Extraordinary Guangdong" in a segment entitled "The new top shot in the gene world."[11] "Who gave him such very rare opportunity?" asked a group of Chinese bioethicists, suspecting the hidden support of influencers in the local or central government.[12] JK kept a running photo diary on his WeChat account of his meetings with high-profile science celebrities and Chinese dignitaries that was worthy of a Kardashian. Deem, Quake, Mello, and more were there for all to see. Lombardi considered JK to be clever but naïve. "I feel bad for the guy," he told me. "He got thrown under the bus by somebody. He wasn't doing this alone. There was some entity who knew exactly what he was doing."[13]

Second, China is famed for its high-tech statewide surveillance. "It's almost as good at surveillance as Russia is at hacking," says George Church, only half-joking. "It's hard to believe the most exceptionally successful surveillance state would not be paying attention to the most amazing story in the history of biology, right?" Church is convinced that somebody in the Chinese government hierarchy was privy to JK's work.[14] JK publicly acknowledged the Chinese Ministry of Science and Technology in his early embryo research. And we know that JK invited Yu Jun, a decorated scientist, BGI co-founder, and member of the Chinese Academy of Sciences, to sit in on at least one of his informed consent sessions with trial volunteers. In China, the needs of the state take precedence over those of the individual: Why should genomics and reproduction be any different? "Emperors have been ruling us for thousands of years," BGI's Wang Jian said. "I know the government is watching us at all times."[15] Would JK's audacious activities be any exception?

Third, China has become a world leader in technology development. China's massive investment in R&D—$300 billion in 2020—has made it a leader in 5G wireless communication, facial identification, artificial intelligence, high-speed rail, and many other areas of technology. In 2019, China successfully landed a lunar rover on the far side of the moon. "It is human nature to explore the unknown world," said the head of China's space program, as China spectacularly laid down another marker in the tightening race for technology supremacy.[16]

Genome editing technology is no exception. Chinese scientists did not develop CRISPR, but they have wasted no time pushing it into the clinic despite concerns of a "Wild East" approach to clinical gene editing.[17] The first patients treated with CRISPR were at the No. 105 Hospital of the People's Liberation Army in Hefei. By the end of 2017, almost one hundred patients in China had been treated with CRISPR genome editing for various cancers including liver, lung, prostate, and blood. There is less emphasis on the sort of fussy regulatory checks demanded by the FDA or its European counterpart. While patients have died, if their deaths are not attributed to CRISPR or the trial protocol, they are not considered adverse events and thus go unreported.[18] By contrast, the first three American cancer patients administered CRISPR gene therapy were not reported until 2020.

Fourth, bioethics is a relatively young scientific discipline, but in China the field didn't begin until the Ministry of Health issued its first national guidelines on medical ethics in 1998.[19] Without enforcement mechanisms and riven with corruption, this did little to deter the harvesting of organs from executed prisoners or the profusion of shady stem cell clinics.

As noted earlier, Chinese researchers published eight of the first ten studies on human embryo editing. Scientists at the Chinese Academy of Sciences reported the first cloned primates[20]—a pair of macaques patriotically named Zhong Zhong and Hua Hua—believing these could be useful animals in medical research. Yi Huso, research director at the Chinese University of Hong Kong Center for Bioethics, observed, "people are saying they can't stop the train of mainland Chinese genetics because it's going too fast."[21] Yi's perspective was eerily prophetic: "We're going to do it, then see what's wrong, then fix it."

In an article entitled "China will always be bad at bioethics," Yangyang Cheng argued that one of the biggest cultural differences between East and West is the use of science to legitimize an authoritarian regime. While Jesse Gelsinger's death set back the field of gene therapy by a decade, if it had occurred in China, "it would most likely have been either covered up or turned into propaganda depicting Gelsinger as a national martyr."[22]

Finally, another contributing factor is China's infamous one-child policy of social engineering and population control, launched by Mao Zedong in 1979. Before it was abandoned in 2015, the policy showcased China's draconian determination to curb population growth and a totalitarian control over women's reproductive rights. This was compounded by the Eugenics and Health Protection Law in 1995, which showed that China's leaders were willing to impose measures not only to reduce overall births but also "new births of inferior quality."[23]

Wang Jian, the billionaire cofounder of BGI, envisions a future society in which ubiquitous genetic testing will help eliminate a plethora of hereditary disorders. "If all the world's children are able to receive genetic testing and prediction for [deafness], [blindness] and related diseases, all these genetic diseases will not exist anymore. How big an impact this will be to humankind!"[24] In her book, *One Child*, Mei Fong quotes the proprietor of an egg donor agency in California whose Chinese clients "almost always want taller, at least 5-foot-5. And they have questions about eyelids." Fong concluded that "an openness to gene editing's worst excesses may prove to be [China's] one-child policy's most unfortunate legacy."[25]

China is most likely to abide by international ethical standards "when its membership in the global political and scientific community depends on it—in other words, when it has no other choice," Cheng writes.[26] For now, ethical debates and commissions dominated by Western scientists and ethicists won't stop it. JK wasn't inclined to wait patiently until every safety concern had been addressed and the world community deigned to give him the green light. As he said, "If it is not me, it will be somebody else."

Benjamin Hurlbut, who corresponded extensively with JK for a spell after his house arrest, also disputes the popular "rogue" narrative. JK's motivations were so familiar they were almost mundane. "Far from 'going rogue' and rejecting the norms and expectations of his professional community, He Jiankui was guided by them." If anything, Hurlbut argues, the "rogue" label belongs not to JK but to "a scientific community that asserts that it alone has the authority to determine what it should and should not do, racing ahead of ethical concerns and public debate.[27]

Hurlbut concludes that JK was following a familiar path that puts "a premium on provocative research, celebrity, national scientific competitiveness, and firsts."[28] Hurlbut writes:

> He was driven by the high-octane milieu of contemporary biotechnology, both in the United States and in China. He internalized ideas that led him to believe that his experiment would elevate his status in the international scientific community, advance his country in the race for scientific and technological

> dominance, and drive scientific progress forward against the headwinds of ethical conservatism and public fear . . .

JK, like many if not most scientists, had aspirations of fame and fortune. He jumped into an area of medical research that was cutting-edge and highly controversial, for which he had neither the technical skills nor maturity to handle. We may never know if he damaged the health of the CRISPR babies. His dreams of scientific immortality were reduced to infamy.

As the scientific community reacted to the CRISPR babies' bombshell, there were ample concerns that someone else might attempt to follow in JK's footsteps. "Who is to say that there won't be one hundred other similar type of experiments that will take place, not only in China but in other medical tourism hotspots?" said Feng Zhang.[29] Around the world, there are hundreds of private, unregulated stem cell clinics luring patients with promises of "tomorrow's treatments today" for everything from chronic back pain to Alzheimer's disease. It would be no surprise if CRISPR clinics, as JK had planned, start to appear.

In testimony for the NAS commission, Sandy Macrae, CEO of Sangamo, said there were two good reasons that no legitimate biotech company would attempt germline editing. First, society wasn't ready for it. Second, "there isn't an obvious disease we want to go for, there isn't a compelling rationale," he said.[30] But Macrae did note that editing a single-cell embryo has one potentially big advantage over trying to edit millions of cells in an adult patient—a reduced chance for an error that could lead to cancer.

In the Spring 2019, as Zhang and others had grimly anticipated, another scientist declared his intentions to perform germline editing. Like JK, Denis Rebrikov, head of the gene-editing laboratory at Russia's largest fertility clinic, the Kulakov National Medical Research Center in Moscow, was little known outside his country. He made his first public proposal to edit human embryos a year before the CRISPR babies were born. Like JK, his initial idea was to target *CCR5*, specifically embryos from HIV-positive women who do not respond to anti-HIV drugs. "I'm crazy enough to do it," he told *Nature.*[31] His preliminary efforts using CRISPR in nonviable human embryos set off alarm bells. "Where is the evidence that he knows how to do this?" asked Lovell-Badge. "Rebrikov needs to hear the concerns of scientists, clinicians, ethicists, and regulators before he falsely gives hope to patients seeking solutions."[32]

But Rebrikov, a former champion in sambo, a Russian martial art, suplexed his critics. "We can't stop progress with words on paper," he said.[33] Russia might not be a free country politically, he said, but it was very free in science. Rebrikov quoted a Russian saying: "If you have success, you are right." If Lulu and Nana were healthy, JK might yet salvage his reputation. And he wouldn't rule out using genome editing for enhancement. "These people who are opposed [to enhancement] *want* to have all these things in their children but only by 'divine providence,' not by science. They are liars or stupid."

Rebrikov changed tack to focus on an inherited form of deafness.[34] Mutations in the gene *GJB2*, which encodes a protein called connexin 26, cause hereditary hearing loss. The incidence is fairly high in Russia, where the mutation is thought to have spread from west to east over thousands of years. The most common *GJB2* mutation is the deletion of a single letter in the sequence. About ten babies are born each year with hereditary deafness resulting from inheriting two dysfunctional copies of *GJB2*. Rebrikov identified several deaf couples in which both partners were homozygous for *GJB2* mutations, meaning that PGT was not an option to produce a hearing-enabled child. If he could use CRISPR to repair one copy of the gene, the child should be able to hear. Rebrikov put the price of germline editing at about 1 million rubles, or $15,500. "I can see the billboard now: 'You choose: a Hyundai Solaris or a Super-Child?'"[35]

Six months after the Hong Kong summit, a small group of Russian medical experts and officials discussed Rebrikov's plans in a closed-door meeting in Moscow with a very special guest—Maria Vorontsova, a pediatric endocrinologist and, more importantly, Vladimir Putin's daughter. A few months later, Rebrikov defended his plans in a meeting at the Russian Academy of Sciences' Institute of Philosophy.[36] One woman seriously considering Rebrikov's trial, known as Yevgenievna, was devastated when her young daughter was declared deaf. Yevgenievna is aware of the risks of genome editing. "Which is better: to hear music or be suffering from cancer? That's the question."[37]

Unlike JK, Rebrikov was fully transparent about his intentions—no secret circle of trust here. He gave interviews to a string of science journals, he had more relevant expertise, and the medical case had more merit than JK's cause. But in July 2019, the WHO director-general, Tedros Ghebreyesus, stated that "regulatory authorities in all countries should not allow any further work in this area until its implications have been properly considered." The Russian Ministry of Health evidently took note, responding that granting Rebrikov permission to undertake clinical germline editing would be premature and irresponsible.

Rebrikov steadfastly quotes a line from Lenin about the Bolsheviks' seizure of power in 1917: "Yesterday was early, tomorrow will be late. Power must be taken today."[38] Speaking of power, the Vorontsova connection raised the tantalizing idea that the Russian president could yet have a say. Putin has toured Rebrikov's institute and publicly demonstrated his familiarity with CRISPR. During a televised town hall at the 19th World Festival of Youth and Students in Sochi, Putin, unscripted, said:

> Mankind has become capable of interfering with the genetic code created by nature or, according to religious people, created by God. What are the practical consequences of that? One can almost imagine that people will be able to create human beings with specific characteristics. This person could be a mathematical genius or a brilliant musician, but also be a soldier—an individual capable of fighting without fear or compassion, without mercy or even pain. You see, humanity can enter, and most likely will enter, in the near future, a very difficult period of its development and existence, a period demanding great responsibility. And what I have just described may be more terrifying than a nuclear bomb. Whenever we do something, and whatever it is that we do, we should never forget about the ethical and moral

> foundations behind our actions. Whatever we do should benefit people. Make them stronger, not destroy them.[39]

Putin is informed and apparently enamored by the possibility of conceiving a few more Tchaikovskys or Rachmaninovs in the next century, though the prospect of super soldiers would have more practical utility. And let's face it, the bit about "ethical and moral foundations" was a nice touch.

Following JK's disappearance and incarceration, the aftershocks continue to ripple through the scientific community. How should scientists respond? Should there be a worldwide ban or moratorium on germline editing research? And who would police the research community to ensure that no violations occurred? In March 2019, a group of eighteen leading scientists marshaled by Eric Lander called for a temporary global moratorium on germline editing research.[40] The signatories included Zhang and Liu, Emmanuelle Charpentier, Keith Joung, bioethicist Françoise Baylis, and Nobel laureate Paul Berg. For Lander, it was his first major pronouncement on genome editing since his notorious "Heroes of CRISPR" review three years earlier.[41]

While there were some disagreements among Lander's coauthors on the merits of heritable editing, there was agreement on the need for a *temporary* moratorium rather than a permanent ban. Under the proposal, nations kept the right to make their own decisions, but committed to suspend any use of clinical germline editing unless certain conditions are met. During the moratorium—Lander thought five years sounded about right—"no clinical uses of germline editing whatsoever are allowed." After that, countries could set their own policies providing that they provided public notice of any germline editing plans; allowed debate of all the medical, technical, and ethical issues during a consultation period; and ensured societal consensus supporting the application.

Future decisions about germline editing should be made by a broad group of stakeholders, "not by scientists, physicians, hospitals or companies, nor the scientific or medical community acting as a whole." And importantly, the moratorium would not hamper clinical research in somatic gene therapy. Lander didn't think an international treaty was plausible: there is no genome editing Thunderbirds task force that can physically stop a rogue actor or government. On the other hand, it would "place major speed bumps in front of the most adventurous plans to re-engineer the human species." The risks of turning a blind eye, potentially hurting patients and eroding public trust, were much worse.

Lander's call was immediately seconded by Francis Collins, who said the community needed a period of reflection and "a substantive debate about benefits and risks that provides opportunities for multiple segments of the world's diverse population to take part."[42] A month later, some sixty researchers and ethicists made a similar plea in a letter to the director of the Department of Health and Human Services. Human genetic manipulation should be unacceptable, they wrote, calling for "a binding global moratorium" until serious scientific, societal, and ethical concerns are addressed.[43] The letter was cosigned by many gene therapy pioneers, including Jean Bennett, Jim Wilson, and Feng Zhang.

Conspicuously absent from joining the moratorium movement was Doudna, even though she'd been invited by Zhang to cosign Lander's article. "This is effectively just rehashing what's been going on for several years," she said. "I don't want to drive others underground with this [moratorium]. I would rather they feel that they can discuss it openly. Gene editing, it's not gone, it's not going away, it's not going to end."[44] George Daley felt that a moratorium raised too many questions, including how long should it last? How is it enforced? And who decides when to rescind it?[45] Church was less diplomatic. Without an enforcement mechanism, he said, "just calling for another moratorium is posturing." Why not crank up the penalties for practicing medicine without FDA approval?[46]

Doudna addressed the issue again on the one-year anniversary of the CRISPR babies. Writing in *Science*, she reiterated that moratoria were not sufficiently strong countermeasures. "The temptation to tinker with the human germ line is not going away," she said. Violators should face loss of funding and publication privileges. "Ensuring responsible use of genome editing will enable CRISPR technology to improve the well-being of millions of people and fulfill its revolutionary potential."[47]

During remarks delivered at Harvard shortly after the commentary came out,[48] Lander assailed the critics with a mixture of disdain and bemusement. His call for a temporary moratorium was "flamingly obvious because in effect we *have* a moratorium" right now. The definition of a moratorium is "temporary prohibition," yet some commentators were "squeamish about even using the dictionary word." Lander thought this was fascinating, as if invoking the *M* word had some magical talismanic properties.

To critics who said a moratorium might be hard to lift? "It lifts!" Lander insisted. For those who said using the *M* word would discourage people from entering the field? Lander wanted to get people's attention, "otherwise people wouldn't have read any further." And for those who claimed his proposal would not have prevented the actions of a rogue actor like JK? Lander pleaded no contest. "It wouldn't—that wasn't the point of it!" Lander insisted. Criminal acts occur: people murder each other despite laws prohibiting it. The proposal was *not* about preventing the actions of certain individuals but about "what countries should choose to do." JK's actions were China's responsibility. In the United States, it's the government's responsibility.* And to folks who might have reservations about the concept of international consultation such as "the Brits," Lander exhaled: "Duh, that's the point! If you don't have the courage of your convictions after that consultation, then you shouldn't be doing it."

JK's experiment was "utterly irresponsible," a "complete screw up" that can happen when work is conducted in complete secrecy. And yet, members of the community knew his intentions and did nothing. "That's not exactly what an upstander does, it's a bystander." Lander wished someone had called the press, but nobody did. "That's really interesting and more disturbing maybe than that he did it." Ultimately, Lander and colleagues felt a moratorium was the most prudent path forward. "I want our kids to be proud that we thought about this carefully," he said.

---

* Since 2016, the U.S. effectively has a ban on human germline editing. An amendment by Representative Robert Aderholt (Alabama) bars the FDA from funding clinical trial applications of gene editing of human embryos.

The call for a moratorium capped a remarkable shift in policy statements. In 2015, the committee of the first ethics summit said germline editing would be "irresponsible" unless and until certain conditions were met. In 2017, the NASEM report concluded it "should be permitted" provided that conditions were met; and in 2018, the Hong Kong summit committee surmised that germline editing was acceptable in principle and needed a translational pathway. For Baylis, a Canadian bioethicist and author who joined Lander's call for a moratorium, that progression was alarming. "There's nothing that's happened that warrants this kind of shift. What's the hurry?"[49]

With JK imprisoned, China has sent a resounding message warning off any other potential genome hotshots. The government required researchers to complete a survey to catalogue gene-editing programs in the country. China belatedly established an ethics commission to plug some of the jurisdictional gaps between the health and science ministries. What effect will these actions have? "It is also almost certain someone will attempt gene editing to make stronger, smarter, more attractive babies. Pandora's box is wide open in China," wrote Mei Fong.[50]

In April 2019, Yang Hui, a young neuroscientist at the Shanghai Institute of Biological Sciences, and Wang Haoyi, a bioethicist at the Chinese Academy of Sciences, published a damning rebuke of JK's work.[51] The authors were "shocked" and "enraged" by JK's "irresponsible conduct" that violated the medical ethics of China and the world. JK had no intention of trying to precisely introduce the Δ32 variant, they alleged, merely to break the *CCR5* gene.

Despite his fierce criticism of JK, Yang Hui was advancing his own research on human embryos using base editing.[52] He didn't mention the JK controversy in the article—it was as if the CRISPR babies had never been born. But in a candid interview, Yang told a Chinese paper that gene editing, like weapons or drugs, could be used for either good or immoral purposes. If we're allowing germline editing, then his goal was to leave no unedited embryo behind. "Even if there is just one embryo left unedited, it will create an ethical problem. An almost 100 percent efficiency is required before the technology can be used on humans," he said.[53]

That day might happen sooner than we think. And while Yang said the production of "superbabies" should be permanently banned, his zeal to follow in JK's footsteps was apparent. Indeed, you could be forgiven for seeing shades of a new Cold War. "We are ahead of the competition in the United States," he said. "We are working with the same spirit as building the first nuclear bomb!"

But if there are groups in China or anywhere else intent on pursuing germline editing, they have an uphill task. "Chromosomal mayhem" is not a phrase widely deployed in *Nature* headlines, but in June 2020 it was not an exaggeration.[54] A trio of bioRxiv preprints from the leading experts on genome editing in human embryos outside China—Niakan, Egli, and Mitalipov—independently reported worrisome "on target" DNA aberrations when using CRISPR to edit specific genes. Niakan's group found that a fraction of embryos displayed unwanted deletions and rearrangements in the vicinity of the targeted gene, while Egli and Mitalipov also reported serious chromosomal damage or rearrangements in some embryos. It was a timely reminder that we still have much to learn about the raw mechanics of DNA repair processes at the earliest stages of embryogenesis before we can safely and responsibly consider genome editing in the human embryo.

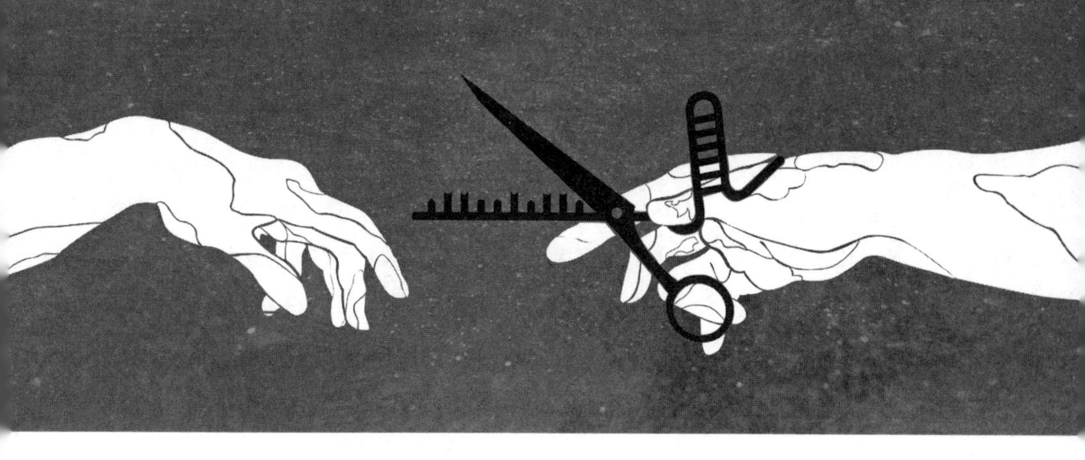

PART IV

"DNA is not just a genetic code. It is in some sense also a moral code."

—Siddhartha Mukherjee

"Soon we must look deep within ourselves and decide what we wish to become. Our childhood having ended, we will hear the true voice of Mephistopheles."

—E. O. Wilson

"We are all mutants, but some of us are more mutant than others."

—Armand Leroi

CHAPTER 20
TO EXTINCTION AND BEYOND

You can't write a book on genome editing without visiting George Church. Or so he tells me. So I make the pilgrimage to the Church lab, nestled in the complex of famous hospitals and institutes that make up Harvard Medical School, a David Ortiz home run from Fenway Park, the proud home of the Boston Red Sox. Church's schedule is crammed with appointments: mine nestles between a check-in with a medical student and a visit from the founders of an Israeli tech company.

I scan the overflowing bookshelves that line an entire wall of Church's office. There are copies of Church's first book, *Regenesis*. There's *Hood*, Luke Timmerman's biography of the inventor of automated DNA sequencing; *One in a Billion*, the story of a landmark clinical case of genome sequencing; and *Woolly*,[1] Ben Mezrich's book about Church's ambitious quest to recreate the woolly mammoth, which is being made into a film. On his desk is a stack of Walter Isaacson bestseller biographies including *Steve Jobs* and *Leonardo Da Vinci*. On top of the pile is *The Innovators*. It occurs to me I'm looking at one. I wonder who can carry off the trademark Church beard in the *Woolly* movie? I'm guessing George Clooney or Jeff Bridges, but all Church will say is that it is his wife's favorite actor.

Church is one of the most imaginative, restless, in-demand scientists alive, despite suffering from narcolepsy. He's warned me more than once to keep panel discussions I'm moderating with him as a guest interesting or he's liable to nod off. In 2019, Church opened a keynote lecture at a gene therapy conference by saying he hoped to make it through his lecture fully conscious. Only he wasn't joking: the day before, he'd fallen asleep—and fallen over—while talking to guests in his office. It's an occupational hazard he mitigates by fasting during the day and never letting his mind let up.

He launches new fields and new biotech companies with merry abandon, pairing "radical technology and radical application" to push the envelope of genetics. He's a pioneer of systems biology and the Personal Genome Project. One of his companies has released what Church fondly calls the "zero-dollar genome," another is developing new gene therapy vectors, yet another is trying to reverse aging. I peek into a room where some of his students are growing brain organoids, a controversial new tool to understand diseases like Alzheimer's and schizophrenia.

If there was a tagline for Church's sprawling operation, it might be: "Read and write DNA without limits" or "life finds a way" from *Jurassic Park.* The genome has become a giant experimental playground for Church. At any given time his lab is populated by dozens of wicked smart students from all corners of the globe, inspired by a legendary scientist who not only encourages but almost demands moonshots and imaginative ideas to turn science fiction into biomedical fact.

Church opens his lectures with his "conflict of interest" slide, a smorgasbord of company and university logos reflecting his incessant entrepreneurial and academic liaisons. Not every venture succeeds, but when they don't it's typically because Church's ideas are too early—it takes times for everyone else to catch up. One of my favorites was a next-generation sequencing outfit in Silicon Valley called Halcyon that could read gene sequences by stretching a DNA molecule like a piece of bubble gum, allowing scientists to visualize the rungs of the double helix under an electron microscope.

He also founded Knome,* the first company to offer personal genome sequencing years before the $1,000 genome became a reality. The first customer, a Swiss biotech executive, paid a staggering $350,000 for the

* Most people pronounced the company "gnome" but Church insisted on calling it "know-me."

privilege. Other early clients included a member of the British royal family and Black Sabbath's Ozzy Osbourne.[2] One of his latest ventures, Nebula Genomics, aims to offer free genomes to customers using something he calls "sponsored sequencing."[3]

He's been a guest on the Stephen Colbert show discussing another futuristic line of research—using DNA as a computer to store reams of digital information. In 2012, Church's team designed a bacterial genome with customized DNA sequence that encoded the full text of *Regenesis*. "Church has been accused of 'playing God,' an accusation abetted by his beard of biblical proportions," Colbert joked in profiling Church for *Time*'s 100 most influential people.[4] The comedian said Church "seems less like God and more like a cross between Darwin and Santa."

Having spent three decades obsessing over the technology for *reading* DNA scripts, Church is increasingly fascinated by writing and editing DNA. He's a leader in the synthetic biology movement and an organization called GP-Write (Genome Project-Write).* But it's not just human genomes that Church is interested in customizing. Church was one of the early developers of CRISPR gene editing, briefly hosting Feng Zhang and publishing one of the first demonstrations of genome editing in human cells in January 2013. Church is always looking to go cheaper, better, faster. Genome editing is no exception. Church jokes that his middle initial (M) stands for "multiplexing."

In *Oryx and Crake*, Margaret Atwood introduced us to the pigoon, a hybrid creature designed for the wealthy elite to supply "an assortment of foolproof human-tissue organs in a transgenic knockout pig host—organs that would transplant smoothly and avoid rejections."[5] Church believes he can revolutionize organ transplantation without going quite so far by providing a safe alternative source of compatible organs from genetically edited pigs. It's been branded "weird science," but if successful, Church's company eGenesis could offer hope to the 115,000 Americans on the organ transplant waiting list.

---

* GP-write was originally named HGP-write, but quickly dropped the H (human) prefix as it made some people uncomfortable to contemplate totally synthetic human genomes.

While about one hundred transplants are performed daily in the United States, twenty patients die waiting for a suitable organ donor. "The closest thing we have to death panels in this country are the decisions made about who gets transplants," Church says. Many people are rejected from an organ transplant because they have infectious diseases or drug addiction with the argument that they wouldn't benefit from a transplant. "But of course they'd benefit. And if you had an abundance of organs, you could do it for everyone."[6]

Church's cofounder and chief technology officer at eGenesis is Luhan Yang, his former grad student and postdoc. Yang grew up in Sichuan Province near Mount Emei, one of the Four Sacred Buddhist Mountains in China—what she calls "Crouching Tiger, Hidden Dragon" territory. After excelling at Peking University, Yang moved to Boston in 2008, working with Prashant Mali on the Church lab's proof of CRISPR gene editing in human cells. Church was soon contacted by physicians at Massachusetts General Hospital (MGH) about improving the prospects of xenotransplantation patients.

eGenesis is located not surprisingly in Kendall Square, one more ambitious biotech company among scores of cool start-ups and established biotech companies. Most aspire for the commercial success of their big pharma neighbors, including Novartis, Pfizer, Amgen, and Biogen. In the company's reception, a flat-panel TV displays a montage of staff holiday photos in Cancun, a present from the company's investors to celebrate the birth of the latest pig litter.

I'm greeted by Wenning Qin, a veteran pharma researcher who after spending most of her career knocking out genes in mice, jumped at the chance to apply her particular set of skills to pigs. Many aspects of the miniature pig's biology make it an ideal organ donor substitute. Pigs have a short four-month gestation and large litters. Pig heart valves and corneas are already used in operations. But two major issues must be addressed before contemplating human transplant of kidneys, livers, and other organs. First, the pig genome contains dozens of embedded viral sequences (PERV, or porcine endogenous retrovirus). While these viruses don't apparently cause any health issues in pigs, unchecked activity in humans could be very dangerous, especially in immunosuppressed organ recipients.

In 2017, eGenesis used CRISPR to successfully remove all sixty-two PERV sequences in the DNA of a pig cell line. The record-setting multiplexing feat made the cover of *Science* under the tasty headline "CRISPR Pigs."[7] Yang's team named the first CRISPR pig Laika, in honor of the stray Soviet dog that became the first live animal in space.[8] Since then, Qin's team has generated a separate line of PERV-free miniature pigs in the United States, named after Greek goddesses: Iris, Hestia, Maia, and Nike, the goddess of victory. Dozens of additional gene edits will be required before clinical trials can commence. The goal is to eliminate, or at least minimize, the risk of immune rejection of a transplanted pig organ. One target gene, for example, codes for an enzyme that produces sugar molecules on the cell surfaces of pigs and other mammals, but not humans.

The precious pig cells are kept frozen in a large liquid nitrogen tank. Any power failure would trigger multiple alarms on staff phones including Qin's. "Would you like to see some pig cells?" she asks. Before I can answer, she lifts a plastic dish containing ninety-six wells out of an incubator and slides it under a microscope. These CRISPRed cells could one day provide a production line of life-saving organs. On a nearby farm, the first piglets with edited immune genes are already trotting around, named after characters in *The Adventures of Tom Sawyer* and *The Last of the Mohicans*.

Back in Boston, pig organs are currently being transplanted into monkeys in preclinical experiments being led by James Markmann, chief of transplant surgery at MGH. Meanwhile, in Hangzhou, China, Yang has taken on a new role as CEO of eGenesis's sister company, Qihan Biotech. Qin tells me the name means "through learning you come to understand" and "a flower ready to blossom."

But it's Church's work on another mammal that has truly captured the attention of the media (and Hollywood): the woolly mammoth, *Mammuthus*

primigenius, the flagship project of the de-extinction revolution. Woolly's backers include venture capitalist Peter Thiel, on a mission to defeat mortality, who donated $100,000 early on in 2015.

The story begins, and may yet resume, in Siberia—about five million square miles of permafrost. Entombed within this deep-soil layer are the remains of animals and plants, a refrigerated repository of an estimated 1,400 gigatons of carbon. If this is released as methane, as global temperatures rise, the compounding effect on the climate would be devastating. The solution proposed by father-and-son team Sergey and Nikita Zimov isn't exactly Jurassic Park but a 4,000-acre pilot project dubbed Pleistocene Park. They plan to restore the ancient ecosystem with mammoths and other animals trampling the insulating snow to enable the arctic frost to penetrate the soil, in turn chilling and compressing the yedoma layers deeper into the ground, postponing thawing and methane release. That's the theory, anyway.

Woolly mammoths died out about 3,000 years ago, ending 100,000 years or more of cohabitation with humans, who probably brought about their demise. Plentiful remains have been found in excellent condition suspended in the frozen tundra. Although the ancient DNA is shattered after millennia in Siberian hibernation, the carefully retrieved fragments are long enough to determine their sequence. Church believes that the woolly mammoth—or a close approximation—can be resurrected, so to speak, by introducing key woolly mammoth genes into the genome of the Asian elephant. The two species are about 0.4 percent different at the DNA level, less than the difference between humans and our closest cousin, the chimpanzee (about a 1 percent difference).

Church made his first trip to meet the Zimovs at Pleistocene Park in August 2018 with a small team including his postdoc Eriona Hysolli. During a grueling fifty-hour journey to the Arctic Circle, the group stopped off in Yakutsk in eastern Siberia. In the lobby of the Polar Star hotel, sporting a CRISPR T-shirt, Church posed for a photograph next to a full-size woolly mammoth replica.[9] Conditions for hiking along the banks of the Kolyma River near Chersky, an arctic town 800 miles west of the Bering Strait, were not ideal. "It's a lovely place if you don't mind having snow flurries and being eaten alive by mosquitoes in the same day," Church said.[10] That's on a good day. "The worst day, it's so cold the mosquitoes cannot live, or there are enough mosquitoes it will literally kill a baby caribou."[11]

Donning protective overalls and gloves, Church dissected six beautiful woolly mammoth specimens with a power drill, extracting DNA from fat, marrow, and muscle. Two genes have already been brought back from extinction so to speak, including the mammoth's hemoglobin gene. While the Zimovs wait for the "elemoths" to arrive, they use a decommissioned tank to remove trees and help other growth. Reindeer, yak, sheep, bison, and horses are already roaming the terrain. Church might need to step on the gas: in 2020, temperatures in a northeastern Siberian town above the arctic circle spiked to a record 100° F.

Beth Shapiro, a paleogeneticist at University of California Santa Cruz, and an HHMI investigator, injects a friendly note of pragmatism. She insists it is not possible to clone a mammoth, which is interesting given that she wrote a book entitled *How to Clone a Mammoth.*[12] Shapiro agrees we could potentially use genome editing to introduce the key base changes, if not all 1.5 million. But then, "we'd have to figure out how to get that individual into a female elephant. Then it would have to be born and raised by elephants. It's hard to imagine that all of those things would happen and we'd end up with something more than just a slightly hairier elephant."[13]

Moreover, Shapiro says, elephants do not fare well in captivity. Until we've figured out how to meet the physical, emotional, and psychological needs of edited elephants, we shouldn't be using them for gene-editing research. She'd rather see the technology used to save endangered elephants, by giving a genetic booster shot if necessary to expand their evolutionary fitness.

Shapiro sees little point in applying heroic measures to bring back extinct species only to place these creatures in a zoo or a park named after a geological epoch. What would we do with resurrected saber-toothed tigers or mastodons? How would millions of passenger pigeons cope with our modern urban environment? But as extinction events in modern times are so often the result of human neglect or suffering, why not apply human technological ingenuity to undo our past mistakes?

A memorial to animal species that have recently gone extinct would include Toughie, the last known Rabbs fringe-limbed tree frog; Sudan, the last male white rhino who was euthanized in 2018 in Kenya; and Lonesome George, the last Pinta Island tortoise from the Galapagos Islands, who died in 2012.[14] The woolly mammoth may be the fanciful face of de-extinction, but gene editing offers a ray of hope for a growing conservation movement.

In 1933, zoologist David Fleay filmed Benjamin, the last Tasmanian tiger in captivity in Hobart, Australia. The grainy black-and-white film captured the caged ferocity of this creature, with the trademark stripes across his lower back and the extraordinary elongated jaws. Three years later, the last of the thylacines was dead. Like Lonesome George and Martha, the last passenger pigeon who died at the Cincinnati Zoo in 1914, Benjamin was the last of his kind—an endling.[15] It is a word, writes Ed Yong, "of soft beauty, heartbreaking solitude, and chilling finality."[16]

Two years before the amphibian in his care at the Atlanta Botanical Gardens died, Mark Mandica, the executive director of the Amphibian Foundation, made a recording of Toughie the frog singing. "He was calling for a mate and there wasn't a mate for him on the entire planet," Mandica said. By the time a species reaches that point, it is just one small, inevitable step to extinction. On Oahu, Hawaii, a trailer provides the last refuge for dozens of species of snails, passing what could be their final days in plastic containers destined to become coffins.

Environmentalist Stewart Brand recounts the passing of Martha, the last passenger pigeon, as if she was a blood relative. Her species was the most abundant bird in the world, with flocks reportedly one mile wide and four hundred miles long, variously described as a biological storm, a feathered tempest, and distant thunder. Yet within little more than a decade, by 1914, five billion birds spanning the North American continent were reduced to zero.* Hunters slaughtered them by the tens of thousands. The fate of the birds might have saved the endangered American bison but other species weren't so lucky. When "Booming Ben," the last heath hen on Martha's Vineyard, died in 1932, it was the lead story in the local newspaper. The

* Martha is at the Smithsonian Institution in Washington, DC, although no longer on display.

editorial was an obituary: "There is no survivor, there is no future, there is no life to be recreated in this form again. We are looking upon the uttermost finality which can be written."[17]

We are in the middle of an existential extinction crisis—the sixth extinction.[18] More than half of all mammalian species have become extinct since 1900. Conservation is a vital tool, as is genetic rescue—increasing the fitness and DNA diversity of species, from California mountain lions to Florida panthers. Genome sequencing is an important tool here for tagging and breeding endangered species—a striking example is a bid to save the endangered Tasmanian devil marsupial, which is threatened by a malignant, orally transmitted cancer. Researchers have released cancer-free devils on a small island off Tasmania in case the main population is unable to stabilize. The endangered black-footed ferret in the Great Plains is at serious risk from bacterial sylvatic (bubonic) plague. Animals bred in captivity can be vaccinated before release, but as proposed by Ryan Phelan and colleagues at the nonprofit Revive & Restore, CRISPR editing offers a means to transfer plague resistance from the domestic ferret to its black-footed cousin.[19]

Meanwhile, similar strategies are needed to save some other iconic American plants and wildlife. The American chestnut tree population has been ravaged by chestnut blight, a disease spread by a Japanese fungus that was first observed in the Bronx Zoo. William Powell's team has engineered a hybrid genetically modified tree that contains a wheat gene that neutralizes the acid produced by the fungus. Powell is petitioning the US government to produce a transgenic forest species,[20] despite opposition from environmentalists who worry about GM trees. But splicing DNA into its nearest living relative isn't always possible. The Stellar's sea cow was hunted to extinction about two hundred years ago. Tempting though it might be to de-extinct this extraordinary creature, Shapiro points out that a baby Stellar's would be larger than the most logical surrogate. Such an effort at de-extinction would end up a bit messy. She also has bad news for fans of the dodo. De-extinct dodo eggs will be just as appetizing to various animals (including humans) as the originals.

When Shapiro and colleagues decoded the first passenger pigeon genomes in 2017, using tissue samples from museum specimens, they found a bizarre pattern—high diversity at the chromosome ends but surprisingly little variation in the middle, a pattern that might have contributed to the species' rapid demise.[21] Her colleague Ben Novak, who has been obsessed with these birds since he was a teenager, is leading "the great comeback." A scheme to revive the passenger pigeon begins with the bird's closest relative, the band-tailed pigeon.

At Monash University in Australia, Novak has taken the first steps, engineering a line of pigeons expressing Cas9, priming their progeny to receive select gene edits based on their passenger pigeon cousins.[22] A best-case scenario suggests that changes in around thirty genes would confer many key traits such as coloring. But editing genes is just the beginning. Novak would have to raise enough birds to coax them to flock;[23] one idea is to raise the first hybrid chicklets using surrogate homing pigeons painted to look like the Real McCoy. He'd also have to reproduce the birds' natural habitat, perhaps the forests of the northeast United States. Novak proposes to calls his prize pigeon *Patagioenas neoectopistes*, or the "new wandering pigeon of America."

"Humans have made a huge hole in nature over the last 10,000 years," says Brand. "We have the ability now, and maybe the moral obligation, to repair some of the damage." Revive & Restore supports projects led by Church, Shapiro, and many others. Hope for the de-extinction movement comes from the story of Celia, the last bucardo mountain goat, which roamed the mountains of Spain. Although Celia died in the wild, some of her cryopreserved ear tissue was used to produce a live animal. It was the first successful de-extinction in history, but unfortunately it died soon after birth from a lung abnormality.[24]

On Hawaii, mosquitoes carrying avian pox and malaria have wiped out more than half of the islands' one hundred species of native birds. Most of the rest are endangered. A species called *Culex quinquefasciatus* was introduced to Hawaii on ships in the early 19th century. The islands' native birds, including the honeycreepers, had no natural resistance to the avian malaria. And with climate change, mosquitoes are able to reach "upslope" to the higher elevations that serve as a natural sanctuary for surviving species. Revive & Restore is contemplating various strategies to reduce the mosquito population, including the sterile insect technique and the introduction of a natural bacterial predator, *Wolbachia*. Potentially the most effective technology involves CRISPR. It is also the most dangerous.

Eradicating diseases like Lyme disease, dengue fever, and especially malaria is a grand challenge on a global scale. And it is one where CRISPR offers a radical solution. What if CRISPR could have an impact on one of the most notorious killers on the planet? Mosquitoes don't have an important role in ecology. They don't pollinate plants or serve as an essential food source for anything. It is unlikely they would be missed, particularly in sub-Saharan Africa. "As the apex predator throughout our odyssey, it appears that her role in our relationship is to act as a countermeasure against uncontrolled human population growth," observes Timothy Winegard.[25]

The solution on offer is called a gene drive; it gives researchers the power to warp the natural Mendelian pattern of inheritance, raising the prospect of halting the spread of devastating infectious diseases. A gene drive is like loaded dice, stacking the odds in favor of a particular copy or version of a gene being passed on to the next generation, rather than leaving it 50:50. Why is that interesting? For decades, biologists have tried to combat the spread of infectious disease or other pests by spraying tons of toxic chemicals or introducing a predatory species, often with dire consequences.

A gene drive offers a much more sophisticated strategy to combat deadly diseases such as malaria, which kills some 650,000 people every year. Scientists would introduce a special DNA element that would act as a sort of poison pill in the *Anopheles gambiae* mosquito. This selfish element can essentially clone itself by inserting a copy into the partner chromosome. The idea was first formulated B.C. (before CRISPR) in 2003 by Austin Burt at Imperial College, London. Burt suggested that a gene drive cassette introduced into 1 percent of the African mosquito population would quickly spread in a chain reaction, affecting 99 percent insects within twenty generations.

Many observers are understandably scared that a gene drive in the wild could go awry, crossing geographic boundaries or spreading into unintended species, threatening the ecological balance of countries across the equator.

Then again, trying to save the lives of more than 400,000 children who perish from malaria each year surely justifies some desperate measures.

Scientists have taken on and defeated malaria on a national scale before. In 1944, the Rockefeller Foundation and the United Nations initiated a program to eradicate malaria-carrying mosquitoes on Sardinia, which claimed 2,000 victims a year. The disease was probably introduced by North African slaves brought to the island after the conquest by the Carthaginians in 502 B.C.E. The peak assault came in the summer of 1948, likened to the Normandy landing, involving 30,000 men who sprayed more than 265 tons of DDT. The campaign eradicated three of the four *zanzare*—endemic species of mosquito and wiped out malaria.[26]

In 2009, a British biotech company, Oxitec, launched a trial using three million genetically infertile male mosquitoes in the Grand Caymans to halt the spread of dengue fever. Following similar trials in Brazil and Malaysia, Oxitec has proposed a release for the Florida Keys, but some residents worry about inadvertent consequences of the modified mosquito release.*

The use of a CRISPR-based gene drive allays some of these concerns as no foreign genes are inserted into the mosquito genome. With CRISPR's ease and precision, researchers have conducted successful gene drives in small lab populations of mosquitoes. But it is one thing to perform a gene drive under the controlled conditions of a basement insectary in London or San Diego. It's another to take this into the real world. The holdback is less technical than social.

Kevin Esvelt leads the Sculpting Evolution group at the MIT Media Lab—an institution, he says, for black sheep who don't fit anywhere else.[27] Esvelt is a leading evangelist in the potential use of CRISPR-Cas9 to develop strategies, including but not limited to gene drives, to combat diseases ferried by ticks and mosquitoes. But with the power of this approach

* Oxitec has dubbed its non-biting mosquitoes Friendly™ but that hasn't convinced all of the affected residents, or the Environmental Protection Agency.

comes tremendous responsibility. Esvelt takes this very seriously: his boyish appearance with sandy hair belies his eloquent intensity.

Esvelt's interest in evolution began with a visit to the Galapagos Islands when he was in the sixth grade. "I wanted to know, how is it that so many marvelous creatures are created? Can we learn how that is done and create equally marvelous things ourselves?" In striving to answer that question, Esvelt has formulated a few ethical objections to the way that natural evolution does things, "the apparent total indifference to animal suffering, to any kind of notion of right or wrong. Evolution is *amoral*. I'm not saying it is immoral, because it is a physical process. But the fact that it does not care about or optimize well-being I view as a fundamental flaw in the universe." That was just the first minute of our interview.

Esvelt's motto then, is "Evolution has no moral compass. We do." At MIT, Esvelt the black sheep wants to develop technology "to continue improving human and environmental well-being" while avoiding traps that could be hazardous to our health. He quotes Charles Darwin: "Man selects for his own good, Nature for that of the being which she tends."

Esvelt obtained his PhD at Harvard working with chemistry professor David Liu on a suitably grand idea: to fast-forward evolution in a tube to optimize the function of proteins and other biomolecules. If you don't understand how a protein works well enough to engineer precise alterations, then generate a billion or more variants, test, pull out the ones that work the best, rinse and repeat. Directed evolution was pioneered by Frances Arnold, who won the Nobel Prize for Chemistry in 2018. But it's a lot of work, Esvelt says, "and I'm a big fan of laziness."

After six years, Esvelt and Liu developed a system called phage-assisted continuous evolution (PACE).* After hundreds of cycles run over a week, they typically produced the engineered variant they're looking for.[28] Esvelt next joined Church's lab. But his early PACE experiments failed because other scientists in the lab were growing large cultures of phage, which kept

* In PACE, only phage that evolve with the desired gene activity can induce production of an essential protein to continue its life cycle. The phage evolve a couple of generations in an hour, generating a billion variants at a time.

infecting his cells. Esvelt turned to the CRISPR-Cas system simply as a means to fend off phage floating in the air.

In 2013, Esvelt joined Prashant Mali, Luhan Yang, and Church in publishing one of the first demonstrations of CRISPR gene editing in human cells. Next he wondered: What if you could teach the cell to do genome editing on its own so that editing could occur during each successive generation? Could there be genes that do this naturally? One possibility was a microbial homing endonuclease that cuts DNA in highly specific sequences and inserts the corresponding gene into the gap. He found Burt's paper from 2003, in which he had tried putting I-SceI from yeast into mosquitoes, copying itself using the cell's natural DNA repair mechanism. "Wow, this guy was a genius," Esvelt remembers thinking. "He thought of this a decade ago!"

Indeed, Burt's radical idea to edit mosquitoes was to harness gene drive systems that occur naturally,[29] probably originating hundreds of millions of years ago. (The cow genome, for instance, is littered with genetic elements from snakes that spread via a gene drive.) Eradicating a few billion *A. gambiae* mosquitoes would have little to no impact on the ecology of the region, while hundreds of mosquito species that do not transmit malaria would be unaffected. It took years to perfect, but in 2011, Burt and Andrea Crisanti finally reported a successful gene drive in mosquitoes in the lab.[30]

Esvelt realized that CRISPR held advantages over endonucleases in constructing a gene drive to eradicate malaria. Prior to the development of CRISPR, nobody had contemplated being able to edit an entire wild species. "The concept was completely absent from science fiction at the time," Esvelt told me. "Literally, no human ever conceived that we might be able to do this. All of a sudden, boom! It looks like we can."

Teaming up with mosquito biologists, in 2014 Esvelt published the concept of a CRISPR-based gene drive.[31] The idea went like this: encode the CRISPR system for making a mutation along with that alteration in the genome. When the mosquito mates with another insect, the offspring inherit one copy of the edited gene along with the CRISPR system that was used to generate that edit. The offspring thus inherit the machinery to cut and replace the wild-type version of the gene. This ensures that the editing is passed down, skewing the normal 50:50 mendelian ratio of inheritance.

About the same time, a group at the University of California, San Diego, led by Ethan Bier and Valentino Gantz, also developed a CRISPR-based gene drive in fruit flies. Working with Anthony James at UC Irvine, they transplanted their CRISPR system into *Anopheles stephensi*, responsible for a fraction of malaria cases in India.[32] Meanwhile, Burt, Crisanti, Tony Nolan, and colleagues reported similar success in *A. gambiae.*[33]

But even as gene drives showed early promise in insectaries, Esvelt worried about the ethical risks of a gene drive potentially running amok and crossing over into other species, as well as the costs of doing nothing. He wrote:

> As one of those who introduced CRISPR-based gene drive to the world, I hold myself morally responsible for any and all consequences that emerge from the technology. In my eyes, if something goes wrong that I might have foreseen, that's on me. If my actions or words inadvertently prevent gene drive from benefiting others, that's on me. If my *failure* to act prevents it from saving lives, that's on me.[34]

The good news is that a CRISPR gene drive is relatively slow, spreading through generations, and easily detectable. "CRISPR is powerful enough that you cannot really build a gene drive that cannot be targeted with CRISPR, meaning whatever one person does, another person can override," says Esvelt. What concerns him more is an accidental gene drive release or unauthorized use, hence the push to ensure that this research was done transparently and responsibly. "You might make a gene drive without even realizing it," he says, which could be introduced into a wild population. A lab accident could devastate the public's trust in science and governance, setting back gene drives the way the Gelsinger tragedy derailed gene therapy. Moreover, what sort of responsibility does the first community that approves a gene drive test have if subsequent applications go awry?[35]

Burt leads Target Malaria, a project funded by the Bill and Melinda Gates Foundation. In a basement insectary located somewhat incongruously in South Kensington, thousands of mosquitoes are housed in cubes of white netting in precisely controlled warm and humid conditions. The male flies suck on sugar water, while the females feast on vials of warm blood. Any rogue mosquito that manages to break quarantine by escaping the double steel doors and electronic security—not to mention an electronic mosquito zapper known affectionately as the Executioner—would then face the inhospitable misery of the dank English climate. What works for malaria could similarly be applied to tackling dengue, yellow fever, Lyme disease, and the Zika epidemic.

It's all very well scheming diagrams and building models about selfish killer genes, but do gene drives work in practice? In 2018, Burt, Crisanti, and Nolan took a giant step in that direction. In their London lair, they crashed caged populations of *A. gambiae* in fewer than eleven generations.[36] The strategy interfered with the insect's sex chromosomes, reducing the proportion of fertile females in each successive generation until the population reached a dead end. Extrapolating from South Kensington to Burkina Faso, which has the third highest number of malaria deaths behind Nigeria and the Democratic Republic of Congo, the strategy could crash a wild mosquito population in about four years. In San Diego, Omar Akbari's team has identified another promising target—a gene that when knocked out, prevents female *Aedes aegypti* mosquitoes, carriers of several viral diseases, from flying. (Males are unaffected.)

Burt says a trial would involve deploying just a few hundred gene-drive mosquitoes in each village. If the social and political concerns can be addressed, Burt reckons that this CRISPR gene drive, coupled with other public health measures such as the use of nets, could eliminate malaria across much of Africa in fifteen years. One low-tech innovation, developed by entomologist (and malaria survivor) Abdoulaye Diabaté, is the Lehmann funnel entry trap, a device that fits to windows and doors from which mosquitoes cannot escape.

As a first step, the National Biosafety Agency in Burkina Faso gave Target Malaria, working with Diabaté, approval for a limited release of a male sterile mosquito strain (not a gene drive). In July 2019, 10,000 fluorescently coated, genetically modified mosquitoes were released in the village of Bana. It was almost literally a drop in the bucket compared to the wild population, but a hugely significant step nonetheless.

But the move has encountered fierce resistance from some Burkinabes. "We refuse to be guinea pigs," says Ali Tapsoba, an anti-GMO campaigner, who fears the irreversibility of a gene drive and his country's lack of resources to deal with it.[37] Mariam Mayet, executive director of the African Centre for Biodiversity, calls Target Malaria a "neocolonial project designed and conceived in the West and telling us what's good for us."[38] It is a common charge from African activists who have campaigned vociferously against the intrusion of giant agbiotech enterprises, notably Monsanto.

"Mosquitoes don't obey national boundaries," says Nnimmo Bassey, a vocal Nigerian environmentalist and critic of the mosquito program in Burkina Faso. Bassey is all for eliminating malaria, but he fears the possibility of CRISPR technology being used for other purposes. "Powerful countries and corporations don't care about people who aren't similar to them," he says. "If you have this power and control in the world when there's no equity, this seemingly nice scientific invention can become extremely dangerous." Bassey favors more mundane low-tech solutions—sanitation, social services.

Of course there are risks associated with releasing gene-edited mosquitoes. Feng Zhang believes that any efforts to unleash gene drives into the wild must include containment measures.* "Some days I feel it would be great to have no mosquitoes at all," he said, but we must be wary of the ecological consequences. "You're removing such a large fraction of the biomass. We need to be careful."[39]

But aren't the dangers of not trying it even greater? "It strikes me as a fake argument to say that something is irreversible," says Church. "There are tons of technologies that are irreversible. But genetics isn't one of them." If something doesn't work properly, he says it can be fixed, as Esvelt and Church have demonstrated in yeast.[40]

Church's Harvard colleague, Amit Chaudhary, grew up in India dreaming of being the next Sachin Tendulkar, the Babe Ruth of cricket, not a CRISPR scientist. His family was poor and there was no escaping

* Neither the Broad Institute nor Caribou will grant a CRISPR license that does not include a rider that says the IP cannot be used for a gene drive.

the mosquitoes, but Choudhary avoided malaria thanks to a Good Knight vaporizer put out in the evening that released insect-repelling pyrethrins.

A chemist by training, Choudhary dares to compare CRISPR to some other transformational discoveries in human history such as fire or the Internet. It comes down to precision control, he says. Humankind was able to control fire. Contrast that with the chaos erupting because of a lack of precision control over the Internet.[41] Gaining control of CRISPR is essential. To that end, Choudhary's group has identified drug-like compounds that can suppress Cas9's ability to cut DNA before it can grasp the relevant sequence.[42]

Choudhary thinks that the vaporizers of his childhood could be repurposed to help regulate gene drives by releasing custom chemicals that would regulate the activity of gene drives. Who needs a helicopter to douse mosquitoes when there is a device that already exists in almost every Indian home?[43] And this could work in Africa too—anywhere dinner is cooked.

At the end of 2018, the UN Convention on Biological Diversity reached a compromise on gene drives, rejecting a moratorium but calling for the informed consent of impacted countries and local communities before contemplating any release.[44] Could a gene drive spread across borders or to other species? Yes, perhaps. But isn't biological warfare against one of the greatest killers of humankind worth the small risk? As Esvelt puts it, "the known harm of malaria greatly outweighs every possible ecological side-effect that has been posited to date, even if all of them occurred at once."[45]

Scientists like Burt, Crisanti, and Esvelt are trying to save the lives of thousands of people each year. Crisanti rejects the criticisms that it would be immoral to attempt to use a gene drive, saying: "What about the moral issue of doing nothing?"[46]

A few years ago, I was sunbathing with my family on a beach in Scituate on Boston's south shore. As we were packing up, a woman approached me. "Hey, do you know what's on your leg?" Sure, I replied. I was sporting a circular rash on my calf, presumably caused by a spider bite. Or so I thought. She shook her head and in a thick Boston accent said: "No. That's a tick bite. You've got Lyme disease."

"How do you know?" I responded disbelievingly.

"I'm an ER nurse, love," she said.

A hasty trip to my GP confirmed the diagnosis, which cleared up with the appropriate antibiotic. I shouldn't have been so skeptical: Lyme disease is particularly common in Massachusetts, where I lived. Picking deer ticks off my brindle beagle-boxer was a daily ritual after hiking in the neighborhood woods.

Tick-borne diseases might not seem to be a public health menace comparable to malaria, but Esvelt makes a convincing case. "The West Coast has earthquakes. The South has hurricanes. The middle of the country has tornadoes. The natural disaster of the Northeast is Lyme disease."[47] Each year, some 300,000 Americans are diagnosed with Lyme disease, signaled by the telltale bullseye rash on their skin. If left untreated, the disease can be debilitating. The disease and other tick-borne diseases are particularly prevalent on the islands of Nantucket and Martha's Vineyard off Cape Cod.

Nantucket is a popular summer retreat for New Englanders and celebrities—and a perfect ecosystem for Lyme disease. Take a large deer population that has few constraints—no wolves, not enough licensed shooting or random car collisions. Lots of deer means lots of ticks, easily evidenced if you simply drag a sheet through the brush. But the ticks' main host is the white-footed mouse, the elusive reservoir of tick-borne disease.

Together with Joanna Buchthal, Esvelt launched the Mice Against Ticks project to engage with a local community to discuss the idea of using CRISPR responsibly to engineer environmental immunity into the mice.[48] If you could release sufficient mice edited with a gene cassette that confers resistance to Lyme disease, this could disrupt the ecological transmission cycle. By encoding antibody genes in the mouse germline that are expressed in newborn mice, Esvelt's team could confer an inherited resistance to disease. Releasing enough genetically vaccinated mice should spread resistance to the next generation and disrupt the tick life cycle.

But first, Esvelt has to win over the residents of Nantucket. He felt a profound ethical obligation to share his research ideas and island residents, insisting that the public must have the final say. To his credit, Esvelt has spoken at public meetings on Nantucket convened by the Boards of Health, and faced vocal apprehension about the release of any genetically modified organisms. The Nantucket steering committee has some well-qualified individuals on it, including Howard Dickler, the former head of the NIAID's infectious disease branch, and John Goldman, editor of an immunology journal. But there are skeptics, too.[49] The consensus was: interested but don't use any foreign DNA in our engineered mice.

On the subject of misery in the world, Darwin wrote: "I cannot persuade myself that a beneficent & omnipotent God would have designedly created the *Ichneumonidæ* with the express intention of their feeding within the living bodies of caterpillars."[50] Beyond malaria and Lyme disease, there are many appalling diseases that could be tackled using the latest genome editing gadgetry. Taming the desert locust would be one.

In similar vein to the *Ichneumonidae*—wasps that paralyze caterpillars, lay their eggs, which hatch and proceed to eat the caterpillars alive from the inside out—consider the New World screwworm. One poor victim was a twelve-year-old girl who returned from a school trip to Colombia and headed straight to the hospital complaining of severe pain in her scalp. After giving the girl morphine, doctors resorted to a novel therapy as described in the medical report: "During her hospital stay, a total of 142 larvae were manually extracted, aided by the application of raw bacon which served as an attractant and petroleum jelly occlusion." Doctors deduced that the episode resulted from a female fly laying eggs in a scalp wound.[51] In another gruesome example, a British woman became infected in her ear canal after walking through a swarm of flies while on holiday in Peru. Back home, doctors found maggots had burrowed more than 1 centimeter through her ear canal. They drowned the maggots with olive oil before finally extracting the parasites.

The screwworm looks like a psychedelic housefly, with large orange eyes and a blue-green body. Its official name, *Cochliomyia hominivorax*, translates as "eater of man." Pregnant "eaters of men" are adept at finding open wounds, sores, and other niches in which to lay eggs. Cattle and livestock also make unsuspecting hosts. Gratifyingly, the screwworm was eradicated across North America in the late 1960s using the sterile insect technique. By irradiating and releasing millions of sterile male flies—a factory in Florida was producing 50 million infertile flies a week—the population was halted in its tracks. During the Reagan administration, an outbreak erupted in Libya due to the import of tainted sheep from South America. The U.S. government stealthily arranged to airdrop millions of sterile flies, technically violating its own sanctions. In 2016, the menace resurfaced in the Florida Keys. After dozens of Key deer deaths, officials employed the same sterile insect technique to squash the screwworm before it could spread onto the mainland.

But this method won't eradicate the screwworm at the source—South American sheep—because of the more rugged terrain. One option is to use the gene drive approach, if the Mercosur countries agree. Esvelt has suggested a more limited "daisy drive" approach, a self-exhausting CRISPR-based gene drive.[52]

An even worse nightmare than accidental outbreaks or unforeseen ecological consequences of a gene drive would be the malicious use of CRISPR as a bioweapon. "Research in genome editing conducted by countries with different regulatory or ethical standards than those of Western countries probably increases the risk of the creation of potentially harmful biological agents or products," the National Security Agency noted drily in its 2016 threat report.[53] CRISPR thus joined the ranks of North Korean nuclear weapons and Syrian chemical weapons.

Adding his voice to the chorus of concern is Bill Gates. "The next epidemic has a good chance of originating on the computer screen of a terrorist intent on using genetic engineering" to create a synthetic smallpox virus or a "super contagious and deadly strain of the flu"—and killing more people than nuclear weapons, Gates told a security conference.[54] His biggest worry: the nefarious use of CRISPR to engineer a new flu strain, combining potent virulence with extreme infectivity. Gates' worry is not frivolous: CRISPR can be used easily without expensive lab equipment or specialized training. CRISPR-based kits available from outfits like the Odin, launched by biohacker Josiah Zayner, sell for less than $500 in some cases. Pathogen-specific kits are "offered up like so many choices at a grocery store," according to a RAND Corporation report.[55]

These are legitimate concerns, but have been overshadowed by recent events. As the world saw in 2020, we don't need DNA-designing despots or basement biohackers playing with CRISPR to cause a pandemic—nature is quite capable on her own.

CHAPTER 21
FARM AID

In the late 1990s, Mark Lynas was a proud environmental activist, a member of a group called Earth First! that would frequently lay waste to fields of genetically modified (GM) crops with machetes in the middle of the night. In 1998, Lynas devised a plan that, if successful, would have made him a household name. Hatched in secret for fear of being bugged, Lynas and his fellow hoodlums decided to kidnap Dolly the sheep.

Created by a Scottish team at the Roslin Institute led by Keith Campbell and Sir Ian Wilmut, the lamb formerly known as 6LL3, cloned from an ovine mammary gland cell, was born in July 1996. Affectionately renamed by a technician, Dolly lived at the Institute just outside Edinburgh. Her birth was kept secret until the results were published to worldwide acclaim and angst six months later. Posing as a researcher, Lynas was granted access to the Institute's library, before hunting for the shed holding the world's most famous sheep. Meanwhile, a female accomplice posed as a lost American tourist outside the institute's perimeter. The audacious plan might have worked, except that Dolly's whereabouts weren't signposted. Nor was it immediately obvious how to tell Dolly apart from the hundreds of other sheep in the facility. "The Roslin scientists had outfoxed us by hiding Dolly in plain sight," Lynas later confessed.[1]

Several years after his Scottish caper, Lynas had an epiphany. The more he researched the science underlying genetically modified organisms (GMOs) and climate change for books such as *Six Degrees*, the more he realized his blinded ignorance. Indeed, the evidence overwhelmingly supports the safety of GMOs. In 2016, the National Academies of Science, Engineering and Medicine published a major report concluding that GMOs did not harm animals nor cause any health problems in humans in the food supply.[2] A group of more than one hundred Nobel laureates called on Greenpeace to end its opposition to GMOs.[3]

Lynas admitted his own mistakes during a keynote speech in 2013 at a major farming conference before a shocked audience.[4] He apologized for ripping up GM crops and helping to demonize a technology with profound environmental benefits. He was immediately accused of being a Monsanto shill and worse. A journalist meeting the reformed Lynas at his home near Oxford remarked that the former ecowarrior was handsome and vaguely fashionable, in the manner of a member of Coldplay whose name you can't quite remember.[5]

Many scientists believe the biggest impact of CRISPR will come not in pharma but in farming. "The most profound thing we'll see in terms of CRISPR's effects on people's everyday lives will be in the agricultural sector," predicts Doudna.[6] "The CRISPR craze has pretty much swept through plant biology," agrees Dan Voytas, a professor at the University of Minnesota and cofounder of Calyxt.[7] In 2017, state-owned ChemChina bought Syngenta, one of the top three agbiotech companies along with Germany's Bayer and Corteva, for $43 billion. China is undertaking a massive effort to improve the quality of many key crops using CRISPR.[8]

Indeed, some commentators have been stressing this point since the birth of the CRISPR revolution in 2012-13. While most of the fanfare centered on CRISPR's potential for treating human disease, some commentators, including British author and politician Matt Ridley, were struck by the implications for crops. Ten thousand years ago, farmers in what is now Turkey used cross-breeding to select a random mutation in wheat plants in the Q gene on chromosome 5A, which rendered the seed head less brittle and the seed husks easier to harvest efficiently.[9]

In 1798, English political economist Thomas Malthus published a famous treatise in which he showed that human population growth was outstripping the increase in agricultural productivity. The growing competition for resources leads inevitably to a Malthusian collapse caused by war, famine, or pestilence. In her book *The Age of Living Machines*, MIT president emerita Susan Hockfield argues that Malthus was wrong because of the repeated invention of new technologies that have increased agricultural productivity. One example was the introduction of four-field crop rotation, which succeeded (you guessed it) three-field crop rotation in the 18th century. Another is the extraordinary story of William Vogt and guano.

Vogt was an ecologist, ornithologist, and environmentalist, profiled in Charles Mann's book *The Wizard and the Prophet.*[10] Vogt (the prophet) discovered a natural resource—mountains of guano, or bird excrement, as birds roosted on the Chincha Islands off the coast of Peru. The nitrogen-rich guano was used for fertilizer, providing a large portion of Peru's national income. In 1948, Vogt wrote about the earth's "carrying capacity" caused by fundamental ecological processes that set limits on what we can do—or as Mann calls it, the first "we're going to hell" book. After studying the cormorants and the weather patterns, Vogt concluded it was not possible to obtain more guano, "to augment the increment of excrement." But as Hockfield points out, the export of guano to Great Britain caused another surge in agricultural productivity.

The wizard adversary to Vogt's prophet was plant geneticist Norman Borlaug, the father of the Green Revolution. In the mid-1950s, Borlaug, an expert at interspecies hybridization, developed semi-dwarf wheat, which probably saved millions of lives after it was introduced to India in 1962, earning him a Nobel Prize. It now makes up 99 percent of all wheat planted around the world.

To speed up the generation of mutations, Lewis Stadler reported the first use of radiation mutagenesis to create novel mutations in plants in 1928. Half a century ago, scientists used a nuclear reactor to shoot gamma rays at barley seeds, inducing a plethora of random mutations in the DNA. One result was "Golden Promise," a high-yielding, low-sodium barley variety popular with (ironically) organic farmers and brewers.

In the late 1970s, Mary-Dell Chilton, a researcher at Washington University in St. Louis, discovered that crown gall disease, a plant tumor, was caused by a bacterium called *Agrobacterium* inserting a sliver of its own DNA into the plant. That prompted the idea that the same bacterium could be used like a gene therapy vector to shuttle desired genes into plants. In January 1983, along with two other researchers, Chilton gave a talk at the annual Miami Winter Symposium, that she called "the symbolic coming of age of genetic engineering."[11] Indeed, Chilton is recognized as a pioneer of agricultural biotechnology and crop improvement. The method wasn't called gene editing, but some did label it gene jockeying. Genes can also be introduced more directly, literally shooting them into plants as DNA-coated tungsten or gold particles with a gene gun.

Two decades ago, scientists at Syngenta inserted gene sequences from maize, encoding four enzymes into rice plants so that they could synthesize vitamin A, thereby creating transgenic "golden" rice. In Bangladesh, about one in five children are vitamin A–deficient. After interminable delays, the Bangladesh authorities are close to approving Golden Rice.

Anti-GMO activists may be aghast to learn that many of their favorite all-natural foods were in fact genetically modified by nature centuries or millennia ago. In 2015, scientists stumbled upon the fact that every domesticated breed of sweet potato contains DNA from the *Agrobacterium*. "All the sweet potatoes we eat are GMOs," says Johns Hopkins' professor Steven Salzberg.[12] To that list, we can add bananas, cranberries, peanuts, walnuts, and two of my favorite beverages—tea and beer (hops).

To develop the new technologies that will feed almost 10 billion by 2050, not just accounting for our ballooning population but helping crops survive a changing climate, Hockfield says we're going to have to invent our way out.[13] CRISPR gave agricultural scientists a new razor-sharp tool in their gene-editing toolbox to complement if not supersede ZFNs and TALENs. The technology can prevent mushrooms browning, produce strawberries with a longer shelf life, and tomatoes that stay longer on the vine. Bing Yang's group at Iowa State University has engineered promoter mutations to generate resistance to bacterial blight in rice.[14] But for all the progress being made in Iowa cornfields, New York greenhouses, and Beijing rice paddies, scientists must also hope for some natural variation in the minds of regulators and politicians, especially in Europe.

For a serendipitous heirloom of our ability to engineer desirable traits in plants, exhibit A comes the Italian Renaissance artist Giovanni Stanchi. A Stanchi masterpiece from the mid-1600s depicts a selection of fruits—peaches, pears, and a watermelon cut open revealing contents that are almost unrecognizable. The flesh is mostly white, with pale red swirls and dark black seeds. It bears almost no resemblance to the succulent red flesh (the placenta) of the modern domesticated watermelon, as breeders selected for the lycopene-rich flesh.[15]

Go back even further to a few thousand years ago, the small fruit from southern Africa would have to be cracked open with a stone. A mere six varieties begat some 1,200 today. But humankind has been doing this since the dawn of agriculture almost 10,000 years ago. The domestication of maize from teosinte in Central America began when farmers in what is now Mexico practiced selective breeding. Corn today looks nothing like the "natural" crop. Nor do peaches or tomatoes or many other common fruits and vegetables.

But selective breeding only goes so far. "Nature hasn't given us enough mutations," says Zach Lippman, a leading plant geneticist at Cold Spring Harbor Laboratory (and an HHMI investigator). Lippman has had a peculiar fascination with tomatoes since he worked on a Connecticut farm as a teenager. Few consumers relish the anemic tomatoes typically available in the local supermarket. Lippman believes gene editing can lead to big improvements.

In 1923, researchers reported a natural mutation in a farmer's field in Florida. Randomly, a rare mutant tomato plant had developed the property of self-pruning. Crossing these plants was predicted to give rise to "a valuable new race of early tomatoes." These compact "determinate" varieties grow three to four months until they mature and ripen. A busy determinate tomato plant can give rise to twenty pounds of fruit and is the preferred plant for ketchup and paste production.

Tomato plants are grown in long rows, collected when they turn a greenish-orange hue, and then hauled into storage until they are gassed with ethylene to start the ripening process. But as Lippman says, nature hasn't given farmers enough mutations to play with, at least not on a practical timescale. Additional mutations could relieve compact growth to make plants a bit larger, thereby increasing fruit yield. Before CRISPR, Lippman had to treat seeds with chemicals that randomly mutagenize the DNA and then manually search for desirable mutations by scouring row after row of tomatoes. He spent four years compiling a toolkit of new mutations, where several mutations worked collectively to build a higher-yield tomato plant. There had to be an easier way.

CRISPR technology is not about introducing foreign DNA, but working with the plant's own DNA and enhancing natural repair processes. For example, a desirable trait in tomatoes is called jointless, in which the stem leading to the fruit lacks a knuckle or joint. Fresh market tomatoes crossbred with the jointless trait enables high-throughput production and less damage during handling. Lippman's group used CRISPR to engineer a jointless line of tomatoes without having to cross different strains, and can apply this to any variety.[16] His lab has also introduced mutations in the promoter of the self-pruning (*SP*) gene to create a sort of genetic rheostat, tuning the degree of inactivation to help growers adapt tomatoes to more northern latitudes with longer days but shorter growing seasons.[17]

Another fruit of interest is the humble groundcherry (also known as a strawberry tomato). A native plant of Central America, the groundcherry is an orphan crop that never made it in the agricultural major leagues. It's drought-resistant and has a "tropically intoxicating" flavor, Lippman says, but they have long branches and the fruit are fussy to grow. Using CRISPR, Lippman seeks to shortcut thousands of years of selective breeding by introducing several gene edits, influencing traits such as plant size and architecture, fruit size, and flower production.[18] Green groundcherries are naturally the size of marbles, but disrupting the *CLAVATA1* gene produces fruit that are 25 percent larger. Moreover, modifying the groundcherry counterpart to the *SP* gene produces more compact plants that are easier to harvest.[19]

CRISPR edits are not GMOs. Whereas CRISPR cuts the genome in a precise spot, it is the cell's natural DNA repair process that stitches the ends back together. The resulting mutation is no different than what might arise using chemical mutagens or X-rays or occur naturally. That is the enlightened verdict of the U.S. Department of Agriculture (USDA), which decided to treat CRISPR edits no differently than other methods of mutagenesis.

But the Europeans disagreed. In 2018, the European Court of Justice (ECJ) ruled that gene-edited crops do fall under GMO guidelines. Criticism rained down from all quarters: "illogical," "absurd," and "catastrophic" were representative reviews. Lynas said the ruling was "like saying doctors can use [a] blunderbuss but not [a] scalpel." Placing CRISPR and GMOs in the same bucket was like "the Catholic Church classifying ducks as fish," lamented Ewan Birney, a prominent British geneticist. Clive Brown, the chief technology officer of Oxford Nanopore, fumed: "If only these twits"—the ECJ—"realized that all of their beloved vegetables and most farm animals are hideous mutants." And British Conservative MP Owen Paterson said the European Union was condemning itself to become "the world museum of farming."[20]

In China, genome-edited crops are currently regulated as GMOs, but discussions with the Chinese government will likely turn things around. "We are hoping for a better solution than in Europe," says Gao Caixia, a plant biologist at the Chinese Academy of Sciences in Beijing.[21]

We have a global food problem. Worldwide, says Mick Watson, a geneticist at the Roslin Institute, there are about a billion people in the world who are obese and a billion people who are hungry. "This should be a problem that's pretty easy to solve, by taking the food away from the obese people."[22] Watson's facetious sense of humor may not be to everyone's taste but it does not diminish a serious message. The growing world population means that over the next fifty years, the world's farmers will have to produce more food than in the past 10,000 years combined. This has ramifications for the CRISPR craze, too. Imagine if we were able to cure all diseases, using CRISPR and other medical innovations. Watson says we wouldn't all live forever—we'd die of starvation.

The commercial potential of genome editing in the plant world has not been lost on the CRISPR community. With few exceptions, fruit and vegetable consumption in the United States has not improved in fifty years. There are some interesting exceptions, which have nothing to do with biotechnology. Introduced in 1986, the baby carrot has led to a dramatic increase in carrot consumption—3.8 billion pounds of the vegetable per year. In 2008, improvements in farming enabled blueberries to be delivered across the country year-round, resulting in a doubling of annual consumption to 600 million tons.

In 2017, Feng Zhang, David Liu, and Keith Joung, three cofounders of Editas, decided to get the band back together to apply their genome-editing prowess to plants. Pairwise Plants secured a five-year deal with Bayer, owner of Monsanto, to develop improved row crops and boost farm productivity, aiming to make foods that are more affordable, convenient, and sustainable.

Gene editing does not introduce a foreign gene, as happens in GMOs—but introduces a specific change to the DNA, usually to a sequence that already exists in nature. Besides speed and specificity, gene editing offers another benefit. Traditional selection leads to a loss of genetic diversity as lines are crossed and back-crossed. Gene editing can introduce traits without backcrossing, preserving or reintroducing lost variation. Early studies have used CRISPR-Cas12a to cut genes in corn and soybeans.

Gene-edited plant products made a high-temperature, low-key debut in the United States in March 2019, as the donut-frying oil at the Minnesota State Fair. Calyxt, a subsidiary of the French biotech company Cellectis, introduced Calyno, a gene-edited high oleic soybean oil. The oil has zero trans fats and 20 percent less saturated fat than its counterpart. Ironically, most of the soybean oil produced in the US is genetically modified, but Calyxt thinks Calyno—modified using TALENs rather than CRISPR—will prove a healthy, more neutrally flavored alternative to olive oil.

Calyxt says its gene-editing approach will give the American people "healthier food ingredients without compromising the taste of what they already love." For cofounder Voytas, that means a household staple: "We'd like a piece of Wonder Bread to meet all your daily requirements of fiber."[23] Thanks to the USDA regulations, Calyno proudly sports a "non-GMO" label, which doesn't sit well with environmental groups that refuse to recognize the distinction between genome editing and transgenic modification. Calyxt's edited soybean plants in South Dakota and Corteva's "waxy" high-starch corn in Iowa—destined for emulsifiers and glue sticks—are just the first seedlings in a forest of gene-edited crops and foods destined for consumers and livestock, from wheat and potatoes to alfalfa plants.

In China, much of Gao Caixia's efforts are on improving wheat, which imposes an extra two degrees of difficulty. The wheat genome is three times larger than the human genome and even larger than corn, soybean, or rice. Moreover, the wheat genome is hexaploid, meaning it has not one pair of chromosomes (a diploid genome) but three pairs. This extra redundancy means gene editors have to work three times as hard to target a particular gene. But Gao's team has engineered wheat lines that are resistant to powdery mildew[24] and herbicide resistance by inactivating the acetolactate synthase gene.[25] She's also engineering tomatoes like Lippman to change the plant's architecture, flowering time, and vitamin C content.

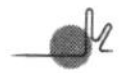

Many genome-editing applications are designed to help farmers increase the yield of their crops. But in some alarming cases, genome editing is the key to the species' very survival. Take orange growers in Florida, the world's second biggest producer behind only Brazil. They have seasonal challenges to overcome: arctic air plunges courtesy of the polar vortex, Atlantic hurricanes tearing up orchards, not to mention political headwinds impacting migrant workers. But the biggest threat is from an invisible source that was only first noticed in the Sunshine State in 2005.

The fruits that supply your freshly squeezed morning orange juice were first cultivated in China some 4,000 years ago, and imported to Europe about five hundred years ago. But a bacterial disease called *huanglongbing* (HLB), also known as yellow dragon disease or citrus greening disease,* has decimated the citrus industry in Florida and may do the same in California.

* The disease was originally called *huanglengbing* ("yellow shoot disease"), but differences in pronunciation resulted in a change to huanglongbing, which was made official in 1995.

It is caused by a bacterium called *Candidatus Liberibacter asiaticus* (CLas). The bacteria are spread by an insect, the Asian citrus psyilld, which feasts on phloem the way mosquitoes gorge on blood.

HLB causes the roots of infected trees to swell and then shrink, depriving the plant of water and nutrients. Meanwhile, the leaves producing sugars via photosynthesis can't transfer them to the rest of the plant because the phloem is blocked. It's like being starved and constipated at the same time.[26] The affected oranges are green, misshapen, and sour, unsuitable for consumption or even concentrate. "It's like AIDS but in citrus," is a common saying in Florida farming circles.[27]

First detected in southern China about one hundred years ago, HLB probably arrived in the United States surreptitiously in the 1990s. Research is hampered by the difficulty in culturing CLas in the laboratory. The economic damage in China, South America, and now Florida is reaching pandemic proportions, with orange groves and fruit production dropping 20–30 percent in the past decade. Tens of millions of trees have been lost worldwide; tens of thousands of jobs and some $5 billion in Florida alone are at risk.

Traditional weapons are ineffective. The insect that carries HLB is hardy with a range that can evade pesticide sprays. Antibiotics are of limited value, as spraying is unable to reach the bacteria hovelling deep inside the orange trees. Many farmers resort to spraying a chemotherapy cocktail of herbicides, pesticides, and fertilizer to combat HLB and citrus canker.

Until recently, it didn't appear that orange trees or any other cultivated citrus crops possessed any natural immunity. Hope grew a few years ago in the form of the Sugar Belle, a cross between a sweet clementine and the Minneola tangelo that is naturally resistant to CLas, growing more phloem to counteract the infection. Another idea is phage therapy, genetically arming a virus that, like CLas, naturally infects the phloem. And then there's CRISPR:[28] one idea is to pump up the promoter that governs activity of a family of plant protease genes to combat the bacteria. But there are challenges in editing polyploid plant species like citrus.[29] The clock is ticking in Southern California as HLB threatens commercial groves.[30]

Southern Gardens Citrus, a subsidiary of U.S. Sugar, is spending millions of dollars on transgenic oranges in a bid to save the entire orange business facing collapse. "We are science geeks," the firm declares proudly on its website. Obviously claims of "100 percent natural" won't fly with the insertion of a transgene. "People are either going to drink transgenic orange juice or they're going to drink apple juice," is how one scientist puts it.[31]

One strategy is to create a "transgenic tree"—a more hostile environment for the bacteria or the insect. Botanist Erik Mirkov has considered scorpion venom, beetle toxin, even a pig gene. But you don't need a PhD to realize that consumers would prefer not to have their orange juice spiked with sarcophagus beetle toxin DNA.[32] The most palatable prospect is an antibacterial gene derived from spinach that encodes a defensin, a hole-puncher protein that punctures the CLas outer membrane. If Southern Gardens can navigate the approval processes, commercial transgenic trees could be planted soon.*

In 1923, Eddie Cantor had a No. 1 hit record with, "Yes! We Have No Bananas," written by Frank Silver and Irving Cohn. Silver based the song on a lament he heard from a Greek vegetable stand proprietor on Long Island. Back then, the tasty variety of banana on sale was known as Gros Michel. But in the early 1900s, plantations in Central and South America were attacked by a fungus called Panama disease. By the 1950s, the Gros Michel had vanished from fruit stands. The fungus, which invades the plants via the roots, is almost impossible to eradicate from contaminated soil. In 2009, author Dan Koeppel, who literally wrote the book on bananas, was in the Democratic Republic of Congo, when he chanced upon someone ferrying bananas across a river. To his astonishment, he recognized them as the vanishingly rare Gros Michels. He peeled back the thick skin and savored his first bite as if sampling a vintage Château Margaux. His verdict

* Sadly Mirkov won't see the fruits of his labors: he died after a short illness in 2018.

was robust, creamy, with notes of . . . "It tastes more like a banana," he said approvingly.[33]

The banana industry was saved from collapse by the Cavendish variety, derived from plants grown in the 1830s at Chatsworth House, an English stately home. The Cavendish is an inferior fruit in most respects but became a commercial mainstay thanks to its resistance to the Panamanian fungus. Or at least it was before "bananageddon." We can't say we weren't warned.

The Cavendish is a monoculture, incapable of evolving because every fruit is a genetic clone. The Panamanian fungus, however, is under no such constraint, and a new strain called Fusarium TR4 (or Tropical Race Four) identified the Cavendish's Achilles' heel. TR4 arose in Taiwan in the 1980s, spread to Australia, and in 2014 jumped to Africa and the Middle East. Five years later, in August 2019, the Colombian Ministry of Agriculture declared a national emergency as TR4 struck in South America, the source of three quarters of the world's banana exports.[34]

There are more than a thousand varieties of banana, but whether consumers, who eat one hundred billion bananas a year, will accept a substitute that doesn't look like a traditional banana is a big question. Maybe they would prefer a fruit that has been genetically modified? A British company, Tropic Biosciences, is using CRISPR to reprogram some of the plant's own RNA interference defenses to target the fungus.[35] The company is also engineering coffee plants that will be genetically depleted of caffeine.

The threats to oranges and bananas illustrate the dilemma that all farmers and agricultural biotech business are now confronting: how to reassure the public that genetically modified, potentially gene-edited, fruits and crops are safe in an age of misinformation, fake news, and a legacy of anti-GMO disinformation. Some manufacturers have brazenly capitalized on this fear and ignorance by slapping "non-GMO" labels on all sorts of foods and fruits—even those that by definition cannot be genetically modified. Take water—an oxygen atom sandwiched between two hydrogen atoms—or salt, the simple union of sodium and chloride ions. These are two of the most natural, ancient compounds on earth. They couldn't be genetically modified if you tried.

Klaas Martens, an organic farming luminary in New York State, is a long-standing opponent of GMOs and the uses of genetic engineering technology such as Roundup for a manageable problem. "When the only tool you have is a hammer, everything turns into a nail," Martens says. But he sees CRISPR as a promising tool that could in some instances be compatible with organic agriculture if it could enhance the natural system.[36]

In Africa and Asia, climate change, disease, and political turmoil pose grave threats to agriculture. The picture in Africa is decidedly mixed. The country furthest along is South Africa, which has been growing GMO maize for a long time. Ruramiso Mashumba, a farmer from Zimbabwe, says there is no other option for his colleagues than genome editing. The effects of climate change, pests, and disease mean farming is not feasible. "The only option is to improve cultivars we have to sustain farming," he says. "At the end of the day, food is key."[37]

A good example is work on cassava, or yuca, a staple tuber crop for some 800 million people across Africa, Asia, and Latin America because its roots are rich in carbohydrates. But this hardy plant also produces a toxin—a chemical related to cyanide—that if insufficiently processed can cause konzo, a motor neuron disease leading to paralysis. CRISPR offers a means to remove the cyanogens by inactivating the genes encoding a pair of enzymes in the cyanogen biosynthetic pathway. Regenerating whole plants harboring these two dormant genes could eliminate konzo. Researchers are also using CRISPR to engineer resistance to the RNA virus that causes cassava brown streak disease.[38] There have been some promising early results, but the virus's capacity to evolve will not make things easy.

Many African nations don't see anything special in CRISPR. In 2012, Kenya summarily banned the import of GMO foods,[39] triggered by the publication of a controversial study by French biologist Gilles-Éric Séralini. Feeding rats a GM maize produced by Monsanto, the French group sensationally reported large tumors in rats.[40] A companion documentary directed by Jean-Paul Jaud called *Tous Cobayes?* (*All of Us Guinea Pigs Now?*) railed at the health risks posed by GM crops and nuclear accidents, raising the unthinkable possibility that a Fukushima-style explosion would not only result in millions of evacuations but also, worse, vineyards in Bordeaux contaminated by radiation.

The Séralini study was retracted by the journal editors two months later,[41] citing flimsy evidence including insufficient animals tested, only to be republished by another journal two years later.[42] Despite widespread renunciation, Séralini remains an influential figure in Kenya. Only in 2019 did the Kenyan government finally give limited approval to farmers to plant GMO cotton.[43] In Uganda, despite extensive debate on a biosafety bill, negative opinions about GMOs on public health prevail. Some opponents argue (falsely) that GMOs result in obesity, as in America.

Nnimmo Bassey, the Nigerian environmentalist, is deeply concerned about climate change, warning that ocean acidification and coastal erosion will breed conflict in his home country. But Bassey doesn't distinguish between the polluters helping to incinerate the planet and industrial corporations seeking to impose genetically edited crops across the continent. Bassey accuses them of cynically taking advantage of poor, hungry Africans purely to gain market access. "They want to bring in new forms of control, new colonial ideas," he says. "In each situation we have alternatives. Food is not just something you swallow. It is life, it is celebration, a cultural activity."[44]

The three "big Ag" players—Bayer/Monsanto, ChemChina/Syngenta, and Corteva, born out of the 2017 Dow-DuPont merger—know they face an uphill battle. Neal Gutterson, Corteva's chief technology officer, told me it's important for African scientists to drive the research, not a bunch of American executives flying in to push their latest technology.[45] There are myriad applications for CRISPR that could result in earlier release onto the market of edited crops, or more important, create bespoke varieties with improved disease resistance or nutritional value.[46] The European market wants an alternative to palm oil, which is currently produced in Malaysia and Indonesia with devastating costs to the ecosystem. With some judicious gene editing, sunflowers can be turned into a palm oil substitute. "The beauty of the technology," Gutterson says, "is when major societal needs can be addressed by emerging technologies."

In *The Happiness Hypothesis*, psychologist Jonathan Haidt introduced the metaphor of the elephant and the rider. The rider is the rational side of our brain, the elephant is the emotional side. While it may appear that the rider is in control, should there be any sort of conflict or disagreement, the elephant is likely to win out. Convincing the public at large, suspicious if not downright hostile to GMOs, that CRISPR and other emerging techniques is safe remains a gargantuan challenge.

If CRISPR is going to help us feed the planet, its impact won't just be felt in the plant kingdom. Livestock and other animals stand to benefit from genome editing to provide disease resistance and other benefits. Selective breeding over recent decades has resulted in some impressive improvements in yield and meat production, averaging around 20–30 percent annually. But with the world population on pace to exceed 9 billion people by 2050, these advances are insufficient at best, trivial at worst.

While genome editing appears the only viable solution to ensuring food security by improving the yield of crops and livestock, regulatory agencies take a different view. In the United States, the regulation of gene editing in plants by the USDA is much less constrained than that of livestock, which is governed by the FDA. When laws don't evolve in keeping with technology, the regulatory agencies are obliged to fit a square peg into a round hole—make the science fit into the existing regulatory framework.

In 2017, the FDA declared that any edited animal DNA would be considered for regulatory purposes to be a drug.[47] This sets up the ludicrous situation where farmers could over time breed a line of dehorned cattle using traditional methods and the FDA would barely bat an eyelid. But expedite the process using CRISPR to engineer the identical genetic tweak and the agency flips out. In Europe, the situation is reversed. Advocates believe that gene editing of food products should be judged on the ends rather than the means. "This technology was developed largely with public funding, and the public should benefit from its intelligent and careful application," argued a group of gene-editing supporters in 2016.[48]

It is quite fitting that the arduous journey of the first GMO fish to market should be that of the salmon. In 1989, the transgenic AquAdvantage salmon was first created in the laboratory by a Massachusetts company, AquaBounty. The gene construct transferred into salmon eggs included a growth hormone gene promoter that allows the fish to grow faster than normal. The modified fish reaches a weight of 500 grams in about 250 days, compared to 400 days for its unmodified sibling. These fast-growing fish, reaching maturity in half the normal time, make land or indoor farming of salmon in 70,000-gallon fiberglass tanks economically feasible. (It's not only healthier for the fish, it abolishes any risk of GMO fish escaping into the wild—especially in landlocked Indiana. Even if they did, the salmon are sterile.)

AquAdvantage salmon were approved for sale in Canada in 2016, a year after the FDA assessed that the transgenic salmon were safe. In May 2019, a shipment of 90,000 eggs left Prince Edward Island, Canada, cleared customs in Chicago, and arrived in Albany, Indiana—1,000 miles from the nearest ocean. After two decades and more than $100 million in regulatory costs, AquAdvantage could finally be sold in the United States. Whether any business will be convinced to sell a big GMO-labeled fish remains to be seen. While AquaBounty touts its supersized fish as sustainable, the situation is not. The enormous promise of CRISPR for food production will be crushed unless gene editing is decoupled from GMOs. Tellingly, AquaBounty has used CRISPR to produce fast-growing tilapia, but opted to produce them in Argentina where the regulatory hurdles were lower.

CRISPR is also coming to livestock, offering a faster, more controlled version of the sort of natural breeding that farmers have been performing for generations. But while the USDA sees it this way, the FDA doesn't, preferring to classify genome editing as a fancier GMO. In livestock, gene editing cows so that females carry the *SRY* male sex-determining gene would ensure an excess of males, which farmers prefer. Another example involves the *polled* gene.

There are more than 270 million dairy cows worldwide producing more than 700 billion liters of milk annually. But within the past 1,000 years, a spontaneous mutation in their cousins, the Angus beef cow, resulted in cows that naturally lacked horns. This *polled* mutation has been selectively bred in Angus cows—hornless animals are easier to house and manage, safer for animals and humans alike.* But most dairy cows do not carry this variant and must have their horns debudded painfully, using a hot iron or chemical cauterization. Many farmers have resisted cross-breeding polled animals, believing that other traits, including milk production, would suffer.

Recombinetics, a Minnesota company, used TALENS to engineer the *polled* mutation into the DNA of dairy cows.[49] The first calf born using this precision breeding method was named Spotigy ("spotty guy") for the black spots where the horns would normally be. At University of California Davis, near Sacramento, Alison Van Eenennaam, a livestock gene-editing evangelist, looks after polled cattle at the so-called Beef Barn. Van Eenennaam was working with a pair of gene-edited polled bulls when she learned that the FDA's guidance would classify gene-edited animals the same as GM animals producing veterinary drugs. "We went from having two bulls that were polled to having two 2,000-pound drugs," Van Eenennaam shrugged.[50]

It got worse. In 2019, FDA bioinformatician Alexis Norris was running computer searches to examine the possibility of gene editing causing off-target effects. She came up empty, but instead, she discovered something awry at the *polled* locus itself: traces of foreign DNA derived from the plasmid vector used to conduct the original editing, including antibiotic genes.[51] Antonio Regalado's headline hit the nail on the head "Gene edited cattle have a major screwup in their DNA."[52]

The episode was not only embarrassing but costly. Recombinetics was on the verge of breeding hornless dairy cows in Brazil, beginning with sperm from Spotigy's half-brother, Buri. Only Buri's calves would be carrying foreign antibiotic genes, making them textbook GMOs. Gene-edited cows from Brazil may be off the table for now, but using CRISPR will allow Recombinetics and others to engineer more precise edits for polled dairy, heat-tolerant beef cattle, and more.

---

* The Celtic *polled* variant is a duplication of 212 bases that replaces a ten-letter stretch in an intergenic region on cow chromosome 1. The resulting hornless trait is inherited in dominant fashion, although the precise mechanism is unknown.

CRISPR will also be an essential tool in the development of disease-resistant livestock. At the University of Missouri, Randall Prather has produced thousands of genetically modified pigs harboring dozens of edited or modified genes. The swine genome is the same size as the human genome and as we saw with eGenesis, well suited to CRISPR gene editing. Prather says, "We alter a handful of [DNA] bases and someone's going to say, 'well, you can't eat that'? I have a hard time with that." He has a right to be exasperated. Every cell division is accompanied by about thirty random mutations, and we don't know whether they're bad or good.

Prather's focus is porcine reproductive and respiratory syndrome (PRRS), caused by a virus that infects white blood cells in the lung leading to viremia. The disease was first detected in the United States in 1987, and in Europe three years later. The economic toll is massive—about $2.5 billion annually in the United States and Europe combined, or more than $6 million a day. Prather and other researchers have identified a cell-surface protein called CD163 as the gatekeeper for PRRS viral entry (analogous to CCR5 and HIV) and thus the prime target for engineering PRRS resistance. With CRISPR technology, the Prather lab was transformed, producing gene-edited piglets in just six months.

To test the CD163 theory, Prather shipped some edited pigs to Bob Rowland at Kansas State University in a blinded experiment. The pigs—edited and controls—were exposed to the virus that causes PRRS while being kept in the same pen. After a month, the lungs were tested for virus. Rowland emailed Prather while on a beach vacation in Florida. "Pigs 40, 43, and 55 remained negative," he said, not knowing those were the gene-edited pigs. Rowland's technicians thought there had been a mistake—they'd never seen pigs resist the virus. Similar results have been reported by groups in China and the Roslin Institute, but Roslin scientist Christine Tait-Burkhard says it will still be a while "before we're eating bacon sandwiches from PRRS-resistant pigs." The technology has been licensed to a British company, Genus, which is seeking regulatory approval.

This is just one lab, one disease. African swine fever (ASF), bovine respiratory disease, pig influenza, and chicken influenza are just a few of the other diseases that CRISPR can help.[53] Between 2018–2019, 150–200 million pigs in China and other parts of Asia became infected with the deadly ASF. Millions of animals have been slaughtered, the price of pork doubling as a result. The disease has reached Europe and is on the verge of entering the United States. At CAS, one of Prather's former trainees, Zhao Jianguo, is leading the charge to develop ASF-resistant hogs, by targeting a gene called RELA. Healready made his mark leading a team effort to render pigs more resistant to cold weather by using CRISPR to knock in a mouse gene called UCP1* that helps them burn more fat.[54]

What we've seen above are some early glimpses of the potential—for that is all it really is right now—of CRISPR to transform agriculture and help feed the planet. The Green Revolution and other agricultural advances of the 20th century were part of an explosion in new technology, much of it arising from the convergence of physics with engineering. Just as scientists decoded the physics parts list in the 20th century, the new century will be driven by the parts list of biology—the seminal discoveries of molecular biology spinning off the double helix and the genomic revolution. In the first two decades of the 21st century, we advanced from a White House celebration of the first human genome sequence for about $2 billion to a genome center like the Broad Institute churning out a human genome every five minutes for less than $1,000.

Whether the CRISPR revolution will match the Green Revolution is impossible to say. I'm not suggesting genome editing is the answer to feeding the planet, but it shouldn't be stymied by overregulation or anti-GM hysteria. "We're not special," says Charles Mann soberly. Like protozoa feasting on unlimited nutrients, we're going to hit the edge of the petri dish. Soon. The wizards of Mann's book believe in GM and ultimately genome-edited crops as part of humankind's solution, while the prophets preach conservation and human connection. But the wizards have failed

---

* Pigs carry a UCP1 gene but it is nonfunctional. In the experiment, the Chinese team knocked in the mouse counterpart.

miserably to persuade the public to embrace GM technology, leaving an uphill road for CRISPR.

Mann believes the two camps have much more in common than they let on. There can be a future that embraces genome-edited crops while recognizing the damage caused by industrialization. One idea is that plant scientists should prioritize trees and tuber crops such as cassava or potatoes, rather than wheat and other cereals.

In a greenhouse on the campus of North Carolina State University, Rodolphe Barrangou walks me through a row of young poplar trees, some twelve feet high, all carrying CRISPR-edited genes. These are the first shoots of TreeCo, "the North Carolina Tree Company," Barrangou's latest commercial venture, launched with fellow faculty member Jack Wang, appropriately set in the wood basket of the world. Poplars are abundant trees used for plywood, furniture, and paper. Genome editing can improve the abundance of pulp and lower waste, with applications ranging from climate resilience to timber to bioenergy. Barrangou isn't planning to feed the world just yet: first, he'd like to become the R&D engine for the lumber and forestry industry. But if not him, then surely someone else.

By the time we hit the edge of the petri dish, we're going to need even better gene-editing tools. Fortunately, in this remarkably innovative arena, they're already coming online.

CHAPTER 22
CRISPR PRIME

In 1960, a tall, gentle man named Victor McKusick, a medical geneticist at Johns Hopkins University, published the first edition of a remarkable catalogue of genetic traits and disorders. *Mendelian Inheritance in Man* became the bible of geneticists around the world, the definitive source of information about human genetic diseases and their underlying mutations. After a dozen editions, the catalogue was moved online. It currently lists more than 7,000 discrete genetic disorders and traits. Of those, McKusick was most closely associated with Marfan syndrome, the genetics of which he first described in 1956. Thirty-five years later, McKusick's colleagues at Hopkins identified the faulty gene alongside two other teams. I invited McKusick to write the accompanying commentary in *Nature.*[1] It was only fitting. In it he discussed the notion that President Abraham Lincoln might have had Marfan syndrome.

If he were still alive, the father of medical genetics would be in awe at the progress we've made in documenting the myriad ways in which our genetic software can be corrupted, not to mention the potential of delivering a patch to fix those errors. The Welsh geneticist Steve Jones wrote that the book of life "has a vocabulary—the genes themselves—a grammar, the way in which the inherited information is arranged, and a literature, the thousands of instructions needed to make a human being."[2] The letters on the pages of our books are subject to a host of different insults—substitutions, deletions and insertions, expansions, duplications, and rearrangements. Building on

McKusick's legacy, the global genetic community has documented mutations in about one third of the total number of genes in the human genome. That number will increase.

Some genetic disorders are caused by the tiniest mutation imaginable—the swap of one letter in the genetic code for another. Sickle-cell disease results from an A shifting to a T in the beta-globin gene. Progeria, a genetic form of premature aging, is caused by a C to T substitution in the lamin A gene. Cystic fibrosis is caused by hundreds of different mutations in a single gene, the most prevalent being the loss of three bases coding for a single amino acid. Conversely, the most common mutation in patients with Tay-Sachs disease is the addition of four bases (TATC) in the beta-hexosaminidase gene. Huntington disease, fragile X mental retardation, and dozens of other disorders arise from the bizarre expansion of a tract of repetitive DNA. Other disorders arise from insertions, duplications, or deletions of longer stretches of DNA, including entire chromosomes such as trisomy 21, or Down syndrome. And there are many more subtle genetic defects, including epigenetic mutations that silence one copy of a gene, depending on which parent supplied the gene.

As genome engineers contemplate the plethora of genes and mutations that need to be corrected to treat or cure genetic disease, they will need a deluxe toolbox that will extend beyond CRISPR-Cas9. For all of the astonishing progress since 2012–13, there are strong signs that the CRISPR toolbox is receiving a major upgrade with a suite of new tools that riff on the original CRISPR gene editing machinery. Cas9 engineers a complete break in the DNA, and while the ability to stitch in the desired sequence or repair is improving rapidly, the process still lacks the requisite precision for most therapeutic applications. The classic CRISPR technology will only work therapeutically in a fraction of genetic diseases.

If you were to draw it up on a whiteboard, the Holy Grail of genome editing would be to develop a technology that can modify a single letter of the genetic code without cleaving the DNA in the process. Almost before the ink was dry on the classic CRISPR papers, researchers were studying the building blocks, seeking to modify and adapt them. That has been the goal of many investigators, but one in particular stands out in his mission to design a truly precise molecular editor. He's a prodigiously talented scientist of Asian descent whose talent was on display in high school before enrolling at Harvard, excelling in his PhD in California, and hitting his prime at the Broad Institute. But this isn't about Feng Zhang.

A decade older than Zhang, there are indeed some striking similarities in David Liu's career. The son of Taiwanese parents, Liu was born and raised in Riverside, California. His mother was a physics professor, his father an engineer. Like Zhang, Liu's scientific talent shone brightly in high school, driven by what he admits was some "immature competitiveness." In 1990, he placed second in the national Westinghouse Science Talent Search competition, and first in his high school.

As a freshman at Harvard University, Liu's interests gravitated toward physics rather than chemistry. But that changed in December 1990 when he traveled to Stockholm as one of five top US students to attend lectures by the newly minted Nobel laureates, including Harvard chemistry professor E. J. Corey. Liu was enthralled by Corey's work on creating new molecules, like assembling Lego blocks. Afterwards, Liu told Corey he wanted to work in his lab. He got that opportunity and eventually graduated top of his class of more than 1,600 students in 1994. Years later, Corey told the *Boston Globe* that Liu was "going to be a superstar."[3]

Liu moved back to California to take up a PhD at Berkeley with Peter Schultz, a talented molecular biologist who was literally rewriting the genetic code. Liu spent the next few years studying methods to expand genetic alphabet—to encode and incorporate synthetic amino acids (beyond the twenty that occur naturally in the body) into proteins. A lecture on his groundbreaking graduate work back at Harvard turned into a de facto faculty interview. Occasionally scientists shine so spectacularly during their PhD that, like a professional basketball team drafting a high school prodigy, a university will offer them a faculty position. Harvard offered Liu a professorship, bypassing the usual four to five years of postdoctoral training. It was too tempting to refuse, but he doesn't recommend others try it. "I had no idea what I was doing," Liu admits.[4]

In autumn 1999, at the ripe old age of twenty-six, Liu joined the ranks of Harvard's illustrious chemistry faculty, with its seven Nobel Prizes since

1964. Having demonstrated the possibilities of performing molecular evolution on the building blocks of life, Liu decided to go for something really big: protein evolution in a test tube. During his first decade at Harvard, Liu's lab made a name in building new technologies for molecular evolution and applying them to treat human disease. The key method is called phage-assisted continuous evolution (PACE), developed by Kevin Esvelt.[5] In 1999, Liu even dabbled in a type of genome editing—an effort to construct a gene activator made up of a DNA (or RNA) triple helix that could be targeted to various sites in the genome to regulate genes or cut DNA. Liu admits the project "utterly failed" but even though he moved productively into other areas of research, his interest in performing chemistry on the genome stuck.[6]

To lure the top students, competing against some of the biggest names in chemistry like Schreiber, Szostak, and Whitesides, Liu promoted his annual lab open house with eye-catching posters. One showed Liu metamorphosizing into Regis Philbin, the host of *Who Wants to Be a Millionaire?* Others showed him in full *Matrix* regalia or dressed up as *Spider-Man* villain Doctor Octopus. The strategy apparently worked: Liu's research took off and in 2005, he was promoted to full professor, just thirty-one years of age. That same year, he joined Doudna as one of about three hundred investigators appointed (and technically employed) by the Howard Hughes Medical Institute, one of the highest echelons in American biomedical research.

For all of Liu's brilliance and commitment in the lab, he strives to maintain a healthy work-life balance. "Chemistry is life, but life is a lot more than chemistry," he says.[7] Some of his hobbies hearken back to his engineering upbringing. In the early 2000s, he built a featherweight airplane that could almost hover indoors, before drones became a phenomenon. He also built a Lego robot called the mousapult, which would entertain his cats by throwing a toy in the direction of a heat signature, using a sensor from a burglar alarm.

Liu's most intense—and lucrative—hobby, first picked up during his student years, was blackjack. His mathematical ability to count cards (a legal activity) became a teachable moment and something bordering on an obsession. Liu started teaching a weekly course for enthusiastic students from which he cultivated a devoted squad of fourteen "blackjack ninjas."

Every few months, the young professor would lead a delegation to Las Vegas and spend the weekend gambling—sometimes running fifteen hours at a time. Liu joked he was just hoping to earn enough to buy his wife a nice pair of earrings, but his posse was known to win "absurd sums of money."[8]

On Sunday nights, Liu took the JetBlue red-eye flight from Las Vegas to Boston, rolling up to teach his morning chemistry lecture pretty tired. He would ask himself why he was flying to casinos to gamble with students.[9] Occasionally though, after a particularly successful trip, he wondered if being a chemistry professor was all it was cracked up to be. Eventually, his hand was forced: the MGM Grand Casino in Las Vegas banned him. But he still carries a laminated card of calculations in his wallet should opportunity knock.

In his office on the third floor of the Broad Institute, besides Liu's own art, skilled photography, and mineral collection, a visitor cannot help but notice the thirty-pound, three-foot Iron Man "Hulkbuster" replica. It's the perfect metaphor: the ultimate shield to protect against the excesses of the Hulk's gamma-ray-induced rampages. Like Tony Stark, Liu has a penchant for inventing cool technologies to shield humans against genetic mutations, stacking the odds and seeing how high he can fly. "He's going to be the godfather of CRISPR 2.0," says Gerald Joyce, director of the Genomics Institute of the Novartis Research Foundation.[10] After listening to a spellbinding lecture from Liu in the Canadian Rockies in early 2020, a scientist sitting next to me whispered in my ear: "He's a genius!"

Growing up in upstate New York, Nicole Gaudelli's love of science and nature was nurtured by her father and grandfather. She loved going to zoos, fishing, growing crystals, and building water rockets. She thought about being a doctor, but her father suggested that she could help many more people by being a research scientist. During her PhD at Johns Hopkins, Gaudelli was captivated by a guest seminar given by Liu talking about PACE and molecular evolution. She decided to apply for a coveted position in Liu's lab for her postdoc.

Shortly after Gaudelli arrived in the lab in 2014, she befriended a new postdoc from southern California who had just earned her PhD from Caltech. Alexis Komor was working on something completely different—a project inspired by months of email exchanges with Liu prior to her arrival. Komor had interviewed with Liu eighteen months before finishing her PhD, hoping to persuade a big-name chemist that she could flourish in his group.

Komor began emailing Liu ideas for her postdoc project ("mutually guided brainstorming" is how Liu puts it). One item was an idea she'd sketched out to fulfill a Caltech graduation requirement: she wanted to evolve a ribonuclease enzyme in the lab so that it could degrade a specific sequence of RNA. Liu liked it but suggested she think about DNA-based editors, in particular the CRISPR-associated nuclease, Cas9. On November 1, 2013, he emailed her: "If you could program a specific A-to-G (for example) change in the human genome, you could really transform genome engineering and possibly human therapeutics."

Komor was excited but confused. "Why is he so crazy about this Cas9 thing?!" she thought.[11] But she kept refining her idea and by the time she arrived in Boston in September 2014, the basic idea of base editing had been born. Ironically, Liu had confused Komor and Gaudelli in the run-up to their arrival, mixing up their respective project ideas. On Komor's first day, Liu introduced her to the rest of the male-dominated lab, and then Gaudelli. "This is Nicole. I kept confusing you. You can see why!" The pair burst out laughing: Gaudelli has dark hair and eyes, unmistakable Italian heritage. Komor is quintessential Californian, blond hair and blue eyes. They became fast friends.

Komor's first six months were uncomfortable, far apart from her husband who was still in California finishing his PhD. She hoped Boston would be a short-term stay to complete a postdoc so she could return to her family and the California sunshine. "Technology development projects are super risky," she told me. "At the beginning I didn't know what I was doing!" Group meetings in which her plans were microdissected by quizzical, sometime skeptical colleagues, were the bane of her existence.

Although Doudna was a classical chemist, the field of CRISPR genome editing had evolved as a largely biological discipline. Komor and Liu brought a different skill set, and it paid off. "Single-stranded DNA is a lot more reactive than when it is double-stranded," Komor says. When Cas9 binds to DNA, it unzips the double helix to expose a stretch of about five bases of single-stranded DNA. Here, then, was a window to perform some cool chemistry. Komor began with cytidine deaminase, an enzyme that converts cytidine (C) to uracil (U), but only works on single-stranded DNA. By tethering the deaminase to an inactive ("dead") form of Cas9, she would create a homing machine to seek out a target DNA sequence and unspool a short stretch of DNA (without cutting the strand) upon which the cytidine deaminase could act.

After about eight months, Komor had a prototype base editor working that could convert a C:G basepair into a U:G mismatch pairing. Now she faced a new problem: the cell's DNA repair system won't tolerate the mismatch, so it tries to restore a natural basepair, like finding the right match in a jigsaw puzzle. Facing this U:G intermediate, Komor needed to tip the odds to favor the solution she wanted—coax the cell to repair the G, which would result in an T:A basepair.* She had to find a trick to complete the base edit, rather than watch the cell's DNA repair process simply undo her good work by fixing the U, reverting the base pairing back to where she'd started.

Komor's first trick was to find a way to block an enzyme that "rips out uracils like nobody's business." So she fused a third component to the base editor—an inhibitor of uracil DNA glycosylase, or the ripper. It shifted the balance a bit but not as much as she hoped. And then one day, she had an epiphany while talking to a colleague in the lab kitchen. "It just came to me," she recalls. "Oh my God, we're working with an endonuclease!" Although she was working with a "dead" form of Cas9, it was still a nuclease that cleaves DNA. By replacing a single amino acid in the enzyme, Komor could restore a "nickase" function that would

---

* In this example, the C base editor targets a C:G basepair and deaminates the C to a U, resulting in a U:G mismatch. The cell's DNA repair processes seek to repair the mismatch in one of two ways: either by switching the U back to a C; or by fixing the G to an A, thus creating a U:A, or T:A basepair. The goal was to push the system toward the latter, resulting in a C:G to T:A substitution.

clip one strand of the double helix. By nicking the G-containing strand (leaving the U intact) she could trigger the cell's DNA repair machinery to fix the G rather than the U nucleotide.

When Komor told Liu about her brilliant idea, he started swearing: he'd wanted to start writing up the paper for fear of being scooped. But Komor's idea was obviously worth trying. "How quickly can you do it?" Komor spent Christmas 2015 back home in California drafting the manuscript, editing it on Christmas Day, and even foregoing her ten-year high school reunion. The *Nature* reviewers initially gave the paper a rough ride on technical grounds, and it was rejected. Komor worked tirelessly to rebut each of the criticisms, while Liu phoned the editor, Angela Eggleston, to appeal. The revised paper was accepted and eventually published in April 2016.[12] When I asked Komor why they chose *Nature*, she laughed. "Where else would we send it?!"

As Komor was developing the first C-to-T base editor (CBE), Gaudelli grew increasingly interested in her friend's research. After much internal debate, she decided to abandon her own project, switching instead to try to develop a novel base editor that could do the reverse reaction—an A-to-G base editor. This would be a more useful setup for medical applications as about half of the known pathogenic mutations in human genes involve mutations of a G to A. (Indeed, there is a high spontaneous mutation rate involving the deamination of cytidines de novo to uracil, resulting in an erroneous T:A basepair.) Developing a system that could reverse this common source of human mutation could have a profound medical benefit. There was one small problem however: Gaudelli didn't have any starting material.

Liu had one unbreakable rule in his lab that had endured more than fifteen years: never start a project by evolving the starting material. But Gaudelli didn't have much choice: there was no natural enzyme that deaminates A to G in DNA.* Undaunted, Gaudelli trained her sights on a bacterial enzyme called tadA, which works on RNA, not DNA. With nothing to lose, Gaudelli performed a slightly crazy experiment—evolution in a test

* Deamination of A actually yields inosine (I), but this is read as G.

tube to try to generate the desired properties.* The first round of evolution yielded a mutation that enabled the altered enzyme to tackle single-stranded DNA instead of RNA as its substrate. The mutation was in the precise location that Gaudelli would have expected. She sent a quick slide to Liu, who started swearing again. "Holy—, this is our smoking gun," he replied.

Several rounds later, Gaudelli had evolved a potent A-base editor, or ABE. She was also able to demonstrate the ability to modify mutations in genes responsible for hereditary diseases including hemochromatosis and sickle-cell disease. Like Komor, Gaudelli's base-editing exploits also earned her a first-author paper in *Nature.*[13] By now, rival journal editors were visiting Liu to solicit hot papers like Gaudelli's. It sailed through peer review over a long weekend. Researchers around the world immediately jumped on the base editing bandwagon.[14]

Looking back, Gaudelli took an almost ludicrous gamble but she pays tribute to the nurturing environment in Liu's lab and her "Hulkbuster" of a boss, who "just makes you feel invincible."[15] She could have had any faculty position she wanted, but she elected to join Beam Therapeutics, a new biotech company Liu cofounded with his comrades in arms, Feng Zhang and Keith Joung. Gaudelli started to think about the friends and family base editing might eventually help. "What if one of those people was my father? My grandfather? What if that was a hypothetical child of mine?"

A friend of Liu's, a pediatric oncologist at Stanford named Agnieszka Czechowicz, came up with the company's name. She texted Liu her suggestion—Beam—which evokes a laser, a precision technology. "It also happens to stand for 'Base Editing And More'," she pointed out.

"What's the 'more'?" Liu asked.

"I'm sure you'll figure it out," she replied.[16]

Several Liu and Zhang postdocs, including Fei Ann Ran, have followed Gaudelli's path to Beam's facility in a building next to Novartis's R&D headquarters in the former Necco candy factory. In 2019, CEO

* Liu later found out that five groups had conducted the same initial experiment, fusing an RNA adenine deaminase, replacing our cytidine deaminase, but all five saw no editing. "No-one else made the crazy-sounding decision that we did to go ahead and evolve one."

John Evans took Beam public, raising a tidy $180 million, which will help them build a new headquarters in the heart of Kendall Square.

In less than five years, base editing had evolved from a speculative postdoc proposal to a pair of landmark *Nature* papers, rapid uptake in labs around the world, and a public biotech company. The creation of base editors is an impressive feat of chemistry. As Liu told me: "These molecular machines have to search the genome for a single target position, open up the DNA, perform chemical surgery directly on a base to rearrange the atoms—then do nothing else [except] defend the edit from the cell's fervent desire to undo them."[17]

The first two base editors offer a means to edit "all the easy mutations," says Komor. Liu anticipates "there'll be a library of base editors and you'll pull out the book that matches exactly what you need." That choice will be influenced by the desired edit, the sequence context, off-target effects, and so on. In March 2020, Liu underscored that prediction: working in collaboration with Jennifer Doudna's group, another postdoc, Michelle Richter, unveiled a new-and-evolved version of the A base editor that was six hundred times more active than Gaudelli's original.[18] Just as with CRISPR-Cas9, base editors are prone to cutting at off-targets. But scientists are working fast to improve their specificity. To keep this in perspective, note that in each of your 10 trillion cells, the genome is constantly mutating. Hundreds of times a day, a C is mutated to a U, which if left unchecked, would become a C-to-T mutation.

It will be years before a base editing drug is available, with many hurdles to climb to get there. Komor, who is now a university professor in her beloved Southern California, sees great promise in not just treating symptoms, but "curing the disease." That's what I'd expect her to say, but Liu's group has already used base editing to correct a mouse model of a rare but devastating genetic disease. Progeria, or Hutchinson-Gilford progeria syndrome, is a dominantly inherited disease caused by a single-base mutation in the laminA gene that results in extreme premature aging. The mutant protein, called progerin, wreaks damage in the aorta and other tissues. Affected children rarely live beyond fifteen years of age.

Liu partnered with NIH director Francis Collins, who years earlier had developed a mouse model of the disease carrying the human progeria mutation. Liu's team delivered the base editor via a pair of AAV vectors, using a molecular Velcro to splice the components together once inside the cell. The ABE corrected the mutation and squashed production of progerin. In the treated mice, cells regained their normal shape, and the aorta was restored to near-normal health. Stunningly, the treated mice look healthy and live longer than the progeria mice. How much longer Liu couldn't exactly say—for the good reason they were still alive. "We're really excited," Liu said, moving forward "carefully but quickly" to advance this revolutionary treatment from mice to boys and girls.[19]

What about more common diseases? Verve Therapeutics, a genome editing start-up, collaborated with Gaudelli's team at Beam to test a one-shot strategy in two monkey models of heart disease. Verve used a lipid nanoparticle to deliver the ABE to the liver of crab-eating macaques to inactivate a pair of known genes that regulate cholesterol. CEO Sek Kathiresan reported a dramatic lowering of "bad" LDL cholesterol and triglycerides in animals targeted at the *PCSK9* and *ANGPTL3* genes, respectively.[20] The results need to be confirmed and extended in humans, which is some years away. But base editing could help realize Kathiresan's dream of a "one-and-done genome editing medicine for heart disease," providing an alternative to chronic statins and reduce the 18 million cardiovascular deaths each year.

From Archimedes in the bath to Isaac Newton's bruised head, there are many legendary aha moments in science history. Perhaps the most bizarre episode belongs to the late Kary Mullis, who recalled the invention of the polymerase chain reaction in a quite extraordinary Nobel lecture. It's too good not to relive here:

> One Friday night I was driving, as was my custom, from Berkeley up to Mendocino where I had a cabin far away from everything off in the woods . . . As I drove through the mountains that night,

> the stalks of the California buckeyes heavily in blossom leaned over into the road. The air was moist and cool and filled with their heady aroma . . . EUREKA!!!! . . . EUREKA again!!!! . . . I stopped the car at mile marker 46.7 on Highway 128. In the glove compartment I found some paper and a pen . . . "Dear Thor!" I exclaimed. I had solved the most annoying problems in DNA chemistry in a single lightning bolt . . . We got to my cabin and I started drawing little diagrams . . . with the aid of a last bottle of good Mendocino County cabernet, I settled into a perplexed semiconsciousness . . . The first successful experiment happened on December 16th. I remember the date. It was the birthday of Cynthia, my former wife . . . There is a general place in your brain, I think, reserved for "melancholy of relationships past." It grows and prospers as life progresses, forcing you finally, against your grain, to listen to country music.[21]

By comparison, Andrew Anzalone's story of the genesis of "prime editing"—fueled by caffeine not cabernet, ambling around the streets of Lower Manhattan rather than speeding through the Napa night—could use a little work in the dramatic license stakes. But his ideas, formed in 2017 before leaving Columbia University for a position in Liu's lab, were crucial in devising a new genome editing platform. A physician-scientist rather than a chemist, Anzalone was inspired by Liu's base editing exploits, but sensed an opportunity to go further. "The base editors were really good for making four possible base changes but they couldn't address the other eight base changes or the small indels," he said.[22*]

In October 2019, Liu unveiled a new genome editing technology developed by Anzalone and other members of the lab that riffed on base editing, expanding the repertoire of potential DNA alterations. "This is the beginning of an aspiration to make any DNA change in any position

* With four bases, each in theory mutable to three other bases, there are a total of 12 possible base substitutions. The C and A base editors developed by Liu's lab account in total for four substitutions (CBE catalyzes C-to-T and G-to-A; the ABE catalyzes the reverse substitutions, A-to-G or T-to-C), known as transitions.

of a living cell or organism," Liu said.[23] I was sitting in the audience of four hundred rapt scientists at the Cold Spring Harbor Laboratory for Liu's first public presentation on prime editing, just ten days before the study was published in (no surprise) *Nature*.[24] "Prime editing is somewhat complicated," Liu admitted. But it works, and the meticulous four-step sequence offers important advantages.

There are more than 75,000 known disease-causing mutations in the human genome—about half of those are point mutations—but most can't be targeted by CRISPR-Cas9 or base editing. While base editing's strength is its ability to make a class of base substitutions known as transitions, they only account for four of the dozen possible base changes. The CBE would in principle fix 14 percent of known point mutations; the ABE accounted for a higher fraction, some 48 percent.

Anzalone wondered if he could build on this technology to engineer any single-base change, transitions and transversions.[25] What he developed is a new system that moves scientists closer to a true search-and-replace function for DNA, regardless of the letter or its location. Naturally, the system would start with the programmable single-guide RNA, which directs Cas9 to the stretch of DNA to be edited. But what if, instead of providing the replacement sequence via a DNA template, he used the same gRNA molecule to supply the edit? The system would have two programmable elements—the target site and the edit itself. Naturally the replacement sequence would have to be converted from the RNA guide to DNA, but luckily there's a very well-known enzyme for that—reverse transcriptase (RT).*

The extended guide RNA, renamed the pegRNA (prime editing-gRNA), specifies the target as well as the desired edit. Similar to the construction of the base editors, Anzalone fused RT to dead Cas9, then figured out a scheme to coax the edited DNA copy into the target sequence. The first step is to engineer the pegRNA and RT enzyme to copy the pegRNA strand into DNA. This results in a flap of DNA that needs to be

* When Anzalone googled "Cas9-RT fusion," he learned that Cas1, which plays a role in capturing viral sequences in the CRISPR defense system, has a natural RT activity. This shows, not surprisingly, how nature uses similar concepts to prime editing but for different purposes.

stitched into the double helix. (Helpfully, cells have enzymes called "flap nucleases" that help in this process.) Finally, the method introduces a nick into the complementary strand, which is then repaired to fully match the edited strand.

Anzalone's approach didn't get off to an auspicious start. When he fused RT to Cas9, he achieved zero editing. But further tests with a batch of RT variants soon resulted in some positive results. If the original CRISPR approach is a molecular scissors and base editing is a more precise pencil eraser, Liu describes prime editing as a word processor, capable of performing a search-and-replace function on any typo in the DNA alphabet. It also carries out some classes of insertions and deletions (indels), including those responsible for the most common form of cystic fibrosis (a three-base deletion) and Tay-Sachs disease (a four-base insertion, respectively).

After submitting their report to *Nature*, Liu and Anzalone had little trouble attending to the three anonymous referees' comments. Figuring out the identity of one of them was easy: few reviewers sprinkle words like "quixotic" in their reports. Urnov effusively recommended publication, citing the increased flexibility on editing sites, no more PAM deserts, no DNA donor, few off-targets, and the early data on correcting DNA in neurons.

In their paper, the Liu group showcased the full range of prime editing's prowess—175 different edits, including 100 point mutations of all possible types; repair of known disease mutations in human cells; insertions and deletions in the forty to eighty base range; and simultaneously deleting two bases while converting a G-to-a-T a few bases away. That's like watching Lionel Messi slalom his way through the opposition defense and around the goalkeeper before tapping the ball into the net. Or as Sharon Begley put it, "the genome equivalent of a pool shark's banking the 9 ball off the 7 and sinking the 1, 5, and 6."[26]

Liu's Cold Spring Harbor unveiling of prime editing was pretty much as close to a drop-the-mic moment as I've witnessed at a science meeting. He had given the meeting organizers a deliberately vague summary of his talk for the program book to preserve the element of surprise. I shook my head in astonishment as he flashed a pie chart showing the categories of human disease mutations, and said calmly that prime editing could in theory address 89 percent of them. Thirty minutes later, Liu wrapped up his talk by acknowledging Anzalone, with a smile: "I'm really looking forward to seeing what Andrew can do in the second year of his postdoc!" Doudna, who was sitting in the front row, later declared, "I literally had chills running down my spine" as she savored the latest power upgrade to the CRISPR toolbox.

The media reaction to prime editing was extraordinary, even overshadowing Google's claim of "quantum supremacy" published the same week. Commentators and journalists gushed about this gorgeous new "CRISPR 3.0" technology. The breakthrough even caught Elon Musk's attention, who retweeted a *New Scientist* story. Urnov was much in demand, obligingly dashing off a different analogy for each reporter who called. For *Scientific American*, prime editing was like a new breed of dog. For *STAT*, it was a new superhero joining the Avengers. For *Genetic Engineering & Biotechnology News*, it was a college sports star preparing to join the professional leagues. "We all hope of course it will be like Alex Morgan or Aaron Rodgers in this regard—and we should know soon."[27]

While Urnov speculated that prime editing could be part of an immunotherapy clinical program within a couple of years, a few commentators took issue with Liu's eyebrow-raising estimate of 89 percent mutations that were potentially fixable. Liu is a fastidious scientist who is not inclined to hype his results—because he doesn't need to. In a subsequent talk a month later in Barcelona, he politely but firmly pushed back. "This is a smart audience," he told 1,500 gene therapy experts. "You know the difference between correcting a mutation and actually treating patients."[28]

But how could prime editing be used therapeutically? With protein and RNA components made up of thousands of atoms, the prime editing molecular machinery is too bulky to fit into the standard AAV vector. But Anzalone was able to use a lentivirus to perform prime editing in mouse cortical neurons. The scarcity of PAM sites is not a big issue because of prime editing's greater flexibility with regard to the edit location. The prime editing window is much longer than traditional Cas9 editing, and the system has a lower rate of off-target effects. The system appeared safer than Cas9, for reasons that make sense: whereas CRISPR-Cas9 has just one base-pairing event (when the gRNA aligns with the target sequence), prime editing has two additional pairing steps—the binding of the pegRNA to the target site and the pairing of the flap to the original site—which offer additional opportunities to reject an off-target sequence. "If any of those three pairing events fail, prime editing can't proceed," Liu said.

As with the earlier CRISPR genome editors, delivery will be a big challenge, but Liu was confident he could deliver prime editors into animals, for example by using a pair of AAVs (as used in the progeria mouse model). On the safety question, Liu stressed that all genome editors have off-target effects—chemical binding is an imperfect process, just as all prescription drugs have some sort of side or off-target effect. "Each platform has complementary strengths. All will have roles in basic research and therapeutics," Liu said, mindful that prime editing could steal some thunder from his earlier platform companies, Editas Medicine and Beam Therapeutics. Indeed, by the time prime editing was announced, Liu's latest company, Prime Medicine, had already been formed on paper, with funding from a Google fund and F-Prime, with some rights licensed to Beam.[29]

Prime editing won't be the last word in genome editing technology. I could mention Homology Medicines, which is patching in a full gene delivered by a virus to genomic targets to treat phenylketonuria without CRISPR. Or an Israeli start-up called TargetGene Biotechnologies, which modestly claims it is developing "the world's best therapeutic genome editing platform." Or Tessera Therapeutics which touts "gene writing" as the route "to cure thousands of diseases at their source."[30] In July 2020, Liu unveiled another impressive riff on base editing, turning a bacterial toxin into a precise gene editor that could be delivered to mitochondria to edit mtDNA. Liu's team used TALEs rather than CRISPR as the guide.[31]

Whether it takes five years or fifty, it seems inevitable that we will be able to engineer bespoke variants into the genome precisely and safely.[32] The prospect of rewiring the genetic code to cure deafness or diabetes, sickle-cell or schizophrenia, is getting closer all the time.[33] But why stop there?

CHAPTER 23
VOLITIONAL EVOLUTION

In 2007, Anne Morriss and her partner decided to start a family. The two women went to a sperm bank and chose a donor based on a few criteria ("sporty" was high on the list). Sperm banks typically only screen donors for a couple of genetic diseases—cystic fibrosis and spinal muscular atrophy—but Morriss had no reason to be alarmed. However, a few days after giving birth to a baby boy, she received a distressing phone call from a Massachusetts public health employee.

"Is your son okay? Is he still alive?"

A stricken Morriss stammered, "Yes, I think so. I just put him down for a nap."

"Can you go and check?"

After birth, Morriss's son had received the standard Guthrie heel-stick test, in which a drop of the baby's blood is screened for a few dozen serious genetic disorders. Against all odds, Morriss and the anonymous sperm donor were both carriers of a rare recessive genetic trait—MCAD deficiency, which affects about 1 in 17,000 people. This gene encodes an enzyme called medium-chain acyl-CoA dehydrogenase that helps the body convert fats into energy. Morriss and her partner immediately modified Alec's diet before any serious problems ensued. A fortuitous phone call had saved their son's life, but hospital bags remain packed by the Morriss front door in case of an emergency.

A motivated Morriss teamed up with Princeton University geneticist Lee Silver (author of *Remaking Eden*) to launch a next-gen diagnostics company called GenePeeks. Using Silver's "Matchright" algorithm, GenePeeks created "digital babies" by virtually matching the DNA of the client with potential sperm donors. The client could then see which donors were predicted to have an increased risk of creating an embryo with one of hundreds of genetic disorders. These individuals could then be excluded from the donor pool, without selecting or freezing spare human embryos.

Silver and Morriss were featured in a story on *60 Minutes*, almost literally their fifteen minutes of fame. In the future, Silver predicted "people will not use sex to reproduce" because it was too dangerous to leave inheritance to chance. I published my first story on GenePeeks in January 2013, the same week as the first demonstrations of CRISPR editing in mammalian cells.[1] Silver and Morriss weren't attempting genome editing, but in their own way were looking to skew the odds in the otherwise random genetic assortment that underlies conception. "Our mission was to empower parents with insight and information that could protect their future children from devastating diseases," Morriss told me. Several hundred couples used GenePeeks's pre-fertilization matchmaking service before the company ran out of money and dissolved. Evidently the world wasn't ready for digital babies. But is it ready for #CRISPRbabies?

We probably won't hear about another gene-edited baby for several years given the widespread support for a moratorium, but this state of affairs won't last forever. The next time someone tries this, whether government approved or secretly in some offshore CRISPR clinic, they will probably succeed. And so we should ask: under what circumstances, if any, might germline editing be justified?

Despite evidence that gene editing in human embryos using CRISPR-Cas9 poses risks of "on target" DNA rearrangements, the pace of research suggests that in a few years, we will have the technical ability to safely perform precision DNA surgery in a human embryo.* Fyodor Urnov suggests

---

* Studies first reported in June 2020 by three eminent groups (one in the UK, two in the US) showed damaging "on target" DNA rearrangements in a fraction of human embryos edited in the lab using CRISPR-Cas9. There is still much we do not fully understand about DNA repair and recombination in embryogenesis.

a thought experiment: Let's say we are part of a group that has raised $1 billion and can draft a biotech dream team, the Avengers of genome engineering. Can we imagine that embryo editing reaches the stage where we can finesse the editing so there are no rearrangements, no off-target effects, no mosaicism? "We could get there quickly," Urnov surmises. "But the $64,000 challenge is: What are we going to do?"[2]

We've been wrestling with this dilemma for decades, since before the recombinant DNA revolution. But there is a renewed urgency since the first reports of human embryo editing. Eric Lander addressed this issue head-on at the first international genome editing conference in 2015. In preimplantation genetic testing (PGT), we already have a method, he argued, to reduce the transmission of disease genes. "The truth is, if we really care about avoiding cases of genetic disease, germline editing is not the first, second, third, or fourth thing we should be thinking about," he said.[3]

The use of PGT has exploded since its development in 1990 by Alan Handyside, Robert Winston, and colleagues in London.[4] Many clinics offer couples the chance to screen their IVF embryos, taking a cellular biopsy after the embryos are about five days old (about 250 cells). Embryologists carefully remove two or three cells from the blastocyst bundle, then amplify and sequence the target gene. This allows them to designate each embryo as being healthy or carrying one or two copies of the mutant gene. Following this scorecard, the couple can choose which embryos to implant and which to put into deep freeze (or perhaps donate for research purposes). For couples who carry a recessive disease gene, there is a one-in-four chance of a child inheriting the disorder. PGT can theoretically eliminate that risk by analyzing the embryos after IVF to identify the healthy embryos for implantation. For dominantly inherited disorders such as Huntington's disease, half of the IVF embryos conceived by a Huntington's patient would on average to possess the disease gene. Over the past two decades, PGT has been performed about 1 million times; about a tenth of those cases are to test for a monogenic disorder.[5]

There are rare cases where PGT won't be helpful. If both members of a couple have a recessive disease such as cystic fibrosis or the deafness gene that Rebrikov is contemplating, one could potentially CRISPR the embryos to fix one or both copies of that gene, restoring it to the healthy (wild type) version. There are also rare situations where a patient has inherited two copies of a dominant disease gene. PGT would be redundant as the patient is guaranteed to pass on the disease to his or her progeny. Only germline editing (fixing both copies of the broken gene) could produce a healthy biological child.

"It's not a very big need, but it's not nothing," Lander observed. "If we truly care about preventing needless genetic disease, we should be empowering genetic diagnostics for families, not editing embryos." Researchers at a fertility clinic in Los Angeles took a stab at estimating what "not nothing" actually entails.[6] The numbers of cases were few and far between, perhaps a few dozen a year in the United States. But IVF is not a trivial procedure: it is expensive, painful, and frequently unsuccessful, producing insufficient healthy embryos to ensure a healthy pregnancy.

As we survey the human genome and identify more of the genetic variants and pathways that underlie more common diseases, the menu of PGT services will inevitably expand, straying beyond the merely medical. In fact, it's already happening. The Ferny Fertility Clinic in New York offers couples a cosmetic gene test to select embryos for a particular eye color. "As has been happening from the beginning of humankind, only mom and dad can 'make' the eye color by combining their own unique genetics into the new child," says the clinic's founder, Jeffrey Steinberg.[7] For a while, the Ferny clinic even offered a discount for people with blue, green, or hazel eyes.

In the future, there may be a way to circumvent editing embryos completely. An alternative approach that is gaining interest is to edit the eggs or sperm prior to fertilization. For example, at the Weill Cornell Medical Center in New York, embryologist Gianpiero Palermo is literally zapping sperm (excess material donated for research) to target the *BRCA2* gene using CRISPR.[8] To coax the CRISPR molecules into the sperm heads, technician June Wang pulses the sperm with a quick electric shock. She places the vial containing 50 million sperm in an electroporation machine and turns it up to 11—1,100 volts, that is. The pulse loosens up the densely packed DNA in the head of the sperm to give Cas9 access to its intended target.

"Before 2020, germ-line engineering to cure severe genetic disease in human embryos will be an established therapeutic option."[9] Lawyer and author Philip Reilly made that prediction in 2000, and while it hasn't quite materialized the way he anticipated, he was correct in believing we would cross the germline threshold, reaching into the genetic fabric in a human embryo to rewrite the book of life. Designing humans—editing humanity—does not seem quite so far-fetched now our species has dared to cross the germline. As evolutionary biologist Mark Pagel wrote before the JK debacle: "The first truly and thoroughly designed humans are more than just the subjects of science fiction: they are on our doorsteps, waiting to be allowed in."[10] Reilly predicts that by 2050, germline editing would be as routine as cosmetic surgery.

At this point, I'm contractually obliged to bring up *Brave New World*, the classic novel published in 1932. As we've seen, discussion of embryo selection, genetic modification, and designer babies inevitably conjures up a reference to Huxley's dystopian vision. It's been that way for decades in discussions about medical involvement in procreation and eugenics, test-tube babies, and Dolly the sheep. As Leon Kass wrote in 2001, "Huxley saw it coming."[11]

But as Derek So has pointed out, *Brave New World* was never intended to be a warning about technologies such as genome editing.[12] Huxley doesn't describe any form of genetic engineering or testing. The upper caste in *Brave New World* were smarter than the remainder not because they were enhanced but because the lower castes were deliberately subjected to impairment. Nor is the novel a good example of parents selecting designer babies. Huxley himself was much more worried about totalitarianism than new reproductive selection technologies. Huxley, like his brother Julian, was a member of the Eugenics Education Society and believed England should enforce mandatory sterilization lest the country devolve into a nation of half-wits.

After giving a talk recently about CRISPR at my children's former high school in Lexington, Massachusetts, a student stumped me by asking if I'd read Margaret Atwood's MaddAddam trilogy, set in a hyper-capitalistic late 21st century. In the first book of the trilogy, *Oryx and Crake*, a brilliant geneticist named Crake usurps natural selection to conceive and create a superspecies adapted to thrive in a post-pandemic society, on a planet ravaged by climate change. They replaced socially normal mating customs with features beneficial to procreation and survival. The Crakers had beautiful skin of many colors resistant to sun damage and able to repel insect bites and infection. They also boasted bovine-like digestive systems requiring only nutrients provided by ubiquitous weeds.

The ability to eat weeds isn't high on anyone's wish list (yet) but some of us do possess extraordinary "superhuman" traits. Take seventy-year-old Jo Cameron, who lives near Loch Ness in Scotland. Her life has been free of pain and anxiety—she barely felt any discomfort giving birth, although has suffered plenty of serious bruises and burns. She didn't appreciate her superpower until undergoing hip surgery in her sixties, managing the pain with just a small dose of acetaminophen. In 2019, researchers found a variant in one of Cameron's genes called FAAH-OUT, which raises levels of anandamide, an endogenous cannabinoid, and makes her almost unable to feel pain.

The first example of a pain-resistance mutation was first described in 2006 in members of a consanguineous (inbred) Pakistani family, incriminating a gene that codes for a sodium ion channel called NaV1.7, involved in the propagation of nerve signals.[13] One boy in particular was a well-known street artist, painlessly walking on hot coals or stabbing his arms with knives. On his fourteenth birthday, he jumped off a house roof and fell to his death. Pain has its purpose enabling humans to comprehend and internalize risk. The English team that made that gene discovery learned about another extraordinary family from Siena, Italy, featuring people who cannot feel extreme pain or temperature. For example, Letizia Marsili shrugged off a nasty crash while skiing, only to discover that evening that she'd sustained a broken shoulder.[14] The Marsilis carry a mutation in a gene called *ZFHX2*, and now have a syndrome named after them. This discovery could eventually turn into a powerful non-addictive pain reliever.

Marvel creator Stan Lee made an entire television series devoted to real superhuman genetic outliers—echolocation, extreme endurance, temperature resistance, mathematical wizards, and people with eidetic or photographic memory.[15] The actor Marilu Henner is the most famous person (although all told there are only a dozen) with total memory recall, a condition called hyperthymesia or highly superior autobiographical memory (HSAM). Scientists are eagerly trying to untangle the neural basis of this extraordinary ability and its possible genetic underpinnings.

Julian Savulescu, a philosopher at Oxford University, can reel off a list of traits he'd like to see engineered into humans that would have made Stan Lee blush. Bat sonar. Hawklike vision. Enhanced memory. Radical life extension. Increased IQ to the point that we become a separate species. Humans have been seeking to enhance the quality of life for years. We add iodine to salt and vitamin D to milk, and calcium to orange juice. We take Ritalin to improve concentration, hormones to improve vitality, and undergo Lasik surgery to dispense with spectacles. We perform IVF, prenatal diagnosis, and PGT, what some term liberal eugenics. "Parents should be allowed [to undertake embryo editing] provided they don't harm their children or other people," Savulescu suggests.

George Church is somewhat agnostic about germline editing but supports using the protective effects of known gene variants to aid human health and longevity. The ends are more important than the means. For years, Church has compiled a set of gene variants that offer potential physical, medical, or behavioral advantages—some call it the transhumanist wish list (see table on the next page). Some of these gene discoveries are already fueling drug discovery advances. But if germline editing was ever offered in the future, these would be some of the first genes on the clinic menu.

CCR5 was on this list long before the JK scandal, as resistance to viral infection is in principle a highly desirable trait. That was true even before the COVID-19 pandemic began.[16] We don't yet know of a gene variant that confers resistance to SARS-CoV-2, but such protective polymorphisms likely exist. The FUT2 receptor is the cellular foothold for the norovirus, which afflicts hospitals and cruise ships with regularity. Immunity to the winter vomiting bug would be nice but knocking out

FUT2 appears to increase risk of Crohn's disease and colon cancer. There are few free lunches in the human gene pool.

Table: A list of gene variants that offer potential medical or other advantages

GENE	MUTATION	EFFECT
CCR5	-/-	HIV resistance
FUT2	-/-	Norovirus resistance
PCSK9, ANGPTL3	-/-	Low coronary disease
APP	A673T/+	Low Alzheimer's
GHR, GH	-/-	Low cancer
SLC30A8	-/+	Low T2 Diabetes
IFIH1	E627X/+	Low T1 Diabetes
LRP5	G171V/+	Extra-strong bones
MSTN	-/-	Lean muscles
SCN9A, FAAH-OUT, ZFXH2	-/-	Insensitivity to pain
ABCC11	-/-	Low odor production
DEC2	-/-	Reduced sleep

What about protection against dementia or premature aging? We've already seen how one version of the apolipoprotein E (APOE) gene on chromosome 19—*APOE4*—is associated with a roughly tenfold increased risk of developing Alzheimer's. Editing the E4 variant to the E2 or E3 form might lower disease risk and is worth exploring. In Colombia, a huge extended family suffers from a rare hereditary form of early-onset Alzheimer's caused by a mutation in the gene for presenilin. About 1,200 family members harboring this mutation are affected—except one.[17] It turns out this seventy-three-year-old woman carries another mutation in the *APOE* gene called APOE-Christchurch, originally discovered in the 1980s by a team in Christchurch, New Zealand.[18] This variant codes for a protective nonstick version of the normal protein, reducing the prevalence of protein aggregates in the brain.

There is also evidence that elevated levels of a protein called Klotho, sometimes dubbed the longevity gene, can improve cognition and protect against Alzheimer's—at least in mice. A Japanese group named the gene after Clotho, daughter of Zeus, and one of the three Fates in Greek mythology. Several biotechnology companies—seemingly driven by Silicon Valley billionaires contemplating their own mortality—are desperately seeking genes that might slow down the aging process.

Other genes that would be prime candidates for future genetic modification are those that govern risk for obesity and cardiovascular disease, diabetes, and hypertension. We know humans will go to extremes to address body weight and heart health, from liposuction and gastric bypass surgery to billions of dollars spent annually on statins and other drugs. While obesity and heart disease are complex traits influenced by the interaction of multiple genes and environmental factors, some rare mutations with a profound influence on body weight are known.

In the mid-1990s, Helen Hobbs, a geneticist at the University of Texas Southwestern Medical Center, set out to identify individuals with rare mutations that might offer protection against heart disease. One of the women she screened was an African American yoga instructor who possessed an enviable cholesterol level: just 14 mg/deciliter, compared to the average of 100 mg/dl. The woman inherited two faulty copies of the gene that encodes PCSK9, a regulator of the LDL receptor. Knocking out PCSK9 increases the number of LDL receptors in the liver that mop up "bad" cholesterol. "Of all the intriguing DNA sequences spat out by the Human Genome Project and its ancillary studies, perhaps none is a more promising candidate to have a rapid, large-scale impact on human health than *PCSK9*," wrote Stephen Hall.[19]

Sure enough, two PCSK9 inhibitors, Praluent and Repatha, were approved by the FDA in 2015. Cardiologists Sekar Kathiresan and Kiran Musunuru (as noted in the previous chapter) cofounded Verve Therapeutics to develop gene-editing approaches to treat patients at high risk of heart disease by mimicking the rare, naturally occurring protective variants seen in genes like *PCSK9* and *ANGPTL3*. Just to be clear, the company adds a disclaimer: "We will not edit embryos, sperm cells, or egg cells."

Another coveted trait is the ability to thrive on just a few hours of sleep. In 2009, Ying-Hui Fu, a geneticist who studies circadian rhythms at the University of California, San Francisco, reported the discovery of a private mutation in *DEC2* in a mother and her daughter, both "natural short sleepers" who need only six hours of sleep a night (with no evident downsides). They awake each morning around 4:30 A.M. alert and ready to start the day. The mutation appears to release the brake on production of a hormone called orexin that is linked to wakefulness.

Earlier we met some rare genetic mutants who carry the scars and injuries that accompany a pain-free existence. No doctor would recommend eliminating pain sensitivity, but such concerns wouldn't deter some hawkish politicians from fantasizing about an elite force of gene-edited unsullied.* This notion has already been raised in Congress, during one of the first hearings on CRISPR. In 2015, Jennifer Doudna was the star witness in a briefing convened by the Research & Technology subcommittee. Brad Sherman, a Democrat congressman from California, remarked that it took just six years from the development of atomic energy to the dropping of the atomic bomb. It might be unethical, he said, but some countries would jump at the chance to create "super soldiers" with enhanced courage, stamina, and strength. He asked the experts if anyone would like to suggest a timeframe to produce such super soldiers? Doudna and her fellow panelists tittered nervously, unsure how to answer an apparently serious question.[20]

Ameliorating pain would have one medical benefit in the context of late-stage cancer. But would it ever be feasible or sensible to edit embryos to provide some sort of cancer vaccination? We give teenagers an HPV vaccine to reduce their risk of cervical and other virally-caused cancers, but could that and more be engineered from birth? One intriguing idea for a genome inoculation is amplifying the number of copies of an essential tumor suppressor gene—p53, the so-called "guardian of the genome." Elephants never forget, so the saying goes; apparently, they never get cancer, either. This makes little sense, for if cancer risk is proportionate to the number of cells (and cell divisions) in an animal, then elephants should be at extreme

* From *Game of Thrones*—unsullied as in resistant to pain, not eunuchs.

risk. Yet across the animal kingdom, the odds of developing cancer show no link to body size—a conundrum known as Peto's paradox, first posed by British epidemiologist, Richard Peto.

In 2012, Vincent Lynch at the University of Chicago discovered surprisingly that the elephant genome carries a whopping twenty copies of *p53*,[21] which happens to be the most frequently mutated gene in cancer.* I've heard speculative proposals that adding a *p53* cassette (say five to ten additional copies) could provide lifelong protection against cancer. Researchers are studying the idea of boosting p53 levels as a form of genetic protection against radiation. America's new Space Force—*Maybe your purpose on this planet isn't on this planet*—won't go far unless scientists can devise a mechanism to protect astronauts from excessive, dangerous amounts of radiation (as would be endured on say a voyage to Mars). Urnov's team at IGI has received funding from the Defense Advanced Research Projects Agency to conduct CRISPR screens to identify gene variants that could help soldiers survive radiation exposure by giving them, in Urnov's words, "a molecular coat of armor."[22]

But who is to say that other genes and fanciful ideas won't prove more realistic, perhaps a suicide mechanism for cancer cells? Is that such a bad genetic modification? "Some people will want to never allow germline genome editing because they think it's bad for humanity," said Robin Lovell-Badge, a vocal critic of He Jiankui's actions. But what he says next might surprise some. "That scares me. I don't like closing and locking doors. Take global warming—we might need to modify ourselves."[23]

Gene cassettes might be normal genes or they could be custom DNA sequences. There is growing excitement around synthetic biology, in which molecular engineers design custom gene circuits that can be tested in our favorite model organisms, yeast or fruit flies or mice. Before the end of this century, we could be installing next-gen DNA circuits in the genomes of the next generation. But before we get too carried away, there's a problem. "Imagine two generations from now: Harry meets Sally," says Lander. "Harry has

* Patients who inherit mutations in *p53* have Li-Fraumeni syndrome, and are at risk of developing a variety of different cancers, indicating p53's critical role in governing cell growth in many different tissues and cell types.

inherited one circuit. Sally has inherited another clever circuit. No one has a clue what will happen when they coexist in their offspring . . . It's complicated."[24]

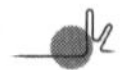

Most discussions about designer babies quickly descend into debates about intelligence and other supposedly desirable physical and behavioral traits. But these utopian fantasies overlook the daunting genetic complexity and heterogeneity of these complex traits. Although Church and others have demonstrated impressive technical virtuosity to edit hundreds of genes simultaneously, the prospect of precisely sculpting scores of specific genes in a human embryo—and achieving the desired outcome—is not feasible at present. But it won't stay that way forever. Before we can answer the question of whether we should contemplate such interventions, we first need to identify the genes required to alter human behavior or personality or cognition. That's by no means straightforward.

What if, in our brave new genetic future, CRISPR clinics decide to add height or mathematical ability or skin color or even intelligence to the menu? This is still the domain of science fiction. These are highly polygenic traits, shaped not by the large effects of solitary genes but by the combined influence of hundreds of genes. Height is a classic example: it is one of the most polygenic traits known, with variants in hundreds of genes associated with a person's stature.

Before the Human Genome Project, we naïvely thought that variants in single genes could account for major mental illnesses and complex behavioral traits. It was a classic case of looking for your lost car keys only under the lamppost. In 1988, *Nature* published a British claim for the mapping of a schizophrenia gene that didn't hold up. *Science* trumped that by publishing evidence for a "gay gene" on the X chromosome. The evidence was perilously thin, collected from fewer than fifty gay couples, and never replicated.

A quarter century later, Benjamin Neale's team at the Broad Institute performed a state-of-the-art genome-wide analysis involving 1 million DNA markers on a database of nearly 500,000 people. The results painted a vastly more complex picture of same-sex behavior, one in which gene variants explained less than half the variance in the trait. The top five gene "hits" made up less than 1 percent of the variance. Nevertheless, even highly polygenic diseases and traits may have simple genetic switches. "Just because it is polygenic doesn't mean it doesn't have a monogenetic solution," Church says. For example, stature is a very polygenic trait, but many patients with short stature can be treated with human growth hormone.

Would a musician want to ensure their child had the mysterious gift of perfect (or absolute) pitch? My father had perfect pitch, which propelled him to a successful career in the West End as musical director of hit shows including *Cabaret* and *Fiddler on the Roof*. As a boy, I'd be ushered backstage after a Saturday matinee to meet a young Judi Dench or Topol. If there's a gene for perfect pitch, I didn't inherit it.* My former colleague Alissa Poh recounted her daily experience living with absolute pitch: her car horn hovers between and E and F, her cell phone rings in A minor, while her refrigerator hums in B-flat.[25] Studies by Jane Gitschier among others support a nature *and* nurture model—the trait is manifest by inheriting an as yet unidentified gene along with early musical training.[26] But perfect pitch doesn't make a musician, nor are all great musicians born with perfect pitch.

In the next fifty to one hundred years, it might become possible to apply genome surgery on artistic or mathematical behavior. And, as Church predicts, once we can do it for a single gene, we will develop safe methods to extend this in parallel, multiplexing edits at multiple genes simultaneously.

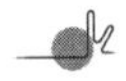

In early 2019, I was invited to attend an unusual conference at the Ditchley Estate, a quintessential English stately home reminiscent of Downton Abbey just outside Oxford, ostensibly to discuss the intersection of gene editing and artificial intelligence (AI). During World War II, Winston Churchill spent weekends there (Checkers, the official retreat of the prime minister, was too recognizable for the Luftwaffe). After a welcoming

* I was a decent singer until my voice broke. At age twelve, I sang at Covent Garden in *Carmen* as a member of the troop of street urchins belting out "Avec la garde montante" in front of the Queen Mother. I would then try and wrestle a melon off Placido Domingo in Act IV.

reception with tea and biscuits, forty of us took our seats at a long boardroom table in the old library. The first speaker to be introduced was Stephen Hsu, a theoretical physicist at Michigan State University. This seemed like a peculiar choice—until Hsu started talking.

Hsu's interest in genetics traces back to his childhood, avidly watching *Star Trek* and pondering Kirk, Khan, and the Eugenics Wars. "If I get to be one of the scientists who makes real some amazing trope from science fiction, that would be the most awesome thing in the world," Hsu told *Radiolab.*[27] Although he gravitated toward physics, Hsu remained fascinated by the link between genetics and intelligence. He was formerly an advisor to BGI's controversial Cognitive Genomics project, since aborted. Now he believes he can apply AI to the prediction of complex polygenic traits including cognitive ability. Once, when asked to give his view of a superior human intelligence, Hsu offered as an example John von Neumann, the 20th-century polymath, developer of game theory, and computer science, who was capable of total recall and a photographic memory. "In my opinion," Hsu says, "genotypes exist that correspond to phenotypes as far beyond von Neumann as he was beyond a normal human."

Hsu cofounded a PGT clinic called Genomic Prediction, located in an unremarkable office park off the New Jersey Turnpike, a short drive from Manhattan. Dressed in a T-shirt and torn genes, Nathan Treff, the company's chief medical officer, met me in his small office decorated with framed posters of Pearl Jam and Iron Man.[28] Genomic Prediction offers the usual menu of PGT services—tests for chromosomal abnormalities and genetic diseases such as CF, Tay-Sachs, and Huntington's disease. But Genomic Prediction goes further, offering couples tests for polygenic conditions including heart disease, obesity, diabetes, and short stature. Also on the menu: low cognitive ability.

Since a landmark paper from researchers at the Wellcome Trust Sanger Institute in 2007, researchers have identified thousands of gene variants that influence our risk for hundreds of complex traits.* We can't point to a

* Genome-wide association studies are performed by surveying a million SNPs on hundreds of thousands of patients and controls. The associations are graphed on an end-to-end map of human chromosomes known as a Manhattan plot for its resemblance to the jagged Big Apple skyline. The tallest peaks indicate the location of the strongest associations.

solitary genetic risk factor for type 2 diabetes or obesity, but we can confidently circle dozens or hundreds of specific DNA variants that influence our susceptibility to these and other disorders.

More recently, government-funded databases such as the UK Biobank have made available full genomic and medical data on some 500,000 (mostly European) volunteers. This allows researchers to run machine learning programs to "train" on the data, looking for genetic predictors for medical and behavioral traits. Kathiresan's group at Harvard Medical School developed polygenic risk scores (PRS) for five complex diseases including heart disease, type 2 diabetes, and breast cancer.[29] Hsu's team extended this work, but he doesn't stop at merely medical disorders. By identifying some 20,000 DNA variants that influence height, Hsu claims he can build an algorithm to calculate a PRS and predict someone's height to plus or minus one inch.[30]

Whereas most investigators are studying PRS in patients, Hsu is courting controversy by insisting he can calculate a PRS before birth in an individual embryo. Hence Genomic Prediction's menu of available polygenic risks includes short stature and low cognitive ability. Hsu scornfully dismisses criticisms, shocked that human geneticists have virtually no idea what their colleagues in livestock or corn breeding are doing.

The genetics of intelligence is a controversial and fraught issue. A recent Canadian study examined the effect of DNA deletions or insertions (copy number variants) in more than 24,000 people. The authors concluded that the number of genes that influenced intelligence was around 10,000—fully half the genes in the human genome.[31] But that hasn't deterred Hsu. Projects like the UK Biobank haven't been conducting IQ tests but have asked volunteers to list their level of education. Using that as a proxy for IQ, Hsu's team mined these data for DNA markers associated with cognitive ability. He says he can predict cognitive ability with a correlation of about 30–40 percent. In the same way that a college dean would look askance at a student who had underperformed on the SAT, Hsu says, "parents may deserve a warning if we find that an embryo has super elevated risk of intellectual disability."[32]

Championing couples' "reproductive liberty," Treff's team has already reported IVF embryo screenings for breast cancer and type 1 diabetes.[33]

Hsu says they can go further to predict which embryo would develop with low cognitive ability—that is, it carries an excess of DNA variants that are predicted to depress IQ below a clinical threshold. The company is not offering clients the option to select for an extra-intelligent embryo—not because the technology isn't there, Hsu says, just that society isn't ready for it. Suppose you want to start a family and you learn that embryo #4 was predicted to be in the top percentile of cognitive ability. What would you do? Would you rather rely on the embryologist judging the health of each embryo by shape and morphology (as happens now) or by analysis of its DNA?

Hsu offers another chilling scenario: What if the Singapore government, as an example, were to invite his company to alert parents if one of their embryos is likely to be well above average intelligence? Hsu can imagine a situation where "I guess in Singapore it is okay but we don't feel like Americans are ready for it."[34] If and when Americans decide they *are* ready for it, I expect Hsu would be happy to oblige. Editing embryos for intelligence remains a fantasy, but ranking embryos via PRS doesn't look so far away.[35]

Hsu's critics argue that selecting embryos based on PRS is risky business, rife with statistical ambiguity, geographic bias, and ethical fragility. "It might be better than a horoscope, but we don't know—but I don't think that Genomic Prediction does either," observes Hank Greely.[36] An Israeli study concluded that the gain afforded by PRS calculations—the difference between the "top ranked" embryo from the average—was about 2.5 cm for height and 2.5 IQ points.[37] That's hardly enough to justify the expense and hassle of IVF. "The prediction is not good enough to individually identify an embryo with a certain characteristic," Kathiresan told me. "There's not a 1:1 correlation between score and outcome. It's a probabilistic model. It's not appropriate in my view to use for embryo selection."[38]

Even if you disagree, geneticist Laura Hercher warns that Hsu's company can't guarantee giving a couple a child that is not going to get sick and die. "If you are buying into that fantasy, you're going to be angry. I hope that it is at [Genomic Prediction] and not at your kid," she says.[39]

Assuming that CRISPR or base or prime editing is safe (or no worse than the mutation rate caused by background radiation) is there any fundamental reason why it should not be allowed on human embryos for whatever applications we deem to be appropriate? In 1997, at the halfway point of the Human Genome Project, UNESCO declared the human genome to be a priceless heirloom. "The human genome underlies the fundamental unity of all members of the human family, as well as the recognition of their inherent dignity and diversity. In a symbolic sense, it is the heritage of humanity."[40]

It's a splendid line, but is the human genome really the heritage of humanity? The sacrosanct property of humankind, to be preserved and protected like a priceless masterpiece? Look but please don't touch? "It seems the equivalent of The Ark of the Covenant," Greely declared. And, as anyone who saw the Indiana Jones film *Raiders of the Lost Ark* knows, "it cannot be allowed fall into the wrong hands."[41] The human genome belongs to all of us, argues Françoise Baylis, but there is no pristine genome exhibit we can put on display as the perfect sequence of the genome. What we have instead are 7.5 billion genomes, all iterations of what went before and their future descendants.

What, then, was the Human Genome Project if not the definitive textbook sequence? The fabled reference genome that President Clinton announced in June 2000 was a patchwork quilt with a dozen anonymous contributors. A decade later, major advances in DNA sequencing technologies made it feasible to have your own genome decoded. In 2010, I attended a conference at MIT where the first two dozen genome pioneers gathered under one roof, including Jim Watson, Harvard's Henry "Skip" Gates, and George Church.[42] A few notables, including Craig Venter and Black Sabbath's Ozzy Osbourne, were absent, but I sensed that this would be the last time that (almost) every person sequenced on the planet was in one room. But nobody was proposing to anoint any individual genome to represent the human species. The genome of a Beckham or Beyoncé is littered with mutations just like yours and mine.

While there is nothing sacrosanct about the human genome, there is widespread opposition to the notion of tampering with the germline, knitting by hand a permanent sequence alteration that would be passed on to future generations. But in a practical sense, editing a human embryo is a much safer proposition than treating a child or adult. Church argues that germline editing offers three intrinsic advantages. First, it is more effective than other delivery systems at reaching all cells in the body. Second, after administering the edit, every future child and descendant would receive the edit free rather than costing millions of dollars for their own somatic gene therapy. And third, germline editing goes through a single cell, whereas somatic therapies impact millions of cells—assuming we can sort out delivering genes to the brain or other hard-to-reach organs—any one of which could become cancerous.[43] "Somatic gene therapy has been hopeless as a therapy, because you've got to get the gene to billions of cells," says Savulescu.[44]

There is no inviolate reason why we should not contemplate germline editing, and indeed there are arguments in favor from a medical and economic standpoint. But that still doesn't answer the question of why or when? "Proceed with caution" is a common refrain whenever a new genetics technology looks poised to transform medicine and assisted reproduction. It was the title of Neil Holtzman's book in the 1980s warning of the potential dangers of prenatal DNA diagnosis. The *Lancet* editor Richard Horton echoed the phrase shortly before the birth of the CRISPR babies.[45]

That's reasonable, but what if deliberately refusing to alter the human genome affects the future just as if we allow germline intervention? "If it ever became possible to eliminate, say, the gene that causes cystic fibrosis, not then to do so would condemn future generations unnecessarily to suffer from a wretched condition," writes Kenan Malik. "There is nothing ethically superior in leaving things be if it is possible to change them for the better."[46]

What if we wanted to correct a devastating gene mutation? Huntington's disease (HD) only affects 1 in 50,000 people, but it is incurable. Children of HD patients have a 50:50 chance of inheriting the disease gene. In the Ken Burns documentary *The Gene*, Jenny Allen makes the fraught decision to take a genetic test that will reveal her destiny. Her mother and two of her siblings have HD. As her doctor reveals that she did not inherit the defective gene, Jenny bursts into tears, waves of joy and relief mingled with survivor's guilt. One day, germline editing could permanently fix this mutation, snipping out the faulty sequence to restore a functioning version of the HD gene. "Why should anyone object if the genome of the 50,000th person is coaxed back to the normal conventional sequence?" Greely asks reasonably.[47]

Savulescu goes further, arguing that parents have a downright obligation to maximize the potential of their children. Eradicating genetic disease is not bad in itself, he says, but our technology won't stop with reverting to healthy forms of genes. We could introduce novel variations that have not been encountered in our species before.

In the summer of 2015, after the initial furor around CRISPR and human embryos, Harvard professor Steven Pinker wrote a strident op-ed in the *Boston Globe*. In a world that promised a biomedical bonanza to improve people's health and longevity, Pinker argued, "the primary moral goal for today's bioethics can be summarized in a single sentence. Get out of the way."[48] While individuals must be protected from harm, a "truly ethical bioethics" should not hold back research in red tape or moratoria, nor should it sow panic about potential future harms or bandy about perverse analogies with Nazi atrocities or science-fiction dystopias like—you guessed it—*Brave New World* or Andrew Niccol's sci-fi film *Gattaca*.

"When science moves faster than moral understanding," Harvard philosopher Michael Sandel wrote in 2004, "men and women struggle to articulate their own unease." The genomic revolution has induced "a kind of moral vertigo."[49] That unease has been triggered numerous times before and after the genetic engineering revolution—the structure of the double helix, the solution of the genetic code, the recombinant DNA revolution, prenatal genetic diagnosis, embryonic stem cells, and the cloning of Dolly. "Test tube baby" was an epithet in many circles but five million IVF babies are an effective riposte to critics of assisted reproductive technology.

With CRIPSR, history is repeating itself, only this time we have the lives of three genetically manipulated human beings weighing on our collective conscience. Lulu, Nana, and the third CRISPR baby did not ask to be genetically modified. "We should all hope and pray that these two little girls are okay," Francis Collins said. "They did nothing to bring this trouble down upon them. They certainly didn't give their consent."[50] True enough, but then again, no embryo or person has ever given informed consent over the circumstances of their conception or the mash-up of genetic material that accompanied fertilization.

Long before CRISPR babies, some argued that engineering genetic enhancements for cognitive or musical talent or athletic ability would steer such children toward a particular destiny, depriving them of free will. But as Sandel noted, this implies that children are naturally free to choose their fate. "None of us chooses his genetic inheritance," Sandel wrote. The alternative to a genetically enhanced child "is not one whose future is unbound by particular talents but one at the mercy of the genetic lottery."[51] Gene editing poses a threat to human dignity. The drive for perfection—mastery of a sport, instrument, or cognitive skill—obscures humanity's achievements. Sandel says the sin of enhancement would be the evasion of training and hard work. He quotes William May, who said, "to appreciate kids as gifts is to accept them as they come."

In the world of sports, fair play is a forgotten virtue as some professional and aspiring athletes take steroids, growth hormone, testosterone, or erythropoietin to steal an advantage on the competition. Governing bodies can detect traces of chemical doping, but genome editing opens up a Pandora's box. "There will be others seeking to fill [JK's'] research shoes and the possibility of hidden funding to attempt to create the perfect athlete," warned Lord Colin Moynihan, speaking in the House of Lords. "Gene editing clearly has huge benefits, such as relieving the burden of heritable diseases. However, it has no place in the sports arms race if we are to protect the integrity of competitive sport."[52]

A bigger concern surrounding genome editing is that it would exacerbate social divisions and inequality. How will access to germline editing be based on need rather than means? The first approved gene therapies are setting record-breaking prices, such as Novartis's staggering $2 million price for Zolgensma. As companies hike the price of generic drugs, some pharma executives, spouting fiduciary responsibility, seem more interested in putting their shareholders ahead of their patients. Genome editing companies developing somatic therapies aren't going to give these precious medicines away as they seek to recoup the vast sums invested in R&D and manufacturing. By contrast, a company developing a germline therapy might be able to offer a more affordable procedure, as the CRISPR machinery would only be administered to a single cell (or gamete).

Equal access to 21st-century medicines is a major concern, says Church, who like Lander, thinks the alternative to genetic therapy is actually genetic counseling. With the cost of genome sequencing dropping toward a paltry $100, "everybody could now get their genome sequenced and avoid a huge fraction of these expensive orphan drugs and gene therapies by genetic counseling."[53] The last thing Church wants is a have-and-have-not society. "When people talk about the ethics of CRISPR, 90 percent of it should be, and probably is, about equal distribution of expensive technology."

In some circles there is a revulsion at the prospect of man-made genetic alterations muddying the gene pool and decreasing human diversity. Would society function better if everyone received an assist on their IQ score? In most countries, high IQ correlates with wealth, health, and overall well-being. But there are ways of addressing that imbalance without resorting to genome surgery in the womb. Why does intelligence, wealth, and job status determine life outcomes so much in the first place? "That is something which should be addressed, rather than reified," says British philosopher Gulzaar Barn.[54]

Physical attractiveness is rarely a disability in life. A CRISPR clinic offering facial prediction would further the notion that a woman's value is derived from her appearance, says Barn, accentuating division and privilege in society. In a world with increasing wealth disparity, this procedure would trickle down from the wealthy. Early efforts to predict facial features from DNA by artist Heather Dewey-Hagborg and Craig Venter[55] met criticism,[56] but they too will improve. In a fair society, everyone should have access (if they wish) to these sorts of genetic endowments, although that's a pipe dream. Barn argues we should be addressing societies' institutions and structures so that life outcomes aren't so dictated by these factors. "We need to consider whether it is right that a small number of unrepresentative, rich funders and scientists are able to implement technologies that have the ability to radically alter society in unprecedented ways."

Those opposed to germline editing argue that we would lose diversity, reinforcing the stigma and discrimination faced by those with disabilities or other genetic conditions. "There is value in human fragility that would be lost if disabilities were made to disappear," says the *Lancet*'s Horton.[57] The eugenic concern about "weeding out" disabilities applies more urgently to PGT, where hundreds of thousands of embryos are screened each year, while sub-optimal embryos are consigned to a state of suspended animation. Western countries routinely screen for Down syndrome (trisomy 21) and other trisomies, the incidence of which correlates with increasing maternal age. In Iceland and Denmark, the number of babies born with Down syndrome annually has been reduced to single digits. Columnist George Will, who's eldest son has Down syndrome, accused Iceland of implementing a "final solution" to the disorder.[58] Meanwhile, some states in the U.S. have passed laws making it illegal to prevent abortion of fetuses diagnosed with trisomy 21.

Genome editing won't change society that much in the near future, but the prospect of genetic enhancement would only accentuate societal differences instead of trying to stem such inequalities. As Barn says, we're all prone to illness and in need of help from others at some point in our lives, which warrants greater investment in public services and support for the less fortunate. "Retaining a more empathetic approach, predicated on the belief that every human life is valuable, is crucial for ensuring a well-functioning society that works for the benefit of all." Philosopher Mike Parker says the best possible life is not necessarily one in which all goes well. Human flourishing involves aspects of both strength and weakness.

Genome editing stokes fears among many disabled people of "society's fear of the deviant." This sense of ableism is "denying us our personhood and our right to exist because we don't fit society's ideals," says Rebecca Cokley, a disability advocate who served in the Obama Administration. Cokley has a form of dwarfism called achondroplasia. She sees her condition as a "rich and diverse culture," a culture she wants to pass onto her children. "We should have that right," she wrote in an op-ed in 2017 in the *Washington Post.* It was titled "Please Don't Edit Me Out."[59]

Ethan Weiss, a physician-scientist at University of California San Francisco, and his wife nicknamed their daughter "Billy Idol" for her fluorescent blond hair. Doctors eventually diagnosed Ruthie with albinism, caused by a mutation in the *OCA2* gene. "I did imagine that genetic engineering could someday help kids who were diagnosed right after birth," Weiss wrote. "But I focused instead on just loving and supporting the child I had, and not the one I wished I had."[60] Tempting though technologies like germline editing might sound, their usage raises concerns that "the world will be less kind, less compassionate, less patient, when or if there are no more children like Ruthie." Weiss insisted that he and his wife were better parents for raising their daughter, but more importantly, "we believe the world is a better place for having kids like Ruthie in it, and we want the world to think hard about whether it really wants to go down a path of engineering a world where there are no Ruthies."

Discussions about whether we dare to place our own designs on the human genetic code inevitably paint a picture of a slippery slope. "Slopes are only slippery if they catch us unaware and we have strayed on to them inadequately equipped," wrote the British philosopher John Harris.[61]

Since the first inkling of gene therapy in the early '70s, we have tended to regard somatic gene therapy as noble and idealistic, life-saving, whereas germline therapy (or editing) is dangerous and immoral. Those who wanted to push for enhancement did so under the banner of eugenics, looking to perfect the human species. But for the past fifty years, as we look down on the proverbial slippery slope, it is as if there is a giant wall halfway down the hill with no back door or underground tunnel. That barrier prevents us sliding to a dystopian "fully synthesized natural world where nothing exists outside human intentionality." So says John Evans, a sociologist at the University of California San Diego, who argues that the somatic/germline distinction is on its last legs. The mid-slope barrier worked well in a 20th-century era where attempts to cure monogenic genetic diseases were in stark contrast to eugenic fantasies of an Aryan race. But that barrier has dissolved as the new genomic century has produced a much deeper understanding of genes and diseases. The debate has evolved from changing the species to changing an individual.

Evans contemplates several other types of barrier as we contemplate our descent. One is a safety barrier, which is essentially in the same position as the germline barrier. This is actually more of a speed bump, which slides down the slope as our skill and precision improves. Another is the biological reality barrier, which says that it will be impossible to edit for perfect pitch or higher intelligence because the genetics are too complex. "This is a loser move," Evans says. In the early '80s, the eminent geneticist Arno Motulsky waved off discussions of germline editing because he insisted it would not be possible for fifty to three hundred years. Finally, Evans says there is the Boundary of Humanity barrier, which demarcates any natural human gene variant from a novel mutation that no human has ever had.

The ethical debate about germline editing will rage on for years if not decades. The reports from august committees convened by the National Academies of Science and the WHO will be valuable, but by no means the last word. I'm not pushing germline editing, nor am I unequivocally opposed on moral, religious, or scientific grounds. I suspect there will come a time when the pros will outweigh the cons, at least in some situations.

Meanwhile, other areas of medicine and surgery are advancing relentlessly. In 2005, a French woman named Isabelle Dinoire who was disfigured when she was bitten by Tanya, her golden retriever, became the first person to undergo a face transplant (the donor had committed suicide). Despite the medical and ethical controversies, dozens of facial transplants have been performed subsequently. Surgeons can correct birth defects such as spina bifida in utero, while the fetus is still in the womb. DNA surgery seems to be the next logical frontier.

Clinical geneticist Helen O'Neill, at University College London, concludes: "No technology is perfect—not IVF nor genome editing—but when combining these and applying them to the most flawed of systems, human biology, we may ask ourselves 'When will good ever be good enough?'"[62]

CHAPTER 24
BASES LOADED

CRISPR genome editing is not yet a decade old but, as I've tried to show in this book, it is poised to transform myriad areas of science, medicine, and agriculture. But there are many scientific, regulatory, and ethical challenges ahead. One setback in a clinical trial could precipitate another decade in the dark ages, as happened to gene therapy. Tim Hunt, Editas's former head of corporate affairs, is refreshingly honest about the business challenges ahead for "cash-eating machines" like Editas commercializing somatic genome editing. "We've raised $500–600 million. We'll need $1–1.5 *billion* before we have a product on the market," he said in early 2020. "CRISPR is often described as fast, cheap, and easy. But making a medicine is not fast, not cheap, and not easy. It's a long journey."[1]

From building a company, managing preclinical research and clinical trials, bankrolling manufacture, enhancing delivery, and navigating regulatory red tape, the road to commercial success is tortuous. And in some cases, genome editing is targeting a very niche market. The numbers of patients receiving some therapies, such as those with orphan diseases, are very small. All of these factors drive up the price of potentially life-saving drugs. Ross Wilson and Dana Carroll, exhorted genome editing companies and regulators "to accept the challenge to make genome editing therapeutics affordable and accessible, which would represent a massive contribution to global health justice."[2]

There may come a time when it makes sense to perform germline editing, but not now, not yet. Maybe in a decade or two, we might consider heritable genome editing to be technically safe, ethically sound, medically justified, and publicly supported. I believe that day will come eventually. Perhaps by 2032, the centennial of *Brave New World*. Or 2053, the one hundredth anniversary of the double helix. Or 2078, when Louise Brown turns one hundred. Or 2100, a century since we first cracked the sequence of the human genome.

No drug comes without side effects; no surgery is completely safe. Genome editing won't be any different, but scientists are well advanced in overcoming those issues. What is ethically appropriate will depend on a deep discussion about values and beliefs involving a host of stakeholders, not just scientists and physicians. Medically justified will be rare for genetic disorders given the prevalence of PGT, and other applications appear frivolous or fictional. Public and government approval is also in question: CRISPR clinics could become an extension of the burgeoning medical tourism industry. Perhaps some nations will sanction genome editing for their citizens, even if the rest of the world isn't ready.

Scientists and physicians, ethicists and lawyers, sociologists and politicians have been debating the merits and dangers of CRISPR since 2015. But every day, babies, children, and adults are being diagnosed with deadly genetic disorders. For these patients and their families, genome editing offers a powerful life-saving medical technology with no bounds. In late 2019, after NIH director Francis Collins gave a public address in a Washington, DC, hotel, a woman in the audience named Neena Nizar navigated her wheelchair to the microphone. Nizar is president of the Jansen's Foundation and one of only two dozen patients in the world (and two adults in the United States) with that rare genetic disease. Patients like Nizar suffer weakened bones and cartilage, and endure multiple rods, pins, and clamps to hold their brittle bones together.

Nizar challenged Collins, demanding to know why more wasn't being done to help patients like her suffering one of thousands of genetic disorders. "Science and medicine have a responsibility to try to find answers" to rare diseases, Collins assured her. But the need for germline editing, for Jansen's or a plethora of other conditions, was "uncompelling to me," because of the availability of PGT to create and test embryos. "So why not just implant those? We don't need a gene-editing solution."[3]

Afterward, Nizar smiled ruefully when I asked her if that was answer she was looking for. Both her sons have inherited the same disease. "It's easy to get on your high horse when you're not in our position. If editing an IVF embryo is the best option to mitigate the pain that a child would otherwise suffer, then give us the choice," she said. "*Let* them say we're playing God."[4] Debates over ethics and eugenics, hyperagency and human flourishing are all well and good. But patients just want a chance for a normal, healthy life. If modern science offers them hope, who is going to dare take that away?

Indeed, there are signs of hope as we enter a new era of hyper-personalized medicine, where patients with ultra-rare gene mutations can receive a customized therapy with FDA approval. At Boston Children's Hospital, pediatric neurologist Tim Yu has developed bespoke drugs for children with Batten disease[5] and ataxia telangiectasia.[6] Yu has also advised a group including Yale University's Monkol Lek to devise a custom CRISPR therapy for Terry Horgan, a 24-year-old patient with Duchenne muscular dystrophy. Horgan's mutation is in the first exon of the dystrophin gene, not part of the group of mutations addressed by drugs currently on the market or in the clinic. Terry's brother Rich is the founder of Cure Rare Disease, a nonprofit spearheading hyper-personalized medicine. Encouraging, but each of these personalized drugs requires about $1–2 million at a minimum.

Like many researchers in the CRISPR spotlight, Jennifer Doudna receives regular emails from patients and their family members, desperately looking for hope. One message (shared publicly by Urnov) was written by a thirty-six-year-old woman whose cheerful salutation—"Hi Dr. Doudna"—belied the plea that followed. "Time is quickly running out for me." She explained she has a single-nucleotide mutation that causes a severe lethal disease. Fixing it should be easy. "I've watched the continued development and the work you've done," she continued. She'd read a magazine profile of Doudna.

"I'm a very good candidate for CRISPR trials. I would be more than happy to be a study participant."[7]

Deftly handling emails from desperate patients is just one of many new challenges that she has had to take on since the world changed in 2012. She is famous now, one of the most recognizable scientists in the world. Time with her students and postdocs is increasingly precious. No longer just the most valuable player in the Doudna lab, she's also the team president, coach, and general manager. She dons a multitude of hats—researcher, teacher, grant writer, mentor, administrator, accountant, evangelist, ethicist, entrepreneur, author, commentator, advisor, and public speaker—sometimes all in the same day. A team of assistants helps her budget and manage lab operations and vet media requests. Every hour of Doudna's day is charted on a laminated daily calendar, juggling group meetings, advisory boards, budget planning, and interviewing the next wave of students and postdocs who will further advance the CRISPR revolution. After another frantic day, she texts her husband to schedule the drive home.

In late 2019, Doudna flew to Washington, DC, for an especially important appointment. Every five to seven years, HHMI summons its 250 investigators to a closed-door review in which they must highlight their research accomplishments and outline their future plans before a panel of a dozen or more members of the scientific elite. After answering questions, a limousine waits to drive her back to the airport while the advisors vote whether to extend support (more than $1 million/year) for another term. For most investigators, it's the most nerve-wracking ninety minutes of their professional lives. Doudna felt a few butterflies as she entered the room but the thought of the institute severing ties with one of the most celebrated scientists in the world was unthinkable.

A few months later, Doudna faced a new and unexpected public health crisis from of all things a virus: the novel coronavirus.[8] As she and her colleagues prepared to shutter their labs for the pandemic, Doudna felt an overwhelming responsibility to help the local community. On March 13, she addressed her IGI colleagues with pronounced fire and emotion and told them it was time to step up. "Folks, I have come to the conclusion that the IGI must rise and take on this pandemic," she said.[9] With a gaping shortfall in COVID-19 testing capacity, Doudna and colleagues decided to turn a 2,500-square-foot space into a COVID-19 test center. The response to a call for volunteers was stunning: hundreds of people volunteered to help in any way they could.

In less than three weeks, dozens of volunteers from Berkeley and industry partners fitted out a new genetic testing lab on the first floor of the IGI capable of running more than 1,000 diagnostics tests every twenty-four hours.[10] A pair of Doudna's top postdocs, Jennifer Hamilton and Lin Shiao, stepped up to become the lab's technical directors. For the first time, as Megan Molteni observed, Shiao felt that "all those years spent moving tiny bits of liquid around might actually directly change someone's life for the better."[11] Urnov rallied the troops with a quote from *Lord of the Rings*:

> "I wish it need not have happened in my time," said Frodo.
> "So do I," said Gandalf, "and so do all who live to see such times.
> But that is not for them to decide. All we have to decide is what
> to do with the time that is given us."

On April 6, 2020, Dori Tieu, a fire prevention inspector from the Berkeley Fire Department, delivered the first samples on ice in a large polystyrene box to a nervous Urnov. For a few weeks or maybe months, CRISPR could wait.

Eighteen months before the pandemic hit pause on his lab's activities, Feng Zhang was asked if there was a chance that his CRISPR editing tour de force in 2013 could end up being the biggest discovery of his career. "I hope not!" Zhang replied, aghast at the prospect that he had peaked already. "It's a very luck position to be in," he continued. "There are many, many more problems we need to solve."[12] From a young age, his parents had told him: "You should make yourself useful." How could he make a difference against COVID-19?

In his meager spare time, Zhang personally set about adapting his SHERLOCK diagnostic method to develop a relatively simple diagnostic test for COVID-19. The test received emergency FDA approval—the first for a CRISPR diagnostic—in May 2020.[13] There's even an at-home version called STOPCovid,* designed to work in about an about for less than $10 (not counting the outlay for a sous vide to substitute for a lab water bath).[14] And he formed a team with the CEO of Pinterest, Ben Silbermann, that took just three weeks to develop and release an app called How We Feel for people to track their personal health and symptoms in real time.[15] Zhang and Silbermann have been friends since they met in high school in Des Moines, Iowa, a lifetime ago.

Other gene editing luminaries also retooled to tackle the COVID-19 crisis. In March 2020, a venture capitalist in Boston named Tom Cahill, who had launched a $125 million fund backed by a small group of billionaires including Peter Thiel and Stephen Pagliuca, co-owner of the Boston Celtics, organized a conference call to discuss COVID-19. Word spread fast. Cahill knew something was amiss when he couldn't dial in because the call was over-subscribed with hundreds of listeners. "There was a sense of desperation among these masters of the universe" about the threat posed by the virus to their families and businesses, said Rob Copeland, who broke the story.[16]

Cahill decided to convene a 21st century Manhattan Project, the Justice League of scientists to vanquish the virus. Team captain was Stuart Schreiber, a distinguished chemist at Harvard, who created the Scientists to Stop COVID-19. The dozen experts included his colleague David Liu and a Nobel laureate, Michael Rosbash, who said he was the least qualified person on the team. Their report on the best prospects to defeat the virus from a pool of two hundred candidates[17] was sent via some well-placed connections to the White House. Schreiber even attached superhero names to his group, including Batman (Ben Cravatt) and Wonder Woman (Akiko Iwasaki). Iron Man is part of a different superhero universe, so Liu is Cyborg—half man, half machine, genius intellect.

Alas, humans don't have their own version of CRISPR superpower to cut down the coronavirus—but perhaps one day they could receive it. At Stanford, Stanley Qi leads an effort to deploy CRISPR-Cas13 in a method called PAC-MAN, designing CRISPR guide RNAs to seek out and destroy coronavirus RNA sequences.[18] After all, why wouldn't the most popular, versatile tool in the biotechnology arsenal be used to vanquish the virus?

---

* "STOP" stands for SHERLOCK Testing in One Pot.

The irony of this emergency call to arms was not lost on Doudna, Zhang, or any of their colleagues. CRISPR evolved to vanquish a particular group of viruses, the bacteriophages. Now a particularly malevolent virus was spreading around the world by feasting on the airways of a different host. "Bacteria have been dealing with viruses forever," Doudna observed. "They've had to come up with creative ways to fight them. And now here we are, humans, in a pandemic facing this challenge."[19]

As we contemplate the future of genome editing, I can't help but think of the remarkable progress geneticists have made over the past fifty to seventy-five years, unraveling the secrets of the gene and the genome like the Cas enzyme unzipping the double helix. Francis Crick and Jim Watson's classic 1953 letter to *Nature* consisted of eight hundred words and one diagram—a beautiful, elegant pencil drawing of the double helix courtesy of Crick's wife Odile. CRISPR has given us the means to modify the DNA code as easily (almost) as a deft flick of Odile's pencil eraser.

Odile Crick was a professional painter with a penchant for nudes. She wasn't as enamored with the double helix breakthrough as her husband. "You were always coming home and saying things like that, so naturally I thought nothing of it," she recalled. Nevertheless, her double helix became not only the most famous scientific drawing of the 20th century but also the universal symbol of mankind's quest to understand, repair, manipulate, and control the code of life—to read, write, and edit DNA.

Odile only drew one other scientific illustration in her life.[20] Her granddaughter Kindra, an accomplished artist in her own right, told me where to look. It appeared in a book by neuroscientist Christof Koch called *The Quest for Consciousness*.[21] The simple drawing was of a woman with shoulder-length hair in a short dark dress running, a static picture depicting motion. To where exactly is left to our imagination.

CRISPR is moving faster than society can keep up. To where is up to all of us.

ACKNOWLEDGMENTS

But for a friendly chat over a pint at the Cold Spring Harbor Laboratory bar, this book might never have happened. My former *Nature* colleague Alex Gann regaled me about his plans to write a book following receipt of a fellowship from an unexpected source. I took note and eighteen months later, submitted a proposal to the John Simon Guggenheim Memorial Foundation to write a book about CRISPR. In 2017, I was delighted to receive a Guggenheim fellowship in science writing, which supplied the initial validation and impetus to pull this book together.

Profound thanks to my colleagues Mary Ann Liebert and Marianne Russell for their unwavering support in launching *The CRISPR Journal* in 2018. This wonderful opportunity propelled me into the CRISPR community and cover many key events and meetings, including the 2018 summit in Hong Kong, which proved a turning point in the story. Thanks also to my colleagues Bill Levine, Sophie Reisz, John Sterling, Chris Anderson, and the rest of the Liebert team.

I've leaned heavily on the superb reporting of a large group of science writers and journalists. They include Antonio Regalado, Sharon Begley, Jon Cohen, Ryan Cross, David Cyranoski, Lisa Jarvis, Julianna LeMieux, Marilynn Marchione, Amy Maxmen, Megan Molteni, Emily Mullin, Michael Specter, Rob Stein, Ed Yong, Sarah Zhang, and Carl Zimmer. And special thanks to Walter Isaacson for his gratifying encouragement.

Several actors in this drama deserve special thanks. Fyodor Urnov's insights, bons mots, and Russian proverbs are sprinkled throughout this book. He's forgotten more about genome editing than I'll ever know, and in the family tradition, really should write a book. Rodolphe Barrangou, with whom I've worked closely for the past three years; he has been a superb ambassador for CRISPR, and supplied me with numerous insights and opinions, not all of which, sadly, could be included. Samira Kiani, Nicholas Shadid, and Cody Sheehy supplied fascinating insights about germline editing and research in China. Kiran Musunuru generously shared a copy of his book before publication. And Jacob Sherkow continues to be the consummate guide to the ongoing patent drama.

Thanks also to Dana Carroll, Emmanuelle Charpentier, George Church, Le Cong, Kevin Esvelt, Ryan Ferrell, Nicole Gaudelli, Michael Gilmore, Philippe Horvath, Martin Jínek, Alexis Komor, David Liu, Steve Lombardi, Luciano Marraffini, Francisco Mojica, Ann Ran, Virginijus Siksnys, Erik Sontheimer, Ross Wilson, and Andrew Wood, for their expertise. I am in awe of their talent and tip my hat to all the heroes, sung and unsung, of the CRISPR revolution. (The full interviews with some of these individuals can be found in my *Guidepost* podcast series, available on all popular podcast platforms.)

Laurie Goodman and Bette Phimister critiqued drafts of the manuscript with unseemly satisfaction, and T.J. Cradick and Tim Hunt provided much valuable feedback. Thanks to Oona Snoyenobos-West, a fiercely talented microbiologist, for insisting that I write a book on CRISPR (and reminding me that microbes aren't "primitive"!). And thanks to Ardy Arianpour, Judy Chen, Pauline Parry, and Amanda Wren for their invaluable encouragement. Any errors that remain (editing or otherwise) are mine alone.

I am perpetually grateful to my agent Jennifer Gates at Aevitas Creative, who steered me to Jessica Case and Pegasus. Jessica's been a true partner in this project and as patient and collegial an editor as I could have wished for. Thanks also to the fabulous Maria Fernandez for typesetting and the rest of the Pegasus team, including Drew Wheeler (copyediting) and Daniel O'Connor (proofreading). Mon Oo Yee supplied her trademark artistic flourish for the cover. Thanks also to the kind folks who supplied photos, including Adam Bolt, Elsie Chen, Eriona Hysolli, Dana Korsen, Lee McGuire, and Hiroshi Nishimasu.

Finally, this book is dedicated to my friend Michael White, who died in Perth, Australia, in 2018. Without Mike, I wouldn't have made my first record (with Colour Me Pop) or written my first book. Finally, my love and gratitude to my children and my wife Susan, without whose love and support (and bonus proof-reading) this book would never have happened.

FURTHER READING

Misha Angrist. *Here is a Human Being: At the Dawn of Personal Genomics.* New York: Harper, 2010.

Margaret Attwood. *Oryx and Crake.* New York: Doubleday, 2003.

Philip Ball. *Unnatural: The Heretical Idea of Making People.* London: Bodley Head, 2011.

———. *How to Grow a Human: Adventures in How We Are Made and Who We Are.* Chicago: University of Chicago Press, 2019.

Francoise Baylis. *Altered Inheritance: CRISPR and the Ethics of Human Genome Editing.* Cambridge, MA: Harvard University Press, 2019.

George Church and Ed Regis. *Regenesis: How Synthetic Biology Will Reinvent Nature and Ourselves.* New York: Basic Books, 2012.

Kevin Davies. *Cracking the Genome: Inside the Race to Unlock Human DNA.* New York: Free Press, 2001.

———. *The $1,000 Genome: The Revolution in DNA Sequencing and the New Era of Personalized Medicine.* New York: Free Press, 2010.

Jennifer A. Doudna and Samuel H. Sternberg. *A Crack in Creation: Gene Editing and the Unthinkable Power to Control Evolution.* Boston: Houghton Mifflin Harcourt, 2017.

John H. Evans. *The Human Gene Editing Debate.* New York: Oxford University Press, 2020.

Mei Fong. *One Child: The Story of China's Most Radical Experiment.* London: Oneworld Publications, 2016.

Jonathan Glover. *What Sort of People Should There Be?* London: Pelican Books, 1984.

Henry T. Greely. *The End of Sex and the Future of Human Reproduction.* Cambridge, MA: Harvard University Press, 2016.

———. *CRISPR People: The Science and Ethics of Editing Humans.* Cambridge, MA: MIT Press, 2021.

Robin Marantz Henig. *Pandora's Baby: How the First Test Tube Babies Sparked the Reproductive Revolution.* Cold Spring Harbor, NY: Cold Spring Harbor Laboratory Press, 2006.

Susan Hockfield. *The Age of Living Machines: How Biology Will Build the Next Technology Revolution.* New York: W. W. Norton, 2019.

J. Benjamin Hurlbut. *Experiments in Democracy: Human Embryo Research and the Politics of Bioethics.* New York: Columbia University Press, 2017.

Aldous Huxley. *Brave New World.* New York: Harper, 2017. [Originally published in 1932]

Steve Jones. *The Language of Genes: Solving the Mysteries of Our Genetic Past, Present and Future.* New York: Anchor Books, 1994.

Horace Freeland Judson. *The Eighth Day of Creation: The Makers of the Revolution in Biology.* Cold Spring Harbor, NY: Cold Spring Harbor Laboratory Press, 1996.

Sam Kean. *The Violinist's Thumb: And Other Lost Tales of Love, War, and Genius, as Written by Our Genetic Code.* New York: Little, Brown and Company, 2012.

Daniel J. Kevles. *In the Name of Eugenics: Genetics and the Uses of Human Heredity.* New York: Knopf, 1985.

Paul Knoepfler. *GMO Sapiens: The Life-Changing Science of Designer Babies.* Hackensack: World Scientific, 2015.

Dan Koeppel. *Banana: The Fate of the Fruit That Changed the World.* New York: Hudson St Press, 2007.

Jim Kozubek. *Modern Prometheus: Editing the Human Genome with Crispr-Cas9.* New York: Cambridge University Press, 2016.

Ricki Lewis. *The Forever Fix: Gene Therapy and the Boy Who Saved It.* New York: St. Martin's Press, 2012.

Peter Little. *Genetic Destinies.* New York: Oxford University Press, 2002.

Mark Lynas. *The Seeds of Science: Why We Got It So Wrong On GMOs*. New York: Bloomsbury Sigma, 2018.

Jeff Lyon and Peter Gorner. *Altered Fates: Gene Therapy and the Retooling of Human Life*. New York: W. W. Norton, 1995.

Kerry Lynn Macintosh. *Enhanced Beings: Human Germline Modification and the Law*. New York: Cambridge University Press, 2018.

John Maddox. *What Remains to Be Discovered: Mapping the Secrets of the Universe, the Origins of Life, and the Future of the Human Race*. New York: Free Press, 1998.

Charles C. Mann. *The Wizard and the Prophet: Two Groundbreaking Scientists and Their Conflicting Visions of the Future of Our Planet*. New York: Picador, 2018.

Jamie Metzl. *Hacking Darwin: Genetic Engineering and the Future of Humanity*. Chicago: Sourcebooks, 2019.

Ben Mezrich. *Woolly: The True Story of the Quest to Revive One of History's Most Iconic Extinct Creatures*. New York: Atria Books, 2017.

Siddhartha Mukherjee. *The Gene: An Intimate History*. New York: Knopf, 2017.

Kiran Musunuru. *The CRISPR Generation: The Story of the World's First Gene-Edited Babies*. Pennsauken, NJ: BookBaby, 2019.

Erik Parens and Josephine Johnston, eds. *Human Flourishing in an Age of Gene Editing*. New York: Oxford University Press, 2019.

Philip R. Reilly. *Abraham Lincoln's DNA and Other Adventures in Genetics*. Cold Spring Harbor, NY: Cold Spring Harbor Lab Pres, 2000.

Matt Ridley. *Genome: An Autobiography of a Species in 23 Chapters*. New York: HarperCollins, 2000.

———. *How Innovation Works: And Why It Flourishes in Freedom*. New York: Harper, 2020.

Michael J. Sandel. *The Case against Perfection: Ethics in the Age of Genetic Engineering*. Cambridge: Belknap Press, 2009.

Beth Shapiro. *How to Clone a Mammoth: The Science of De-Extinction*. Princeton, NJ: Princeton University Press, 2015.

Lee M. Silver. *Remaking Eden: How Genetic Engineering and Cloning Will Transform the American Family*. New York: Ecco, 2007.

Gregory Stock. *Redesigning Humans: Our Inevitable Genetic Future*. Boston: Houghton Mifflin, 2002.

Larry Thompson. *Correcting the Code: Inventing the Genetic Cure for the Human Body*. New York: Simon & Schuster, 1994.

Luke Timmerman. *Hood: Trailblazer of the Genomics Age*. Seattle: Bandera Press, 2016.

James D. Watson. *The Annotated and Illustrated Double Helix*. New York: Simon & Schuster, 2012.

James D. Watson, Andrew Berry, and Kevin Davies. *DNA: The Story of the Genetic Revolution*. New York: Knopf, 2017.

Timothy C. Winegard. *The Mosquito: A Human History of Our Deadliest Predator*. New York: Dutton, 2019.

Carl Zimmer. *She Has Her Mother's Laugh: The Powers, Perversions, and Potential of Heredity*. New York: Dutton, 2018.

ONLINE RESOURCES

Addgene: Educational Tools and Resources. www.addgene.org/crispr/
Many useful educational tools available at the non-profit repository.

The CRISPR Journal. www.crisprjournal.com
Peer-review journal dedicated to CRISPR and genome editing research (published by Mary Ann Liebert Inc.).

Guidepost: A podcast series from The CRISPR Journal. home.liebertpub.com/lpages/crispr-guidepost-podcast/215/
In-depth interviews with leading practitioners in the world of CRISPR and genome editing.

HHMI BioInteractive: CRISPR-Cas9 Mechanism & Applications. www.biointeractive.org/classroom-resources/crispr-cas-9-mechanism-applications.
Outstanding web animation of CRISPR-Cas9 gene targeting.

Human Nature film (2019). wondercollaborative.org/human-nature-documentary-film/
Superb full-length documentary directed by Adam Bolt.

Innovative Genomics Institute: Education. innovativegenomics.org/education/
A variety of educational tools and engaging digital resources.

Synthego. The Bench blog. www.synthego.com/blog
Useful source of interviews, blog posts and educational materials from California biotech company.

ENDNOTES

Prologue

1. Jon Cohen, "What now for human genome editing?," *Science* 362, (2018): 1090–1092. http://science.sciencemag.org/content/362/6419/1090.
2. Antonio Regalado, "Exclusive: Chinese scientists are creating CRISPR babies," *MIT Technology Review*, November 25, 2018, https://www.technologyreview.com/s/612458/exclusive-chinese-scientists-are-creating-crispr-babies/.
3. Marilynn Marchione, "Chinese researcher claims first gene-edited babies," Associated Press, November 26, 2018, https://www.apnews.com/4997bb7aa36c45449b488e19ac83e86d.
4. Sui-Lee Wee, "Chinese Scientist Who Genetically Edited Babies Gets 3 Years in Prison," *New York Times* December 31, 2019, https://www.nytimes.com/2019/12/30/business/china-scientist-genetic-baby-prison.html.
5. "The era of human gene-editing may have begun. Why that is worrying," *Economist*, December 1, 2018, https://www.economist.com/leaders/2018/12/01/the-era-of-human-gene-editing-may-have-begun-why-that-is-worrying.
6. Elizabeth Pennisi, "The CRISPR Craze," *Science* 341, (2013): 833–836, https://science.sciencemag.org/content/341/6148/833.
7. "Editing Humanity," *Economist*, August 22, 2015, https://www.economist.com/leaders/2015/08/22/editing-humanity.
8. Fraser Nelson, "The return of eugenics," *The Spectator*, April 2016, https://www.spectator.co.uk/article/the-return-of-eugenics.
9. Antonio Regalado, "Who Owns the Biggest Biotech Discovery of the Century?," *MIT Technology Review*, December 4, 2014, https://www.technologyreview.com/s/532796/who-owns-the-biggest-biotech-discovery-of-the-century/.
10. Amy Maxmen, "The Genesis Engine," *WIRED*, August 2015, https://www.wired.com/2015/07/crispr-dna-editing-2/.
11. Kevin Davies, "Nature, genetics and the Niven factor," *Nature Genetics* 39, (2007): 805–806, https://www.nature.com/articles/ng0707-805.
12. Kevin Davies and Michael White, *Breakthrough: The Race to Find the Breast Cancer Gene* (New York: John Wiley & Sons, 1995).
13. Adam Liptak, "Justices, 9–0, Bar Patenting Human Genes," *New York Times*, June 13, 2013, https://www.nytimes.com/2013/06/14/us/supreme-court-rules-human-genes-may-not-be-patented.html.
14. Kevin Davies, *Cracking the Genome* (New York: Free Press, 2001).

15 Kevin Davies, *The $1,000 Genome* (New York: Free Press, 2010).
16 Fastest genetic diagnosis, Guinness World Records, February 3, 2018, https://www.guinnessworldrecords.com/world-records/413563-fastest-genome-sequencing/.
17 Julianna LeMieux, "MGI Delivers the $100 Genome at AGBT Conference," *Genetic Engineering and Biotechnology News,* February 26, 2020, https://www.genengnews.com/news/mgi-delivers-the-100-genome-at-agbt-conference/.
18 Gina Kolata, "Who Needs Hard Drives? Scientists Store Film Clip in DNA," *New York Times,* July 12, 2017, https://www.nytimes.com/2017/07/12/science/film-clip-stored-in-dna.html.
19 Y. Shao et al., "Creating a functional single-chromosome yeast," *Nature* 560, (2018): 331–335.
20 Matthew Warren, "Four new DNA letters double life's alphabet," *Nature,* February 21, 2019, https://www.nature.com/articles/d41586-019-00650-8.
21 James D. Watson, Andrew Berry, and Kevin Davies, *DNA: The Story of the Genetic Revolution* (New York: Knopf, 2017).
22 Rob Stein, "In a 1st, Doctors in U.S. Use CRISPR Tool To Treat Patient With Genetic Disorder," *NPR,* July 29, 2019, https://www.npr.org/sections/health-shots/2019/07/29/744826505/sickle-cell-patient-reveals-why-she-is-volunteering-for-landmark-gene-editing-st.
23 Michael Specter, "How the DNA Revolution Is Changing Us," *National Geographic,* August 2016, https://www.nationalgeographic.com/magazine/2016/08/dna-crispr-gene-editing-science-ethics/.

Chapter 1: The CRISPR Craze

1 Bill Whitaker, "CRISPR: The gene-editing tool revolutionizing biomedical research," *60 Minutes,* April 29, 2018, https://www.cbsnews.com/news/crispr-the-gene-editing-tool-revolutionizing-biomedical-research/.
2 William Kaelin, "Why we can't cure cancer with a moonshot," *Washington Post,* February 11, 2020, https://www.washingtonpost.com/opinions/the-problem-with-trying-to-cure-cancer-with-a-moonshot/2020/02/11/87632bba-2d84-11ea-9b60-817cc18cf173_story.html.
3 Lesley Goldberg, "Jennifer Lopez Sets Futuristic Bio-Terror Drama at NBC (Exclusive)," *Hollywood Reporter,* October 18, 2016, https://www.hollywoodreporter.com/live-feed/jennifer-lopez-sets-futuristic-bio-939509.
4 Neal Baer, "Covid-19 is scary. Could a rogue scientist use CRISPR to conjure another pandemic?," *STAT,* March 26, 2020, https://www.statnews.com/2020/03/26/could-rogue-scientist-use-crispr-create-pandemic/.
5 Walter Isaacson, "Should the rich be allowed to buy the best genes?," *Air Mail,* July 27, 2019, https://airmail.news/issues/2019-7-27/should-the-rich-be-allowed-to-buy-the-best-genes.
6 Mary-Claire King, "Emmanuelle Charpentier and Jennifer Doudna: Creators of Gene-Editing Technology," *Time,* April 16, 2015, https://time.com/collection-post/3822554/emmanuelle-charpentier-jennifer-doudna-2015-time-100/.
7 Jean-Eric Paquet, Kavli banquet speech, September 4, 2018, http://kavliprize.org/events-and-features/video-2018-kavli-prize-banquet.
8 Leah Sherwood, "Genome editing pioneer and Hilo High graduate Jennifer Doudna speaks at UH Hilo about her discovery: CRISPR technology," *UH Hilo Stories,* September 19, 2018, https://hilo.hawaii.edu/news/stories/2018/09/19/genome-editing-pioneer-and-hilo-high-graduate-jennifer-doudna-speaks-at-uh-hilo-about-her-discovery-crispr-technology/.
9 Katie Hasson, "Senate HELP Committee holds hearing on gene editing technology," *Center for Genetics and Society,* November 15, 2017, https://www.geneticsandsociety.org/biopolitical-times/senate-help-committee-holds-hearing-gene-editing-technology.

10 U.S. Senate Committee on Health, Education, Labor & Pensions, "Gene Editing Technology: Innovation and Impact," November 14, 2017, https://www.help.senate.gov/hearings/gene-editing-technology-innovation-and-impact.
11 Pope Francis, "Address of His Holiness Pope Francis to participants at the International Conference organized by the Pontifical Council for Culture on Regenerative Medicine," April 28, 2018, http://w2.vatican.va/content/francesco/en/speeches/2018/april/documents/papa-francesco_20180428_conferenza-pcc.html.
12 C. Brokowski, "Do CRISPR Germline Ethics Statements Cut It?," *CRISPR Journal* 1, (2018): 115–125, https://www.liebertpub.com/doi/10.1089/crispr.2017.0024.
13 April Glaser and Will Oremus, "Tomorrow's Children, Edited," *Slate*, November 28, 2018, https://slate.com/technology/2018/11/if-then-podcast-antonio-regalado-crispr-human-gene-editing-china.html.
14 Francis Collins, "Experts debate: Are we playing with fire when we edit human genes?," *STAT*, November 17, 2015, https://www.statnews.com/2015/11/17/gene-editing-embryo-crispr/#Collins.
15 E. S. Lander et al., "Adopt a moratorium on heritable genome editing," *Nature*, March 13, 2019, https://www.nature.com/articles/d41586-019-00726-5.
16 Rachel Cocker, "This Harvard scientist wants your DNA to wipe out inherited diseases—should you hand it over?," *Telegraph*, March 16, 2019, https://www.telegraph.co.uk/global-health/science-and-disease/harvard-scientist-wants-dna-wipe-inherited-diseases-should/.
17 Sarah Marsh, "Essays Reveal Stephen Hawking Predicted Race of 'Superhumans'," *Guardian*, October 14, 2018, https://www.theguardian.com/science/2018/oct/14/stephen-hawking-predicted-new-race-of-superhumans-essays-reveal.
18 Rob Stein, "First U.S. Patients Treated With CRISPR As Human Gene-Editing Trials Get Underway," *NPR*, April 16, 2019, https://www.npr.org/sections/health-shots/2019/04/16/712402435/first-u-s-patients-treated-with-crispr-as-gene-editing-human-trials-get-underway.

Chapter 2: A Cut Above

1 White House, "Announcing the Completion of the First Survey of the Entire Human Genome at the White House," YouTube video, 40:32, last viewed June 26, 2020, https://www.youtube.com/watch?v=Y_8XRkb-wbY.
2 Nicholas Wade, "Genetic Code of Human Life Is Cracked by Scientists," *New York Times*, June 27, 2000, http://movies2.nytimes.com/library/national/science/062700sci-genome.html.
3 Kevin Davies, "Deanna Church on the Reference Genome Past, Present and Future," *Bio-IT World*, April 22, 2013, http://www.bio-itworld.com/2013/4/22/church-on-reference-genomes-past-present-future.html.
4 R. Chen and A. J. Butte, "The reference human genome demonstrates high risk of type 1 diabetes and other disorders," *Pacific Symposium on Biocomputing*, 2011 (2010): 231–242, https://www.worldscientific.com/doi/abs/10.1142/9789814335058_0025.
5 John Maddox, *What Remains To Be Discovered* (New York: Free Press, 1999).
6 Fyodor D. Urnov, "Genome Editing B.C. (Before CRISPR): Lessons from the 'Old Testament,'" *CRISPR Journal* 1, (2018): 115–125, https://www.liebertpub.com/doi/10.1089/crispr.2018.29007.fyu.
7 Shirley Tilghman, in *The Gene*, PBS, 2020, https://www.pbs.org/kenburns/the-gene/.
8 Rebecca Robbins, "The best and worst analogies for CRISPR, ranked," *STAT*, December 8, 2017, https://www.statnews.com/2017/12/08/crispr-analogies-ranked/.
9 Lina Dahlberg and Anna Groat Carmona, "CRISPR-Cas Technology In and Out of the Classroom," *CRISPR Journal* 1, (2018): 107–114, https://www.liebertpub.com/doi/10.1089/crispr.2018.0007.

10 C. LaManna and R. Barrangou, "Enabling the Rise of a CRISPR World," *CRISPR Journal* 1, (2018): 205–208, https://www.liebertpub.com/doi/10.1089/crispr.2018.0022.
11 K. Davies and R. Barrangou, "MasterChef at Work: An Interview with Rodolphe Barrangou," *CRISPR Journal* 1, (2018): 219–222, https://www.liebertpub.com/doi/full/10.1089/crispr.2018.29015.int?url_ver=Z39.88-2003&rfr_id=ori:rid:crossref.org&rfr_dat=cr_pub%20%200pubmed.
12 Hank Greely, quoted in Mark Shwartz, "Target, Delete, Repair," *Stanford Medicine*, Winter 2018, https://stanmed.stanford.edu/2018winter/CRISPR-for-gene-editing-is-revolutionary-but-it-comes-with-risks.html.
13 Luciano Marraffini, "CRISPR Frontiers" (discussion, New York Academy of Sciences, February 24, 2020).
14 S. Wiles, "Monday micro—200 million light years of viruses?!," *Infectious Thoughts* August 5, 2014, https://sciblogs.co.nz/infectious-thoughts/2014/08/05/monday-micro-200-million-light-years-of-viruses/.
15 S. Klompe and S. H. Sternberg, "Harnessing A Billion Years of Experimentation: The Ongoing Exploration and Exploitation of CRISPR-Cas Immune Systems," *CRISPR Journal* 1, (2018): 141-158.
16 Fyodor Urnov in *Human Nature* (2019), https://wondercollaborative.org/human-nature-documentary-film/.
17 CSHL Leading Strand, "CSHL Keynote, Dr Blake Wiedenheft, Montana State University," YouTube video, 21:21, last viewed June 26, 2020, https://www.youtube.com/watch?v=2x5VoReHV_4&t=.
18 F. Jiang and J. A. Doudna, "CRISPR-Cas9 Structures and Mechanisms," *Annual Review of Biophysics* 46, (2017): 505–529, https://www.annualreviews.org/doi/full/10.1146/annurev-biophys-062215-010822.
19 HHMI BioInteractive, "CRISPR-Cas9 Mechanism & Applications," https://www.biointeractive.org/classroom-resources/crispr-cas-9-mechanism-applications.
20 M. Shibata et al., "Real-space and real-time dynamics of CRISPR-Cas9 visualized by high-speed atomic force microscopy," *Nature Communications* 8, (2017): 1430, https://www.nature.com/articles/s41467-017-01466-8.
21 D. Lawson Jones et al., "Kinetics of dCas9 target search in *Escherichia coli*," *Science* 357, (2017): 1420–1424, https://science.sciencemag.org/content/357/6358/1420?.
22 Andrew Wood, phone interview, August 28, 2019.
23 Rodolphe Barrangou, "CRISPR-Cas: From Bacterial Adaptive Immunity to a Genome Editing Revolution," *XBio,* September 2019, https://explorebiology.org/summary/genetics/crispr-cas:-from-bacterial-adaptive-immunity-to-a-genome-editing-revolution
24 S. Hwang and K. L. Maxwell, "Meet the Anti-CRISPRs: Widespread Protein Inhibitors of CRISPR-Cas Systems," *CRISPR Journal* 2, (2019): 23–30, https://www.liebertpub.com/doi/full/10.1089/crispr.2018.0052.
25 M. Adli, "The CRISPR tool kit for genome editing and beyond." *Nature Communications* 9, (2018): 1911, https://www.nature.com/articles/s41467-018-04252-2.
26 P. T. Harrison and S. Hart, "A beginner's guide to gene editing," *Experimental Physiology* 103, (2018): 439–448, https://physoc.onlinelibrary.wiley.com/doi/full/10.1113/EP086047.
27 Jennifer Doudna, Keystone Symposium, Banff, Canada, February 9, 2020.

Chapter 3: We Can Be Heroes

1 Fyodor Urnov, "Genome Engineering," Keystone Symposium, Victoria Island, Canada, February 21, 2019.
2 Francisco Mojica, interview, Santa Pola, Spain, May 1, 2018.

3 Ed Yong, "The Unique Merger That Made You (and Ewe, and Yew)," *Nautilus*, February 6, 2014, http://nautil.us/issue/10/mergers--acquisitions/the-unique-merger-that-made-you-and-ewe-and-yew.

4 Manuel Ansede, "Francis Mojica, de las salinas a la quiniela del Nobel," *El País*, May 18, 2017, https://elpais.com/elpais/2017/05/18/eps/1495058731_149505.html.

5 F. J. M. Mojica et al., Transcription at different salinities of Haloferax mediterranei sequences adjacent to partially modified PstI sites. *Molecular Microbiology* 9, (1993): 613–621.

6 Y. Ishino et al., "Nucleotide sequence of the iap gene, responsible for alkaline phosphatase isozyme conversion in Escherichia coli, and identification of the gene product," *Journal of Bacteriology* 169, (1987): 5429–5433, https://jb.asm.org/content/jb/169/12/5429.full.pdf.

7 F. J. M. Mojica et al., "Long stretches of short tandem repeats are present in the largest replicons of the *Archaea Haloferax mediterranei* and *Haloferax volcanii* and could be involved in replicon partitioning," *Molecular Microbiology* 17, (1995): 85–93, DOI: 10.1111/j.1365-2958.1995.mmi_17010085.x.

8 Clara Rodriguez Fernandez, "Interview with Francis Mojica, the Spanish scientist that [sic] discovered CRISPR," *Labiotech*, November 13, 2017, https://labiotech.eu/francis-mojica-crispr-interview/.

9 B. Masepohl et al., "Long tandemly repeated repetitive (LTRR) sequences in the filamentous cyanobacterium Anabaena sp. PCC 7120," *Biochimica et Biophysica Acta* 1307, (1996): 26–30, https://www.sciencedirect.com/science/article/abs/pii/0167478196000401.

10 K. S. Makarova et al., "A DNA repair system specific for thermophilic Archaea and bacteria predicted by genomic context analysis," *Nucleic Acids Research* 30, (2002): 482–496, https://www.ncbi.nlm.nih.gov/pmc/articles/PMC99818/.

11 K. Davies and F. Mojica, "Crazy About CRISPR: An Interview with Francisco Mojica," *CRISPR Journal* 1, (2018): 29–33, https://www.liebertpub.com/doi/10.1089/crispr.2017.28999.int.

12 Ibid.

13 Molly Campbell, "Francisco Mojica: The Modest Microbiologist Who Discovered and Named CRISPR," *Technology Networks,* October 14, 2019, https://www.technologynetworks.com/genomics/articles/francis-mojica-the-modest-microbiologist-who-discovered-and-named-crispr-325093.

14 K. Davies and F. Mojica, "Crazy About CRISPR: An Interview with Francisco Mojica," *CRISPR Journal* 1, (2018): 29–33, https://www.liebertpub.com/doi/10.1089/crispr.2017.28999.int.

15 César Díez-Villaseñor, email, October 28, 2017.

16 F. J. M. Mojica et al., "Intervening Sequences of Regularly Spaced Prokaryotic Repeats Derive From Foreign Genetic Elements," *Journal of Molecular Evolution* 60, (2005): 174–182, https://link.springer.com/article/10.1007%2Fs00239-004-0046-3.

17 F. J. M. Mojica and F. Rodriguez-Valera, "The discovery of CRISPR in archaea and bacteria," *FEBS Journal* 283, (2016): 3162–3169, https://febs.onlinelibrary.wiley.com/doi/full/10.1111/febs.13766.

18 C. Pourcel et al., "CRISPR elements in *Yersinia pestis* acquire new repeats by preferential uptake of bacteriophage DNA, and provide additional tools for evolutionary studies," *Microbiology* 151, (2005): 653–663, https://mic.microbiologyresearch.org/content/journal/micro/10.1099/mic.0.27437-0.

19 A. Bolotin et al., "Clustered regularly interspaced short palindrome repeats (CRISPRs) have spacers of extrachromosomal origin," *Microbiology* 151, (2005): 2551–2661.

20 Philippe Horvath, interview, Vilnius, Lithuania, June 21, 2018.

21 K. Davies and R. Barrangou, "MasterChef at Work: An Interview with Rodolphe Barrangou," *CRISPR Journal* 1, 219–222 (2018), https://www.liebertpub.com/doi/10.1089/crispr.2018.29015.int.

22 K. Davies and S. Moineau, "The Phage Whisperer: An Interview with Sylvain Moineau," *CRISPR Journal* 1, (2018): 363–366, https://www.liebertpub.com/doi/10.1089/crispr.2018.29037.kda.

23 R. Barrangou et al., "CRISPR Provides Acquired Resistance Against Viruses in Prokaryotes," *Science* 315, (2007): 1709-1712, DOI: 10.1126/science.1138140.

24 Philippe Horvath, "New Hot Papers—2008," *Science Watch*, July 2008, http://archive.sciencewatch.com/dr/nhp/2008/pdf/08julnhpHorvath.pdf.

Chapter 4: "Thelma and Louise"

1 Jennifer Kahn, "The CRISPR Quandary," *New York Times Magazine*, November 9, 2015, https://www.nytimes.com/2015/11/15/magazine/the-crispr-quandary.html.

2 Colin Tudge, *The Engineer in the Garden* (New York: Hill and Wang, 1994).

3 Melissa Marino, "Biography of Jennifer A. Doudna," *Proceedings of the National Academy of Sciences* 101, (2004): 16987–16989, https://www.pnas.org/content/101/49/16987.

4 Vic Myer, Keystone symposium, Banff, Canada, February 9, 2020.

5 K. Makarova et al., "A putative RNA-interference-based immune system in prokaryotes: computational analysis of the predicted enzymatic machinery, functional analogies with eukaryotic RNAi, and hypothetical mechanisms of action," *Biology Direct* 1, (2006): 7, https://www.ncbi.nlm.nih.gov/pmc/articles/PMC1462988/.

6 Jennifer Doudna, "Jennifer Doudna on the future of gene editing," *Berkeley News*, April 10, 2019, https://news.berkeley.edu/2019/04/10/berkeley-talks-transcript-jennifer-doudna-future-of-gene-editing/.

7 K. D. Seed et al., "A bacteriophage encodes its own CRISPR/Cas adaptive response to evade host innate immunity," *Nature* 494, (2013): 489–491, https://www.ncbi.nlm.nih.gov/pmc/articles/PMC3587790/.

8 B. Al-Shayeb et al., "Clades of huge phages from across Earth's ecosystems," *Nature* 578, (2013): 425–431, https://www.nature.com/articles/s41586-020-2007-4.

9 Jill Banfield, in *Human Nature*, 2019, https://wondercollaborative.org/human-nature-documentary-film/.

10 Ross Wilson, interview, San Francisco, March 13, 2019.

11 B. Wiedenheft et al., "Structural basis for DNase activity of a conserved protein implicated in CRISPR-mediated genome defense," *Structure* 17, (2009): 904–912, https://doi.org/10.1016/j.str.2009.03.019.

12 Lisa Jarvis, "A day in the life of Jennifer Doudna," *Chemical & Engineering News*, March 8, 2020, https://cen.acs.org/biological-chemistry/gene-editing/A-day-with-Jennifer-Doudna-Trying-to-keep-up-with-one-of-the-world-most-sought-after-scientists/98/i9.

13 Press release, "Genentech announces vice president appointment in research," January 21, 2009, https://www.gene.com/media/press-releases/11787/2009-01-21/genentech-announces-vice-president-appoi.

14 M. Jínek and J. A. Doudna, "A three-dimensional view of the molecular machinery of RNA interference," *Nature* 457, (2009): 405–412, https://www.nature.com/articles/nature07755.

15 Katrin Koller, "You should always have something crazy cooking on the back burner," *BaseLaunch*, October 17, 2017, https://www.baselaunch.ch/you-should-always-have-something--crazy-cooking-on-the-back-burner-2/.

16 H. Deveau et al., "Phage Response to CRISPR-Encoded Resistance in *Streptococcus thermophilus*," *Journal of Bacteriology* 190, (2008): 1390–1400, https://www.ncbi.nlm.nih.gov/pmc/articles/PMC2238228/.

17 F. J. M. Mojica et al., "Short Motif Sequences Determine the Targets of the Prokaryotic CRISPR Defence System," *Microbiology* 155, (2009): 733740.

18 A. F. Andersson and J. F. Banfield, "Virus Population Dynamics and Acquired Virus Resistance in Natural Microbial Communities," *Science* 320, (2008): 1047–1050, https://science.sciencemag.org/content/320/5879/1047.abstract.

19 S. J. J. Brouns et al., "Small CRISPR RNAs guide antiviral defense in prokaryotes," *Science* 321, (2008): 960–964, https://www.ncbi.nlm.nih.gov/pmc/articles/PMC5898235/.

20 Mark van der Meijs, "'Without John van der Oost, CRISPR-Cas would never have become this big,'" *Resource,* June 19, 2019, https://resource.wur.nl/en/science/show/Without-John-van-der-Oost-CRISPR-Cas-would-never-have-become-this-big-.htm.

21 L. A. Marraffini and E. J. Sontheimer, "CRISPR interference limits horizontal gene transfer in Staphylococci by targeting DNA," *Science* 322, (2008): 1843–1845, https://www.ncbi.nlm.nih.gov/pmc/articles/PMC2695655/.

22 Will Doss, "The CRISPR Revolution," *Northwestern Medicine,* February 16, 2018, https://magazine.nm.org/2018/02/16/the-crispr-revolution/.

23 K. Davies and S. Moineau, "The Phage Whisperer: An Interview with Sylvain Moineau," *CRISPR Journal* 1, (2018): 363–366, https://www.liebertpub.com/doi/10.1089/crispr.2018.29037.kda.

24 J. E. Garneau et al., "The CRISPR/Cas bacterial immune system cleaves bacteriophage and plasmid DNA," *Nature* 468, (2010): 67–71, https://www.nature.com/articles/nature09523.

25 Pauline Freour, "Emmanuelle Charpentier: 'Des qu'on manipule le vivant, il y a un risqué de derive,'" *Le Figaro,* March 22, 2016, http://sante.lefigaro.fr/actualite/2016/03/22/24766-emmanuelle-charpentier-quon-manipule-vivant-il-y-risque-derive.

26 Alison Abbott, "The quiet revolutionary: How the co-discovery of CRISPR explosively changed Emmanuelle Charpentier's life," *Nature,* April 27, 2016, https://www.nature.com/news/the-quiet-revolutionary-how-the-co-discovery-of-crispr-explosively-changed-emmanuelle-charpentier-s-life-1.19814.

27 Florence Rosier, "Emmanuelle Charpentier, le 'charmant petit monstre' du génie génétique," *Le Monde,* January 9, 2015, https://www.lemonde.fr/sciences/article/2018/05/31/emmanuelle-charpentier-le-charmant-petit-monstre-du-genie-genetique_4559167_1650684.html.

28 N. Herzberg, "Les nouvelles icones de la biologie," *Le Monde,* August 1, 2018, https://www.lemonde.fr/festival/article/2016/08/01/la-piste-aux-etoiles_4977125_4415198.html.

29 The annual lecture is given in honor of the son of Columbia University chemistry professor Stephen Lippard, who died age seven of a neurological disease. Emmanuelle Charpentier, "The 44th Annual Andrew Mark Lippard Memorial Lecture," (lecture, Columbia University, New York, September 26, 2018), http://www.columbianeurology.org/44th-annual-andrew-mark-lippard-memorial-lecture.

30 Jacques Monod, *Chance and Necessity* (New York: Vintage Books, 1972).

31 E. Charpentier, "The Kavli Prize: An autobiography by: Emmanuelle Charpentier," 2018, http://kavliprize.org/sites/default/files/%25nid%25/autobiagraphies_attachments/Emmanuelle%20Charpentier_autobiography.pdf.

32 K. Davies and E. Charpentier, "Finding Her Niche: An Interview with Emmanuelle Charpentier," *CRISPR Journal* 2, (2019):17–22, https://www.liebertpub.com/doi/10.1089/crispr.2019.29042.kda

33 Ibid.

Chapter 5: DNA Surgery

1 E. Deltcheva et al., "CRISPR RNA maturation by *trans*-encoded small RNA and host factor RNase III," *Nature* 471, (2011): 602–607, https://www.ncbi.nlm.nih.gov/pmc/articles/PMC3070239/.

2 Jennifer Kahn, "The Crispr Quandary," *New York Times Magazine*, November 9, 2015, https://www.nytimes.com/2015/11/15/magazine/the-crispr-quandary.html.

3 K. Davies and M. Jínek, "The CRISPR-RNA World: An Interview with Martin Jínek," *CRISPR Journal* 3, (2020): 68–72, https://www.liebertpub.com/doi/10.1089/crispr.2020.29091.mji.

4 Jennifer Doudna, "Why genome editing will change our lives," *Financial Times*, March 14, 2018, https://www.ft.com/content/582d382c-2647-11e8-b27e-cc62a39d57a0.

5 M. Jínek et al., "A programmable dual-RNA—guided DNA endonuclease in adaptive bacterial immunity," *Science* 337, (2012): 816–821, https://science.sciencemag.org/content/337/6096/816/tab-article-info.

6 DOE/Lawrence Berkeley National Laboratory, "Programmable DNA scissors found for bacterial immune system," *Science Daily*, June 28, 2012, https://www.sciencedaily.com/releases/2012/06/120628193020.htm.

7 A. Pollack, "A powerful new way to edit DNA," *New York Times*, March 3, 2014, https://www.nytimes.com/2014/03/04/health/a-powerful-new-way-to-edit-dna.html.

8 S. M. Lee, "Editing DNA could be genetic medicine breakthrough," *San Francisco Chronicle*, September 7, 2014, https://www.sfchronicle.com/technology/article/Editing-DNA-could-be-genetic-medicine-breakthrough-5740320.php.

9 S. J. J. Brouns, "A Swiss Army Knife of Immunity," *Science* 337, (2012): 808–809, https://science.sciencemag.org/content/337/6096/808.

10 Fyodor Urnov in *Human Nature* (2019), https://wondercollaborative.org/human-nature-documentary-film/.

11 R. Barrangou, "RNA-mediated programmable DNA cleavage," *Nature Biotechnology* 30, (2012): 836–868, https://www.nature.com/articles/nbt.2357.

12 Rodolphe Barrangou, interview, Victoria, Canada, February 21, 2019.

13 D. Carroll, "A CRISPR Approach to Gene Targeting," *Molecular Therapy* 20, (2012): 1656–1660, https://www.cell.com/molecular-therapy-family/molecular-therapy/fulltext/S1525-0016(16)32156-6.

14 R. Sapranauskas et al., "The *Streptococcus thermophilus* CRISPR/Cas system provides immunity in *Escherichia coli*," *Nucleic Acids Research* 39, (2011): 9275–9282, doi.org/10.1093/nar/gkr606.

15 "*Cool*," vol. 1, number 1, July 30, 1990, http://cell.com/pb/assets/raw/journals/research/cell/cell-timeline-40/spoof.pdf.

16 K. Davies and V. Siksnys, "From Restriction Enzymes to CRISPR: An Interview with Virginijus Siksnys," *CRISPR Journal* 1, (2018): 137–140, https://www.liebertpub.com/doi/10.1089/crispr.2018.29008.vis.

17 G. Gasiunas et al., "Cas9–crRNA ribonucleoprotein complex mediates specific DNA cleavage for adaptive immunity in bacteria," *Proceedings of the National Academy of Sciences* 109, (2012): E2579–E2586, https://www.pnas.org/content/109/39/E2579?iss=39.

18 Sarah Zhang, "The Battle Over Genome Editing Gets Science All Wrong," *WIRED*, April 18, 2015, https://www.wired.com/2015/10/battle-genome-editing-gets-science-wrong/.

19 R. Dahm, "Friedrich Miescher and the Discovery of DNA," *Developmental Biology* 278, (2005); 274–288, https://doi.org/10.1016/j.ydbio.2004.11.028.

20 Stuart Firestein, "Fundamentally Newsworthy," *The Edge.org*, 2016, " Youtube, https://www.edge.org/response-detail/26718.

21 TEDx Talks ,"O(ú)pravy lidské DNA | Martin Jínek | TEDx Třinec," YouTube video, 21:07, last viewed June 26, 2020, https://www.youtube.com/watch?v=d7kPcjD3PUU.

22 Jin-Soo Kim, email to Doudna and Charpentier, October 3, 2012. PTAB Interference 106048. UC exhibit 1558, https://acts.uspto.gov/ifiling/PublicView.jsp?identifier=106048.

23 George Church, email to Doudna, November 14, 2012. PTAB Interference 106048. UC exhibit 1559, https://acts.uspto.gov/ifiling/PublicView.jsp?identifier=106048.

24 Lisa Jarvis, "A day in the life of Jennifer Doudna," *Chemical & Engineering News*, March 8, 2020, https://cen.acs.org/biological-chemistry/gene-editing/A-day-with-Jennifer-Doudna-Trying-to-keep-up-with-one-of-the-world-most-sought-after-scientists/98/i9.

25 M. Jínek et al., "RNA programmed genome editing in human cells," *eLife* 2, (2013): e00471. https://elifesciences.org/articles/00471.

26 Feng Zhang, email to Doudna, January 2, 2013, PTAB Interference 106048, UC exhibit 1620, https://acts.uspto.gov/ifiling/PublicView.jsp?identifier=106048.

Chapter 6: Field of Dreams

1 M. Boguski, "A Molecular Biologist Visits *Jurassic Park*," *BioTechniques* 12, (1992): 668–669, http://markboguski.net/docs/publications/BioTechniques-1992.pdf.

2 Alice Park, "The editor of life's building blocks," *Time*, October 6, 2016, https://time.com/4518815/feng-zhang-next-generation-leaders/.

3 Carey Goldberg, "CRISPR Wizard Feng Zhang: The Making Of A Sunny Science Superstar," *WBUR*, April 26, 2018, https://www.wbur.org/commonhealth/2018/04/26/feng-zhang-crispr-profile.

4 Ingfei Chen, "The beam of light that flips a switch that turns on the brain," *New York Times*, August 14, 2007, https://www.nytimes.com/2007/08/14/science/14brai.html.

5 John Colapinto, "Lighting the Brain," *New Yorker*, May 18, 2015, https://www.newyorker.com/magazine/2015/05/18/lighting-the-brain.

6 Kerry Grens, "Feng Zhang: The Midas of Methods," *The Scientist*, August 1, 2014, https://www.the-scientist.com/?articles.view/articleNo/40582/title/Feng-Zhang--The-Midas-of-Methods/.

7 F. Zhang, L. Cong et al., "Efficient construction of sequence-specific TAL effectors for modulating mammalian transcription," *Nature Biotechnology* 29, (2011): 149–153, https://www.nature.com/articles/nbt.1775.

8 Le Cong, phone interview, July 18, 2019.

9 Michael Gilmore, email, July 7, 2019.

10 K. L. Palmer and M. S. Gilmore, "Multidrug-resistant enterococci lack CRISPR-*cas*," *mBio* (2010): 1:e00227–10, https://mbio.asm.org/content/1/4/e00227-10/article-info.

11 P. Horvath and R. Barrangou, "CRISPR/Cas, the Immune System of Bacteria and Archaea," *Science* 327, (2010): 167–170, https://science.sciencemag.org/content/327/5962/167.long.

12 J. E. Garneau et al., "The CRISPR/Cas Bacterial Immune System Cleaves Bacteriophage and Plasmid DNA," *Nature* 468, (2010): 67–71, https://www.nature.com/articles/nature09523.

13 Broad Institute, https://www.broadinstitute.org/files/news/pdfs/BroadPriorityStatement.pdf.

14 MIT McGovern Institute, "Meet Feng Zhang," YouTube video, 4:12, last viewed January 3, 2020, https://www.youtube.com/watch?v=EjolOzkYNlk&t=.

15 Luciana Marraffini, personal communication, September 23, 2019.

16 D. Altshuler and D. Cowan, "Isogenic Human Pluripotent Stem Cell-Based Models of Human Disease Mutations," Grant ID: R01-DK-097768, https://commonfund.nih.gov/TRA/recipients12.

17 Sharon Begley, "Meet one of the world's most groundbreaking scientists. He's 34." *STAT*, November 6, 2015, https://www.statnews.com/2015/11/06/hollywood-inspired-scientist-rewrite-code-life/.

18 Amy Maxmen, "Easy DNA editing will remake the world. Buckle up," *WIRED*, December 2015, https://www.wired.com/2015/07/crispr-dna-editing-2/.

19 Le Cong, phone interview, July 18, 2019.

20 Fei Ann Ran, interview, Boston, August 2, 2019.

21 TEDx Talks, "Inspired by nature: harnessing tools from microbes to engineer biology | Fei Ann Ran | TEDxVienna," YouTube video, 15:55, last viewed November 10, 2019, https://www.youtube.com/watch?v=dcD0G1-BPGE.

22 Fei Ann Ran, interview, Boston, August 2, 2019.

23 George Church, interview, Boston, August 2, 2019.

24 K. Davies and K. Esvelt, "Gene Drives, White-Footed Mice, and Black Sheep: An Interview with Kevin Esvelt," *CRISPR Journal* 1, (2018): 319–324, https://www.liebertpub.com/doi/10.1089/crispr.2018.29031.kda.

25 Rodolphe Barrangou, interview, Victoria, Canada, February 21, 2019.

26 L. Cong, F. A. Ran et al., "Multiplex genome engineering using CRISPR/Cas systems," *Science* 339, (2013): 819–823, http://science.sciencemag.org/content/339/6121/819.full.

27 P. Mali et al., "RNA-guided human genome engineering via Cas9," *Science* 339, (2013): 823–826, https://www.ncbi.nlm.nih.gov/pmc/articles/PMC3712628/.

28 S. W. Cho et al., "Targeted genome engineering in human cells with the Cas9 RNA-guided endonuclease," *Nature Biotechnology* 31, (2013): 230–232, https://www.nature.com/articles/nbt.2507.

29 W. Y. Hwang et al., "Efficient genome editing in zebrafish using a CRISPR-Cas system," *Nature Biotechnology* 31, (2013): 277–279, https://www.ncbi.nlm.nih.gov/pmc/articles/PMC3686313/.

30 W. Jiang et al., "CRISPR-assisted editing of bacterial genomes," *Nature Biotechnology* 31, (2013): 233–239, https://www.ncbi.nlm.nih.gov/pmc/articles/PMC3748948/.

31 Hannah Devlin, "Jennifer Doudna: 'I have to be true to who I am as a scientist,'" *Guardian*, July 2, 2017, https://www.theguardian.com/science/2017/jul/02/jennifer-doudna-crispr-i-have-to-be-true-to-who-i-am-as-a-scientist-interview-crack-in-creation.

32 Matt Ridley, "Editing Our Genes, One Letter at a Time," *Wall Street Journal*, January 11, 2013, https://www.wsj.com/articles/SB10001424127887323482504578227661405130902.

33 Matthew Herper, "This protein could change biotech forever," *Forbes*, March 19, 2013, https://www.forbes.com/sites/matthewherper/2013/03/19/the-protein-that-could-change-biotech-forever/.

34 K. Karczewski, "Progress in genomics according to bingo: 2013 edition," *Genome Biology* 14, (2013): 143, https://genomebiology.biomedcentral.com/articles/10.1186/gb4148.

35 Q. Ding et al., "Enhanced Efficiency of Human Pluripotent Stem Cell Genome Editing through Replacing TALENs with CRISPRs," *Cell Stem Cell* 12, (2013): 393–394, https://www.cell.com/cell-stem-cell/fulltext/S1934-5909(13)00101-X

36 T. J. Cradick, online interview, June 30, 2020.

37 Jon Cohen, "The Birth of CRISPR Inc.," *Science* 355, (2017): 680–684, https://science.sciencemag.org/content/355/6326/680.summary..

38 R. Coontz, "*Science*'s top ten breakthroughs of 2013," *Science*, December 19, 2013, https://www.sciencemag.org/news/2013/12/sciences-top-10-breakthroughs-2013.

39 John Travis, "Breakthrough of the Year: CRISPR makes the cut," *Science,* December 17, 2015, https://www.sciencemag.org/news/2015/12/and-science-s-2015-breakthrough-year.

Chapter 7: Prize Fight

1 Patrick Gillooly, "Lander named to Obama's science team," *MIT News*, December 22, 2008, http://news.mit.edu/2008/lander-named-obamas-science-team.

2 Veronique Greenwood and Valerie Ross, "How Feng Zhang modified a cell's genome on the fly," *Popular Science*, October 23, 2013, https://www.popsci.com/science/article/2013-09/feng-zhang/.

3 Eric S. Lander, "The heroes of CRISPR," *Cell* 164, (2016): 18–28, https://www.cell.com/cell/pdf/S0092-8674(15)01705-5.pdf.

4 Michael Eisen, "The villain of CRISPR," *It is NOT Junk* (blog), January 25, 2016, http://www.michaeleisen.org/blog/?p=1825.

5 Nathaniel Comfort, "A Whig history of CRISPR," *Genotopia,* January 18, 2016, https://genotopia.scienceblog.com/573/a-whig-history-of-crispr/.

6 Jennifer Doudna, "PubMed Commons," 2016, posted by Richard Sever, https://twitter.com/search?q=pubmed%20commons%20charpentier&src=typd.

7 Bob Grant, "Credit for CRISPR: A Conversation with George Church," *The Scientist,* December 29, 2015, https://www.the-scientist.com/news-analysis/credit-for-crispr-a-conversation-with-george-church-34306.

8 George Church, interview, Boston, August 2, 2019.

9 Samuel Sternberg, "Humans By Design: Covering the Genome Editing Revolution," New York University, March 6, 2018, https://journalism.nyu.edu/about-us/event/2018-spring/humans-by-design-covering-the-gene-editing-revolution/.

10 Jennifer A. Doudna and Samuel H. Sternberg, *A Crack in Creation* (Boston: Houghton Mifflin Harcourt, 2017).

11 CanadaGairdnerAwards, "Jennifer Doudna-2016 Canada Gairdner Awards Gala," YouTube video, 5:31, last viewed May 1, 2020, https://www.youtube.com/watch?v=jLOSMcQ2iec&t=.

12 Rodolphe Barrangou, interview, Victoria, Canada, February 21, 2019.

13 Ibid.

14 Albany Med, "Albany Medical Center Prize in Medicine and Biomedical Research panel discussion," YouTube video, 1:23:27, last viewed May 10, 2020, https://www.youtube.com/watch?v=oNlukM56bsY.

15 Clara Rodriguez Fernandez, "Francis Mojica, the Spanish Scientist Who Discovered CRISPR," *Labiotech,* April 8, 2019, https://labiotech.eu/interviews/francis-mojica-crispr-interview/.

16 Jennifer A. Doudna, "The promise and challenge of therapeutic genome editing," *Nature* 578, (2020): 229–236, https://www.nature.com/articles/s41586-020-1978-5?proof=true.

17 L. S. Qi et al., "Repurposing CRISPR as an RNA-guided platform for sequence-specific control of gene expression," *Cell* 152, (2013): 1173–1184, https://www.cell.com/fulltext/S0092-8674(13)00211-0.

18 H. A. Rees and D. R. Liu, "Base editing: Precision Chemistry on the Genome and Transcriptome of Living Cells," *Nature Reviews Genetics* 19, (2018): 770–788, https://pubmed.ncbi.nlm.nih.gov/30323312/.

19 B. L. Oakes et al., "CRISPR-Cas9 Circular Permutants as Programmable Scaffolds for Genome Modification," *Cell* 176, (2019): 254–267, https://pubmed.ncbi.nlm.nih.gov/30633905/.

20 D. Burstein et al., "New CRISPR–Cas systems from uncultivated microbes," *Nature* 542, 2017: 237–241.

21 B. Zetsche et al., "Cpf1 Is a Single RNA-guided Endonuclease of a Class 2 CRISPR-Cas System," *Cell* 163, (2015): 759–771, https://www.ncbi.nlm.nih.gov/pmc/articles/PMC4638220/.

22 J. S. Chen et al., "CRISPR-Cas12a target binding unleashes indiscriminate single-strand DNase activity," *Science* 360, (2018): 436–439, https://science.sciencemag.org/content/360/6387/436.

23 J. P. Broughton et al., "CRISPR-Cas12-based detection of SARS-CoV-2," *Nature Biotechnology,* 2020, https://www.nature.com/articles/s41587-020-0513-4.

24 J. S. Gootenberg et al., "Nucleic acid detection with CRISPR-Cas13a/C2c2," *Science* 356, (2017): 438–442, https://science.sciencemag.org/content/356/6336/438.

25 John Carreyrou, *Bad Blood* (New York: Knopf, 2018).

26 M. B. Nourse et al., "Engineering of a miniaturized, robotic clinical laboratory," *Bioeng. Transl. Med.* 3, 58–70 (2018), https://www.ncbi.nlm.nih.gov/pmc/articles/PMC5773944/.

27 O. O. Abudayyeh et al., "Nucleic acid detection of plant genes using CRISPR-Cas13," *CRISPR Journal* 2, (2019): 165-171, https://www.liebertpub.com/doi/10.1089/CRISPR.2019.0011.
28 Elie Dolgin, "The kill-switch for CRISPR that could make gene-editing safer," *Nature* January 15, 2020, https://www.nature.com/articles/d41586-020-00053-0.
29 S. E. Klompe et al., "Transposon-encoded CRISPR–Cas systems direct RNA-guided DNAintegration," *Nature* 571, (2019): 219–225, https://www.nature.com/articles/s41586-019-1323-z.

Chapter 8: Genome Editing B.C.

1 Fyodor Urnov, TRI-CON, San Francisco, February 15, 2018.
2 Editorial, "Method of the Year 2011," *Nature Methods* 9, (2012): 1, https://doi.org/10.1038/nmeth.1852.
3 Mario R. Cappechi, "The making of a scientist," *HHMI Bulletin*, May 1997, https://healthcare.utah.edu/capecchi/HHMI.pdf.
4 Ibid.
5 F. D. Urnov, "Genome Editing B.C. (Before CRISPR): Lasting Lessons from the 'Old Testament,'" *CRISPR Journal* 1, (2018): 34–46, https://www.liebertpub.com/doi/10.1089/crispr.2018.29007.fyu.
6 Fyodor Urnov, interview, Florence, Italy, June 27, 2018.
7 http://sangamoncountyhistory.org/wp/?p=1410.
8 Ed Lanphier, interview, Ross, California, March 4, 2019.
9 S. Hacein-Bey-Abina et al., "Sustained Correction of X-Linked Severe Combined Immunodeficiency by Ex Vivo Gene Therapy," *New England Journal of Medicine* 346, (2002): 1185–1193, https://www.nejm.org/doi/full/10.1056/NEJMoa012616.
10 Douglas Birch, "Hamilton Smith's second chance; Scientist's journey; He won the Nobel, but lost his way. Could he put his family together? And could he crack one of life's great puzzles?," *Baltimore Sun*, April 11, 1999, https://www.baltimoresun.com/news/bs-xpm-1999-04-11-9904120283-story.html.
11 Ibid.
12 Y. G. Kim, J. Cha, and S. Chandrasegaran, "Hybrid restriction enzymes: zinc finger fusions to Fok I cleavage domain," *Proceedings of the National Academy of Sciences USA* 93, (1996): 1156–1160, https://www.ncbi.nlm.nih.gov/pmc/articles/PMC40048/.
13 S. Chandrasegaran and J. Smith, "Chimeric Restriction Enzymes: What is Next?," *Biological Chemistry* 380, (1999): 841–848, https://www.ncbi.nlm.nih.gov/pmc/articles/PMC4033837/.
14 M. Bibikova et al., "Targeted Chromosomal Cleavage and Mutagenesis in Drosophila Using Zinc-Finger Nucleases," *Genetics* 161, (2002): 1169–1175, https://www.genetics.org/content/161/3/1169.long.
15 M. Bibikova et al., "Stimulation of Homologous Recombination through Targeted Cleavage by Chimeric Nucleases," *Molecular and Cellular Biology* 21, (2001): 289–297, https://mcb.asm.org/content/21/1/289.
16 K. Davies and D. Carroll, "Giving Genome Editing the Fingers: An Interview with Dana Carroll," *CRISPR Journal* 2, (2019): 157–162, https://www.liebertpub.com/doi/10.1089/crispr.2019.29058.dca.
17 M. H. Porteus and D. Baltimore, "Chimeric nucleases stimulate gene targeting in human cells," *Science* 5620, (2003): 763, http://science.sciencemag.org/content/300/5620/763.
18 F. D. Urnov et al., "Highly Efficient Endogenous Human Gene Correction Using Designed Zinc-Finger Nucleases," *Nature* 435, (2005): 646–651, https://www.nature.com/articles/nature03556.

19 S. Jaffe, "Giving Genetic Disease the Finger," *WIRED*, July 5, 2005, https://www.wired.com/2005/07/giving-genetic-disease-the-finger/.

20 K. Kandavelou et al., "'Magic' scissors for genome surgery," *Nature Biotechnology* 23, (2005): 686–687, https://www.nature.com/articles/nbt0605-686.

21 B. J. Doranz et al., "A dual-tropic primary HIV-1 isolate that uses fusin and the beta-chemokine receptors CKR-5, CKR-3, and CKR-2b as fusion cofactors," *Cell* 85, (1996): 1148–58, https://www.cell.com/cell/fulltext/S0092-8674(00)81314-8.

22 M. Parmentier, "CCR5 an HIV infection, a view from Brussels," *Frontiers in Immunology* 6, (2015): 295, https://www.ncbi.nlm.nih.gov/pmc/articles/PMC4459230/.

23 M. Samson et al., "Resistance to HIV-1 infection in caucasian individuals bearing mutant alleles of the CCR-5 chemokine receptor gene," *Nature* 382, (1996): 722–725, https://www.nature.com/articles/382722a0.

24 Stephen J. O'Brien, *Tears of the Cheetah* (New York: Thomas Dunne, 2003).

25 T. R. Brown, "I am the Berlin patient: a personal reflection," *AIDS Research and Human Retroviruses* 31, (2015): 2–3, https://www.ncbi.nlm.nih.gov/pmc/articles/PMC4287108/.

26 G. Hutter et al., "Long-term control of HIV by CCR5 delta32/delta32 stem-cell transplantation," *New England Journal of Medicine* 360, (2009): 692–698, https://www.nejm.org/doi/full/10.1056/NEJMoa0802905.

27 F. Urnov, "AWESOME interview with Dr. Fyodor Urnov," *The Sangamo Domain* (blog), December 12, 2015, http://sangamodomain.blogspot.com/2015/12/awesome-interview-with-dr-fyodor-urnov.html.

28 P. Tebas et al., "Gene editing of CCR5 in autologous CD4 T cells of persons infected with HIV," *New England Journal of Medicine* 370, (2014): 901–910, https://www.nejm.org/doi/full/10.1056/nejmoa1300662.

29 Emily Mullin, "Back to the Future: Pre-CRISPR Systems are Driving Therapies to the Clinic," *Genetic Engineering & Biotechnology News,* February 7, 2019, https://www.genengnews.com/insights/back-to-the-future-pre-crispr-systems-are-driving-therapies-to-the-clinic/.

30 CNBC, "Sangamo Biosciences CEO Edward Lanphier | Mad Money | CNBC," YouTube video, 8:19, March 7, 2019, https://www.youtube.com/watch?v=c3dT1sH1PNM.

31 Brian Madeux quoted in M. Marchione, "US scientists try 1st gene editing in the body," AP News, November 15, 2017, https://apnews.com/4ae98919b52e43d8a8960e0e260feb0a.

32 C. Hunter, "A rare disease in two brothers," *Journal of the Royal Society of Medicine* 10, (1917): 104–116, https://www.ncbi.nlm.nih.gov/pmc/articles/PMC2018097/pdf/procrsmed00727-0110.pdf.

33 Sandy Macrae, Genome Editing summit, National Academy of Sciences, Washington, DC, August 13, 2019, https://vimeo.com/showcase/6229550/video/354892447.

34 B. Zeitler et al., "Allele-specific transcriptional repression of mutant *HTT* for the treatment of Huntington's disease," *Nature Medicine* 25, (2019): 1131–1142, https://www.nature.com/articles/s41591-019-0478-3.

35 Ed Rebar, "Genome Engineering," (lecture, Keystone Symposium, Victoria, Canada, February 2019).

Chapter 9: Deliverance or Disaster

1 Anon, "Mount Hope geneticists get more milk from cows by selective breeding," *Life* 5, 50–53 (1938).

2 J. A. Wolff and J. Lederberg, "An early history of gene transfer and therapy," *Human Gene Therapy* 5, (1994): 469–480, https://www.liebertpub.com/doi/abs/10.1089/hum.1994.5.4-469.

3 Derek So, "The Use and Misuse of *Brave New World* in the CRISPR Debate," *CRISPR Journal* 2, (2019): 316–323, https://www.liebertpub.com/doi/10.1089/crispr.2019.0046.

4 Jack Williamson, *Dragon's Island* (New York: Simon & Schuster, 1951).
5 Francis Crick, letter to Michael Crick, March 15, 1953, Wellcome Library, https://wellcomelibrary.org/item/b1948799x.
6 J. D. Watson, correspondence: Letter to Max Delbruck, March 12, 1953, http://scarc.library.oregonstate.edu/coll/pauling/dna/corr/corr432.1-watson-delbruck-19530312-transcript.html.
7 Brenda Maddox, "DNA's double helix: 60 years since life's deep molecular secret was discovered," *Guardian,* February 22, 2013, https://www.theguardian.com/science/2013/feb/22/watson-crick-dna-60th-anniversary-double-helix.
8 J. D. Watson and F. H. C. Crick, "Molecular Structure of Nucleic Acids: A Structure for Deoxyribose Nucleic Acid," *Nature* 171, (1953):737–738, http://dosequis.colorado.edu/Courses/MethodsLogic/papers/WatsonCrick1953.pdf.
9 Brenda Maddox, "DNA's double helix: 60 years since life's deep molecular secret was discovered," *Guardian,* February 22, 2013, https://www.theguardian.com/science/2013/feb/22/watson-crick-dna-60th-anniversary-double-helix.
10 Symposium held at Ohio Wesleyan University, Delaware, on April 6, 1963. The proceedings were published in *The Control of Human Heredity and Evolution* (1965).
11 S. E. Luria, *The Control of Human Heredity and Evolution*, ed. T. M. Sonneborn (New York: Macmillan, 1965).
12 E. M. Witkin, "Remembering Rollin Hotchkiss (1911–2004)," *Genetics* 170, (2005): 1443–1447, https://www.ncbi.nlm.nih.gov/pmc/articles/PMC1449782/.
13 R. Hotchkiss, *The Control of Human Heredity and Evolution,* ed. T. M. Sonneborn (New York: Macmillan, 1965).
14 R. Sinsheimer, "The End of the Beginning," (lecture, Caltech, October 26, 1966). Excerpt in *Human Nature* (2019).
15 R. Sinsheimer, "The Prospect of Designed Genetic Change," *American Scientist* XXXII, (1969): 8–13, http://calteches.library.caltech.edu/2718/1/genetic.pdf.
16 M. Nirenberg, "Will society be prepared?," *Science* 157, (1967): 633, https://science.sciencemag.org/content/157/3789/633.
17 J. Lederberg, "Molecular biology, eugenics and euphenics," *Nature* 198, (1963): 428–429, https://www.nature.com/articles/198428a0.pdf.
18 J. A. Wolff and J. Lederberg, "An early history of gene transfer and therapy," *Human Gene Therapy* 5, (1994): 469–480, https://www.liebertpub.com/doi/abs/10.1089/hum.1994.5.4-469.
19 J. Lederberg, "DNA breakthrough points way to therapy by virus," *Washington Post,* January 13, 1968, https://profiles.nlm.nih.gov/ps/access/BBABSP.pdf.
20 J. Lederberg, in J. A. Wolff and J. Lederberg, "An early history of gene transfer and therapy," *Human Gene Therapy* 5, (1994): 469–480, https://www.liebertpub.com/doi/abs/10.1089/hum.1994.5.4-469.
21 N. A. Wivel and W. F. Anderson, "Human Gene Therapy: Public Policy and Regulatory Issues," in *The Development of Human Gene Therapy*, ed. T. Friedmann (Cold Spring Harbor, NY: Cold Spring Harbor Laboratory Press, 1999), pp 671–689.

Chapter 10: The Rise and Fall of Gene Therapy

1 S. Rogers, "Shope papilloma virus: A passenger in man and its significance to the potential control of the host genome," *Nature* 212, (1966): 1220–1222, https://www.nature.com/articles/2121220a0.
2 T. Friedmann and R. Roblin, "Gene therapy for human genetic disease?," *Science* 175, (1972): 949–955, https://science.sciencemag.org/content/175/4025/949.long.

3 JapanPrize, "2015 Japan Prize Commemorative Lecture: Dr. Theodore Friedmann & Prof. Alan Fischer," YouTube video, 1:07:56, last viewed June 7, 2020, https://www.youtube.com/watch?v=Z5SLxpPLxcw&t=136s.

4 David Baltimore, "Limiting science: A biologist's perspective," *Daedalus*, 107, (1978): 37–45, https://www.jstor.org/stable/20024543?seq=1.

5 M. J. Cline, "Perspectives for gene therapy: inserting new genetic information into mammalian cells by physical techniques and viral vectors," *Pharmacology & Therapeutics* 29, (1985): 69–92.

6 P. Jacobs, "Doctor tried gene therapy on 2 humans," *Washington Post,* October 8, 1980, https://www.washingtonpost.com/archive/politics/1980/10/08/doctor-tried-gene-therapy-on-2-humans/c95d4b44-3e5c-4a48-904c-4bbefe52391b/?utm_term=.b867c1010e22.

7 D. Bartels, "Gene therapy: scientific advances and socio-ethical considerations," *Journal of Social & Biological Structures* 9, (1986) 105–113, https://psycnet.apa.org/record/1987-32279-001.

8 E. Beutler, "The Cline affair," *Molecular Therapy* 4, (2001): 396–397. https://www.cell.com/action/showPdf?pii=S1525-0016%2801%2990486-1.

9 R. Williamson, "Gene therapy," *Nature* 298, (1982): 416–418, https://www.nature.com/articles/298416a0.

10 W. F. Anderson, "Prospects for human gene therapy," *Science* 226, (1984): 401–409, https://science.sciencemag.org/content/226/4673/401.long.

11 P. Gorner and J. Lyon, "Altered Fates," *Chicago Tribune*, March 7, 1986, https://www.chicagotribune.com/news/ct-xpm-1986-03-07-8601170568-story.html.

12 Jonathan Gardner, "New estimate puts cost to develop a new drug at $1B, adding to long-running debate," *BioPharma Dive,* March 3, 2020, https://www.biopharmadive.com/news/new-drug-cost-research-development-market-jama-study/573381/.

13 Larry Thompson, "Human gene therapy debuts at NIH," *Washington Post,* September 15, 1990, https://www.washingtonpost.com/archive/politics/1990/09/15/human-gene-therapy-debuts-at-nih/f98ffb56-aa7f-4529-a5f6-c57cb845a7cd/?utm_term=.82d9fd43ca22.

14 Jeff Lyon and Peter Gomer, *Altered Fates: Gene Therapy and the Retooling of Human Life* (New York: W. W. Norton & Co., 1996).

15 Robin Marantz Henig, "Dr. Anderson's gene machine," *New York Times*, March 31, 1999, https://www.nytimes.com/1991/03/31/magazine/dr-anderson-s-gene-machine.html.

16 Peter Gorner, "Doctors begin world's first gene therapy," *Chicago Tribune*, September 15, 1990, http://articles.chicagotribune.com/1990-09-15/news/9003170543_1_dr-w-*french*-anderson-gene-therapy-adenosine-deaminase.

17 L. M. Muul et al., "Persistence and expression of the adenosine deaminase gene for 12 years and immune reaction to gene transfer components: long-term results of the first clinical gene therapy trial," *Blood* 101, (2003): 2563–2569, https://pubmed.ncbi.nlm.nih.gov/12456496/.

18 J. M. Wilson, "Recollections from a Pioneer Who Provided the Foundation for the Success of Gene Therapy in Treating Severe Combined Immune Deficiencies," *Human Gene Therapy* 27, (2016): 53–56, https://www.liebertpub.com/doi/10.1089/humc.2016.29013.int.

19 T. Friedmann, "A brief history of gene therapy," *Nature Genetics* 2, (1992): 93–98, https://www.nature.com/articles/ng1092-93.

20 Natalie Angier, "Gene experiment to reverse inherited disease is working," *New York Times*, April 1, 1994, https://www.nytimes.com/1994/04/01/us/gene-experiment-to-reverse-inherited-disease-is-working.html.

21 S. H. Orkin and A. G. Motulsky, "Report and recommendations of the panel to assess the NIH investment in research on gene therapy," *OSP*, December 7, 1995, https://osp.od.nih.gov/wp-content/uploads/2014/11/Orkin_Motulsky_Report.pdf.

22 T. Friedmann, "Preface," in *The Development of Human Gene Therapy* (Cold Spring Harbor, NY: Cold Spring Harbor Laboratory Press, 1999).

23 Paul Gelsinger, "Jesse's Intent," http://www.jesse-gelsinger.com/jesses-intent.html.

24 Rick Weiss and Deborah Nelson, "Teen dies undergoing experimental gene therapy," *Washington Post*, September 29, 1999, https://www.washingtonpost.com/wp-srv/WPcap/1999-09/29/060r-092999-idx.html?noredirect=on.

25 Sheryl Gay Stolberg, "The biotech death of Jesse Gelsinger," *New York Times,* November 28, 1999, https://www.nytimes.com/1999/11/28/magazine/the-biotech-death-of-jesse-gelsinger.html.

26 S. E. Raper et al., "Fatal systemic inflammatory response syndrome in a ornithine transcarbamylase deficient patient following adnoviral gene transfer," *Molecular Genetics and Metabolism* 80, (2003): 148–158, https://pubmed.ncbi.nlm.nih.gov/14567964/.

27 Peter Little, *Genetic Destinies* (Oxford, UK: Oxford University Press, 2002).

28 Siddhartha Mukherjee, *The Gene: An Intimate History* (New York: Simon & Schuster, 2017).

29 *The Gene*, PBS 2020, https://www.pbs.org/show/gene/.

30 H. F. Judson, "The Glimmering Promise of Gene Therapy." *MIT Technology Review*, November 1, 2006, https://www.technologyreview.com/s/406797/the-glimmering-promise-of-gene-therapy/.

31 Ibid.

32 K. Davies, "From the Cultural Revolution to the Gene Therapy Revolution: An Interview with Guangping Gao, PhD," *Human Gene Therapy,* February 14, 2020, https://www.liebertpub.com/doi/abs/10.1089/hum.2020.29109.int.

33 Ryan Cross, "The redemption of James Wilson." *Chemical & Engineering News,* September 2019, https://cen.acs.org/business/The-redemption-of-James-Wilson-gene-therapy-pioneer/97/i36.

34 David Schaffer, TRI-CON, San Francisco, March 2, 2019.

35 Julianna LeMieux, "Going Viral: The Next Generation of AAV Vectors," *Genetic Engineering & Biotechnology News,* September 3, 2019, https://www.genengnews.com/insights/going-viral-the-next-generation-of-aav-vectors/.

36 Editorial, "Gene therapy deserves a fresh chance," *Nature* 461, (2009): 1173, https://www.nature.com/articles/4611173a.

37 Melinda Wenner, "Gene therapy: An interview with an unfortunate pioneer," *Scientific American,* September 1, 2009, https://www.scientificamerican.com/article/gene-therapy-an-interview/?redirect=1.

38 Carl Zimmer, "Gene therapy emerges from disgrace to be the next big thing, again," *WIRED*, August 13, 2013, https://www.wired.com/2013/08/the-fall-and-rise-of-gene-therapy-2/.

39 JapanPrize, "2015 Japan Prize Commemorative Lecture: Dr. Theodore Friedmann & Prof. Alain Fischer," YouTube video, 1:07:56, last viewed June 30, 2020, https://youtu.be/Z5SLxpPLxcwe.

Chapter 11: Overnight Success

1 Ricki Lewis, "Luxturna: A giant step forward for blindness gene therapy—a conversation with Dr. Kathy High," *DNA Science Blog*, July 20, 2017, https://blogs.plos.org/dnascience/2017/07/20/luxturna-a-giant-step-forward-for-blindness-gene-therapy-a-conversation-with-dr-kathy-high/.

2 David Dobbs, "Why there's new hope about ending blindness," *National Geographic,* September 2016, https://www.nationalgeographic.com/magazine/2016/09/blindness-treatment-medical-science-cures/.

3 ASGCT, "Seeing the Light with Retinal Gene Therapy: From Fantasy to Reality—Jean Bennett," YouTube video, 45:37, last viewed June 30, 2020, https://youtu.be/jDdFmBxNfUE.

4 J. M. Wilson, "Interview with Jean Bennett, MD, PhD," *Human Gene Therapy* 29, (2018): 7–9, https://doi.org/10.1089/humc.2018.29032.int.

5 A. W. Taylor, "Ocular immune privilege," *Eye* 23, (2009): 1885–1889, https://www.nature.com/articles/eye2008382.

6 Jean Bennett, "My Career Path for Developing Gene Therapy for Blinding Diseases: The Importance of Mentors, Collaborators, and Opportunities," *Human Gene Therapy* 25, (2014): 663–670, https://www.ncbi.nlm.nih.gov/pmc/articles/PMC4137328/.

7 J. M. Wilson, "Interview with Jean Bennett, MD, PhD," *Human Gene Therapy* 29, (2018): 7–9, https://doi.org/10.1089/humc.2018.29032.int.

8 A. Maguire et al., "Safety and efficacy of gene transfer for Leber's Congenital Amaurosis," *New England Journal of Medicine* 358, (2008): 2240–2248, https://www.ncbi.nlm.nih.gov/pmc/articles/PMC2829748/.

9 J. Bennett et al., "Safety and durability of effect of contralateral-eye administration of AAV2 gene therapy in patients with childhood-onset blindness caused by *RPE65* mutations: a follow-on phase 1 trial," *Lancet* 388, (2016): 661–672, https://www.ncbi.nlm.nih.gov/pmc/articles/PMC5351775/.

10 Sharon Begley, "Out of prison, the 'father of gene therapy' faces a harsh reality: a tarnished legacy and an ankle monitor," *STAT*, July 23, 2018, https://www.statnews.com/2018/07/23/w-french-anderson-father-of-gene-therapy/.

11 David Schaffer, TRI-CON conference, San Francisco, March 2, 2019.

12 S. Cannon, "Sickle cell disease advocate & precision medicine leader remembered," *Black Doctor*, April 8, 2016, https://blackdoctor.org/my-story-any-day-without-pain-is-a-good-day/.

13 M. Friend, "Shakir Cannon, patient advocate," *CRISPR Journal* 1, (2018): 24–25, https://www.liebertpub.com/doi/10.1089/crispr.2018.29005.mfr.

14 D. Shriner and C. N. Rotimi, "Whole-genome-sequence-based haplotypes reveal single origin of the sickle allele during the Holocene Wet Phase," *American Journal of Human Genetics* 102, (2018): 547–556, https://www.cell.com/ajhg/fulltext/S0002-9297(18)30048-X.

15 T. L. Savitt and M. F. Goldberg, "Herrick's 1910 case report of sickle cell anemia," *Journal of the American Medical Association* 261, (1989): 266–271, https://doi.org/10.1001/jama.1989.03420020120042.

16 Editorial, "Sickle cell anemia, a race specific disease," *Journal of the American Medical Association* 133, (1947): 33–34, https://jamanetwork.com/journals/jama/article-abstract/290758.

17 A. C. Allison, "Two lessons from the interface of genetics and medicine," *Genetics* 166, (2004): 1591–1599, http://www.genetics.org/content/genetics/166/4/1591.full.pdf.

18 V. M. Ingram, "Gene mutations in human haemoglobin: the chemical difference between normal and sickle cell haemoglobin," *Nature* 180, (1957): 326–328, https://www.nature.com/articles/180326a0.

19 J. Lapook, "Could gene therapy cure sickle cell anemia?," *60 Minutes*, March 10, 2019, https://www.cbsnews.com/news/could-gene-therapy-cure-sickle-cell-anemia-60-minutes/.

20 J-A. Ribeil et al., "Gene therapy in a patient with sickle cell disease," *New England Journal of Medicine* 376, (2017): 848–855, https://www.nejm.org/doi/full/10.1056/NEJMoa1609677.

21 J. Lapook, "Could gene therapy cure sickle cell anemia?," *60 Minutes*, March 10, 2019, https://www.cbsnews.com/news/could-gene-therapy-cure-sickle-cell-anemia-60-minutes/.

22 J. Watson, "Sickling in negro newborns: Its possible relationship to fetal hemoglobin," *American Journal of Medicine* 5, (1948): 159–160, https://doi.org/10.1016/0002-9343(48)90029-1.

23 R. Daggy et al., "Health and disease in Saudi Arabia: The Aramco experience, 1940s–1990s," University of California, 1998, http://texts.cdlib.org/view?docId=kt8m3nb5g6&doc.view=entire_text.

24 V. G. Sankaran et al., "Human fetal hemoglobin expression is regulated by the developmental stage-specific repressor *BCL11A*," *Science* 322, (2008): 1839–1842, http://science.sciencemag.org/content/322/5909/1839.long.

25 K. Weintraub, "New gene therapy shows promise for patients with sickle cell disease," *WBUR*, March 8, 2019, https://www.wbur.org/commonhealth/2019/03/08/gene-therapy-sickle-cell.

26 Sharon Begley, "NIH and Gates Foundation launch effort to bring genetic cures for HIV, sickle cell disease to world's poor," *STAT*, October 23, 2019, https://www.statnews.com/2019/10/23/nih-gates-foundation-genetic-cures-hiv-sickle-cell/.

27 Sean Nolan, "Unite to Cure," The Vatican, April 26, 2018.

28 K. Foust et al., "Intravascular AAV9 preferentially targets neonatal-neurons and adult-astrocytes in CNS," *Nature Biotechnology* 27, (2009): 59–65, https://www.ncbi.nlm.nih.gov/pmc/articles/PMC2895694/.

29 J. R. Mendell et al., "Single-dose gene-replacement therapy for spinal muscular atrophy," *New Engl. J. Med.* 377, (2017): 1713–1722, https://www.nejm.org/doi/full/10.1056/NEJMoa1706198.

30 Larry Luxner, "With Zolgensma's approval, debate shifts to pricing and availability of world's costliest drug," *SMA News Today*, May 29, 2019, https://smanewstoday.com/2019/05/29/zolgensma-approval-shifts-debate-pricing-availability-worlds-costliest-drug/1.

31 Nathan Yates, "I have spinal muscular atrophy. Critics of the $2 million new gene therapy are missing the point," *STAT*, May 31, 2019, https://www.statnews.com/2019/05/31/spinal-muscular-atrophy-zolgensma-price-critics/.

32 E. Mamcarz et al., "Lentiviral Gene Therapy Combined with Low-Dose Busulfan in Infants with SCID-X1," *New England Journal of Medicine* 380, (2019); 1525–1534, https://www.nejm.org/doi/full/10.1056/NEJMoa1815408.

33 M. Cortez, "'Bubble boys' cured in medical breakthrough using gene therapy," *Bloomberg* April 17, 2019, https://www.bloomberg.com/news/articles/2019-04-17/-bubble-boys-cured-in-medical-breakthrough-using-gene-therapy.

34 V. Montazerhodjat, D. M. Weinstock, and A. W. Lo, "Buying cures versus renting health: Financing health care with consumer loans," *Science Translational Medicine* 8, (2016): 327, https://stm.sciencemag.org/content/8/327/327ps6.

35 George Church, interview, Boston, August 3, 2019.

36 Sarah Boseley, "Dismay at lottery for $2.1m drug to treat children with muscle-wasting disease," *Guardian*, December 20, 2019, https://www.theguardian.com/society/2019/dec/20/lottery-prize-zolgensma-drug-zolgensma-children-muscle-wasting-disease.

37 F. W. Twort, "An investigation on the nature of ultra-microscopic viruses," *Lancet* 186, (1915): 1241–1243.

38 A. Dublanchet, "The epic of phage therapy," *Canadian Journal of Infectious Diseases and Medical Microbiology* 18, (2007): 15–18, https://www.ncbi.nlm.nih.gov/pmc/articles/PMC2542892/.

39 Richard Martin, "How ravenous Soviet viruses will save the world," *WIRED*, October 1, 2003, https://www.wired.com/2003/10/phages/.

40 M. Rosen, "Phage therapy treats patient with drug-resistant bacterial infection," Howard Hughes Medical Institute, May 8, 2019, https://www.hhmi.org/news/phage-therapy-treats-patient-with-drug-resistant-bacterial-infection.

41 Julianna LeMieux, "Phage therapy win: Mycobacterium infection halted," *Genetic Engineering & Biotechnology News*, May 8, 2019, https://www.genengnews.com/insights/phage-therapy-win-mycobacterium-infection-halted/.

42 Ben Fidler, "How an Ohio kids' hospital quietly became ground zero for gene therapy," *Xconomy*, April 15, 2019, https://xconomy.com/national/2019/04/15/how-an-ohio-kids-hospital-quietly-became-ground-zero-for-gene-therapy/.

43 Ryan Cross, "The redemption of James Wilson," *Chemical & Engineering News,* September 2019, https://cen.acs.org/business/The-redemption-of-James-Wilson-gene-therapy-pioneer/97/i36.
44 Ibid.
45 C. Hinderer et al., "Severe Toxicity in Nonhuman Primates and Piglets Following High-Dose Intravenous Administration of an Adeno-Associated Virus Vector Expressing Human SMN," *Human Gene Therapy* 29, (2018): 285–298.
46 Ben Fidler, "Two patients die in now-halted study of Audentes gene therapy," *BioPharma Dive*, June 26, 2020, https://www.biopharmadive.com/news/audentes-gene-therapy-patient-deaths/580670/
47 Nicole Paulk, "Gene Therapy: It's Time to Talk about High-Dose AAV," *Genetic Engineering & Biotechnology News*, July 7, 2020, https://www.genengnews.com/topics/genome-editing/gene-therapy-its-time-to-talk-about-high-dose-aav/.

Chapter 12: Fix You

1 Rebecca Robbins, "Billionaire Sean Parker is nerding out on cancer research. Science has never seen anyone quite like him," *STAT,* July 9, 2019, https://www.statnews.com/2019/07/09/sean-parker-cancer-research-science/.
2 F. Baylis and M. McLeod, "First-in-human Phase 1 CRISPR gene editing cancer trials: Are we ready?," *Current Gene Therapy* 17, (2017): 309–319, https://www.ncbi.nlm.nih.gov/pmc/articles/PMC5769084/.
3 Preetika Rana, Amy Dockser Marcus, and Wenxin Fan, "China, Unhampered by Rules, Races Ahead in Gene-Editing Trials," *Wall Street Journal,* January 21, 2018, https://www.wsj.com/articles/china-unhampered-by-rules-races-ahead-in-gene-editing-trials-1516562360.
4 E. A. Stadtmauer et al., "CRISPR-engineered T cells in patients with refractory cancer," *Science* 367, (2020): eaba7365, https://science.sciencemag.org/content/367/6481/eaba7365.
5 Carl June, Keystone Symposium, Banff, Canada, February 9, 2020.
6 G. E. Martyn et al., "Natural regulatory mutations elevate the fetal globin gene via disruption of BCL11A or ZBTB7A binding," *Nature Genetics* 50, (2018): 498–503, https://www.nature.com/articles/s41588-018-0085-0.
7 M. H. Porteus, "A New Class of Medicines through DNA Editing," *New England Journal of Medicine* 380, (2019): 947–959, https://www.nejm.org/doi/full/10.1056/NEJMra1800729.
8 Matthew Porteus, International Human Genome Editing Summit, Hong Kong, November 29, 2018.
9 David Sanchez in *Human Nature* (2019), https://wondercollaborative.org/human-nature-documentary-film/.
10 Rob Stein, "In a 1st, doctors In U.S. use CRISPR tool to treat patient with genetic disorder," *NPR,* July 29, 2019, https://www.npr.org/sections/health-shots/2019/07/29/744826505/sickle-cell-patient-reveals-why-she-is-volunteering-for-landmark-gene-editing-st.
11 Rob Stein, "A Patient Hopes Gene-Editing Can Help With Pain Of Sickle Cell Disease," *NPR,* October 10, 2019, https://www.npr.org/sections/health-shots/2019/10/10/766765780/after-a-life-of-painful-sickle-cell-disease-a-patient-hopes-gene-editing-can-hel.
12 Rob Stein, "A Young Mississippi Woman's Journey Through a Pioneering Gene-Editing Experiment," *NPR,* December 25, 2019, https://www.npr.org/sections/health-shots/2019/12/25/784395525/a-young-mississippi-womans-journey-through-a-pioneering-gene-editing-experiment.
13 Rob Stein, "A Year In, 1st Patient To Get Gene Editing For Sickle Cell Disease is Thriving," *NPR,* June 23, 2020, https://www.npr.org/sections/health-shots/2020/06/23/877543610/a-year-in-1st-patient-to-get-gene-editing-for-sickle-cell-disease-is-thriving.

14 K. Davies and E. Charpentier, "Finding Her Niche: An Interview with Emmanuelle Charpentier," *CRISPR Journal* 2, (2019): 17–22, https://doi.org/10.1089/crispr.2019.29042.kda.
15 Bill Gates, "Gene Editing for Good," *Foreign Affairs,* May/June 2018, https://www.foreignaffairs.com/articles/2018-04-10/gene-editing-good.
16 Kevin Davies, "Avila Therapeutics Targets the Covalent Proteome," *Bio-IT World,* January 28, 2010, http://www.bio-itworld.com/2010/01/28/avila.html.
17 M. L. Maeder et al., "Development of a gene-editing approach to restore vision loss in Leber congenital amaurosis type 10," *Nature Medicine* 25, (2019): 229–233, https://www.nature.com/articles/s41591-018-0327-9.
18 Marilynn Marchione, "Doctors try 1st CRISPR editing in the body for blindness," AP News, March 4, 2020, https://apnews.com/17fcd6ae57d39d06b72ca40fe7cee461.
19 Rob Wright, "A CEO's most formative leadership experience," *Life Science Leader,* February 25, 2019, https://www.lifescienceleader.com/doc/a-ceo-s-most-formative-leadership-experience-0001.
20 Rob Wright, "John Leonard's latest adventure—readying Intellia Therapeutics for the long haul," *Life Science Leader,* March 1, 2019, https://www.lifescienceleader.com/doc/john-leonard-s-latest-adventure-readying-intellia-therapeutics-for-the-long-haul-0001.
21 Alliance for Regenerative Medicine, "2017 Annual Dinner," YouTube video, 1:10:07, last viewed May 2, 2020, https://www.youtube.com/watch?v=wPKyr092HIE.
22 Amy Dockser Marcus, "A year of brutal training for racing in the Andes," *Wall Street Journal,* February 10, 2017, https://www.wsj.com/articles/a-year-of-brutal-training-for-racing-in-the-andes-1486814400.
23 Rodolphe Barrangou, interview, Victoria, Canada, February 21, 2019.
24 U. A. Neil, "A conversation with Eric Olson," *Journal of Clinical Investment* 127, (2017): 403–404, https://www.ncbi.nlm.nih.gov/pmc/articles/PMC5272177/.
25 L. Amoasii et al., "Gene editing restores dystrophin expression in a canine model of Duchenne muscular dystrophy," *Science* 362, (2018): 86–91, https://science.sciencemag.org/content/362/6410/86.
26 V. Iyer et al., "No unexpected CRIPR-Cas9 off-target activity revealed by trio sequencing of gene-edited mice," *PLOS Genetics,* July 9, 2018, https://doi.org/10.1371/journal.pgen.1007503.
27 J. H. Hu et al., "Chemical Biology Approaches to Genome Editing: Understanding, Controlling and Delivering Programmable Nucleases," *Cell Chemical Biology* 23, (2016): 57–73, https://doi.org/10.1016/j.chembio.2015.12.009.
28 M. Kosicki et al., "Repair of double-strand breaks induced by CRISPR-Cas9 leads to large deletions and complex rearrangements," *Nature Biotechnology* 36, (2018): 765–771, https://www.nature.com/articles/nbt.4192.
29 C. T. Charlesworth et al., "Identification of preexisting adaptive immunity to Cas9 proteins in humans," *Nature Medicine* 25, (2019): 249–254, https://www.nature.com/articles/s41591-018-0326-x.
30 A. Mehta and O. M. Merkel, "Immunogenicity of Cas9 Protein," *Journal of Pharmaceutical Sciences* 109, (2020): 62–67, https://pubmed.ncbi.nlm.nih.gov/31589876/.
31 T. Ho and D. P. Lane, "Guardian of Genome Editing," *CRISPR Journal* 1, (2018): 258–260, https://www.liebertpub.com/doi/10.1089/crispr.2018.29021.dal.

Chapter 13: Patent Pending

1 Jennifer Doudna, "How CRISPR lets us edit our DNA," TED, September 2015, https://www.ted.com/talks/jennifer_doudna_how_crispr_lets_us_edit_our_dna.

2 Jacob Sherkow, "How much is a CRISPR patent license worth?," *Forbes*, February 21, 2017, https://www.forbes.com/sites/jacobsherkow/2017/02/21/how-much-is-a-crispr-patent-license-worth/#184e0a0d6b77.

3 M. Jínek, J. A. Doudna, E. Charpentier, and K. Chyliński, "Methods and Composition for RNA-directed site-specific DNA modification," USPTO application, submitted May 25, 2012.

4 Jacob Sherkow, "The CRISPR patent landscape: Past, present, and future," *CRISPR Journal* 1, (2018): 9–11, https://www.liebertpub.com/doi/abs/10.1089/crispr.2018.0044.

5 Erik Sontheimer, NIH grant application, January 2009. Personal communication.

6 Jon Cohen, "How the battle lines over CRISPR were drawn," *Science,* February 15, 2017, https://www.sciencemag.org/news/2017/02/how-battle-lines-over-crispr-were-drawn.

7 "Broad Institute Priority Statement," Exhibit A, May 2016, https://www.broadinstitute.org/files/news/pdfs/BroadPriorityStatement.pdf.

8 Broad Communications, "For journalists: Statement and background on the CRISPR patent process," January 16, 2020, https://www.broadinstitute.org/crispr/journalists-statement-and-background-crispr-patent-process.

9 Ibid.

10 Joe Stanganelli, "Interference: A CRISPR Patent Dispute Roadmap," *Bio-IT World,* January 9, 2017, http://www.bio-itworld.com/2017/1/9/interference-a-crispr-patent-dispute-roadmap.aspx.

11 Sarah Zhang, "An Outdated Law Will Decide the CRISPR Patent Dispute," *Atlantic,* December 7, 2016, https://www.theatlantic.com/science/archive/2016/12/crispr-patent-hearing/509747/.

12 Dana Carroll, "A CRISPR Approach to Gene Targeting," *Molecular Therapy* 20, (2012): 1658–1660, https://www.ncbi.nlm.nih.gov/pmc/articles/PMC3437577/.

13 Lin Shuailiang, email to Doudna, February 28, 2015, https://s3.amazonaws.com/files.technologyreview.com/p/pub/docs/CRISPR-email.pdf.

14 Antonio Regalado, "Patent Office hands win in CRISPR battle to Broad Institute," *MIT Technology Review,* February 16, 2017, https://www.technologyreview.com/s/603662/patent-office-hands-win-in-crispr-battle-to-broad-institute/.

15 K. Davies, "The Evolving Law of CRISPR: Interview with Professor Jacob Sherkow of New York Law School," *Biotechnology Law Report* 37, (2018): 126–130, https://www.liebertpub.com/doi/10.1089/blr.2018.29068.kd.

16 Andrew Pollack, "Harvard and M.I.T. scientists win gene-editing patent fight," *New York Times,* February 15, 2017, https://www.nytimes.com/2017/02/15/science/broad-institute-harvard-mit-gene-editing-patent.html.

17 Sharon Begley, "University of California appeals CRISPR patent setback," *STAT,* April 13, 2017, https://www.statnews.com/2017/04/13/crispr-patent-uc-appeal/.

18 Chuck Stanley, "Berkeley Defends Bid to Overrule PTAB in CRISPR row," *Law360,* April 30, 2018, https://www.law360.com/ip/articles/1038721.

19 Michael Barnes, "The CRISPR revolution," *Catalyst* 9, (2014): 18–20, https://issuu.com/ucb-catalyst/docs/catalyst-sp14?e=8644787/8584210.

20 Jacob Sherkow, "The CRISPR-Cas9 Patent Appeal: Where Do We Go From Here?," *CRISPR Journal* 1, (2018): 309–311, https://www.liebertpub.com/doi/abs/10.1089/crispr.2018.0044.

21 Richard Harris, "East Coast Scientists Win Patent Case Over Medical Research Technology," *NPR*, September 10, 2018, https://www.npr.org/2018/09/10/646422497/east-coast-scientists-win-patent-case-over-medical-research-technology.

22 Sharon Begley, "University of California to be granted long-sought CRISPR patent, possibly reviving dispute with the Broad Institute," *STAT,* February 8, 2019, https://www.statnews.com/2019/02/08/the-university-of-california-gets-its-key-crispr-patent/.

23 Sharon Begley, "Patent office reopens major CRISPR battle between Broad Institute and Univ. of California," *STAT*, June 25, 2019, https://www.statnews.com/2019/06/25/crispr-patents-interference/.

24 H. T. Greely, "CRISPR, Patents, and Nobel Prizes," *Los Angeles Review of Books*, August 23, 2017, https://lareviewofbooks.org/article/crispr-patents-and-nobel-prizes/#!.

25 Christie Rizk, "CRISPR Patent Fight Turns Ugly as UC Accuses Broad Researchers of Lying About Claims," *Genomeweb*, August 1, 2019, https://www.genomeweb.com/business-news/crispr-patent-fight-turns-ugly-uc-accuses-broad-researchers-lying-about-claims#.Xvv3VkVKg2w.

26 Kerry Grens, "That Other CRISPR Patent Dispute," *The Scientist*, August 31, 2016, https://www.the-scientist.com/daily-news/that-other-crispr-patent-dispute-32952.

27 Press release, "The Rockefeller University and Broad Institute of MIT and Harvard nnounce update to CRISPR-Cas9 portfolio filed by Broad," Broad Institute, January 15, 2018, https://www.broadinstitute.org/news/rockefeller-university-and-broad-institute-mit-and-harvard-announce-update-crispr-cas9.

28 Jef Akst, "UC Berkeley receives CRISPR patent in Europe," *The Scientist*, March 24, 2017, https://www.the-scientist.com/?articles.view/articleNo/48987/title/UC-Berkeley-Receives-CRISPR-Patent-in-Europe/.

29 Alex Philippidis, "Rejecting Broad Institute Opposition, EPO Affirms CRISPR Patent Issued to Charpentier, UC, and U. Vienna," *Genetic Engineering & Biotechnology News*, February 10, 2020, https://www.genengnews.com/news/rejecting-broad-institute-opposition-epo-affirms-crispr-patent-issued-to-charpentier-uc-and-u-vienna/.

30 Hannah Kuchler, "Jennifer Doudna, CRISPR scientist, on the ethics of editing humans," *Financial Times*, January 31, 2020, https://www.ft.com/content/6d063e48-4359-11ea-abea-0c7a29cd66fe.

31 Sharon Begley, "Fight for coveted CRISPR patents gets knottier, as MilliporeSigma makes new claims," *STAT Plus*, July 22, 2019, https://www.statnews.com/2019/07/22/milliporesigma-crispr-patent-denial-challenge-university-california/.

32 Joe Stanganelli, "Interference: A CRISPR Patent Dispute Roadmap," *Bio-IT World*, January 9, 2017, http://www.bio-itworld.com/2017/1/9/interference-a-crispr-patent-dispute-roadmap.aspx.

33 Jacob Sherkow, online interview, May 5, 2020.

Chapter 14: #CRISPRbabies

1 Marilynn Marchione, "AP Exclusive: US scientists try 1st gene editing in the body," AP News, November 15, 2017, https://apnews.com/4ae98919b52e43d8a8960e0e260feb0a.

2 Aging Reversed, "AP—CRISPR babies in China," YouTube video, 2:50, November 26, 2018, https://youtu.be/qUiNG1iW4Ww.

3 Jill Adams, "A conversation with Marilynn Marchione," *Open Notebook*, June 25, 2019, https://www.theopennotebook.com/2019/06/25/storygram-marilynn-marchiones-chinese-researcher-claims-first-gene-edited-babies/#qanda.

4 Marilynn Marchione, email, July 25, 2019.

5 K. Musunuru, *The Beagle Has Landed*, December 7, 2018, https://beaglelanded.com/podcasts/kiran-musunuru/.

6 Antonio Regalado, Precision Medicine & Society Conference, Columbia University, April 24, 2019.

7 Hank Greely, TRI-CON, San Francisco, March 3, 2020.

8 M. Araki and T. Ishii, "International regulatory landscape and integration of corrective genome editing into in vitro fertilization," *Reproductive Biology and Endocrinology* 12, (2014): 108, https://doi.org/10.1186/1477-7827-12-108.

9 Cody Sheehy, "A Unique Partnership: Code of the Wild," *CRISPR Journal* 1, (2018): 135–136, https://www.liebertpub.com/doi/10.1089/crispr.2018.29009.csh.

10 Antonio Regalado, "Years before CRISPR babies, this man was the first to edit human embryos," *MIT Technology Review,* December 11, 2018, https://www.technologyreview.com/s/612554/years-before-crispr-babies-this-man-was-the-first-to-edit-human-embryos/.

11 E. Lanphier et al., "Don't Edit the Human Germline," *Nature,* March 12, 2015, https://www.nature.com/news/don-t-edit-the-human-germ-line-1.17111.

12 P. Liang et al., "CRISPR/Cas9-mediated gene editing in human tripronuclear zygotes," *Protein & Cell* 6, (2015): 363–372, https://www.ncbi.nlm.nih.gov/pmc/articles/PMC4417674/.

13 Samira Kiani, phone interview, January 29, 2019.

14 He Jiankui, "Evaluation of the safety and efficacy of gene editing with human embryo CCR5 gene," Chinese Clinical Trial registry, November 8, 2018 (withdrawn November 30, 2018), http://www.chictr.org.cn/showprojen.aspx?proj=32758.

15 Nie Jing-Bao, "He Jiankui's genetic misadventure: Why him? Why China?," *Hastings Center,* December 5, 2018, https://www.thehastingscenter.org/jiankuis-genetic-misadventure-china/. (Additional translation courtesy of Nicholas Shadid.)

16 Antonio Regalado, "Exclusive: Chinese scientists are creating CRISPR babies," *MIT Technology Review*, November 25, 2018, https://www.technologyreview.com/s/612458/exclusive-chinese-scientists-are-creating-crispr-babies/.

17 Marilynn Marchione, "Chinese researcher claims first gene-edited babies," Associated Press, November 26, 2018, https://www.apnews.com/4997bb7aa36c45449b488e19ac83e86d.

18 The He Lab, "About Lulu and Nana: Twin Girls Born Healthy After Gene Surgery As Single-Cell Embryos," YouTube video, 4:43, last viewed June 30, 2020, https://www.youtube.com/watch?v=th0vnOmFltc.

19 Living MacTavish, "In Conversation With Scientist and CRISPR Pioneer Jennifer Doudna," YouTube video, 47:30, November 30, 2017, https://www.youtube.com/watch?v=YVoPRSPEpvU&list=PLdT7Y4C6bsoSUdt2PB1NQlVQgKULTARXk&index=47&t=1898s.

20 Paul Knoepfler, TEDx Vienna, October 2015, https://www.ted.com/talks/paul_knoepfler_the_ethical_dilemma_of_designer_babies?language=en.

21 University of California Television, "Jennifer Doudna in Conversation with Joe Palca," YouTube video, 58:54, last viewed June 30, 2020, https://youtu.be/0nMbNPb3CLc.

22 Pam Belluck, "How to stop rogue gene-editing of human embryos?," *New York Times,* January 23, 2019, https://www.nytimes.com/2019/01/23/health/gene-editing-babies-crispr.html.

Chapter 15: The Boy from Xinhua

1 Yangyang Cheng, "Brave new world with Chinese characteristics," *Bulletin of the Atomic Scientists*, December 21, 2018, https://thebulletin.org/2018/12/brave-new-world-with-chinese-characteristics/.

2 Zoe Low, "China's gene editing Frankenstein He Jiankui, dubbed 'mad genius' by colleagues, had early dreams of becoming Chinese Einstein," *South China Morning Post,* November 27, 2018, https://www.scmp.com/news/china/society/article/2175267/chinas-gene-editing-frankenstein-dubbed-mad-genius-colleagues-had.

3 Yangyang Cheng, "Brave new world with Chinese characteristics," *Bulletin of the Atomic Scientists*, December 21, 2018, https://thebulletin.org/2018/12/brave-new-world-with-chinese-characteristics/.

4 J. He and M. W. Deem, "Heterogeneous diversity of spacers within CRISPR (clustered regularly interspaced short palindromic repeats)," *Physical Review Letters* 105, (2010): 128102, https://journals.aps.org/prl/abstract/10.1103/PhysRevLett.105.128102#fulltext.

5 Kevin Davies, *The $1,000 Genome* (New York: Free Press, 2010).

6 Patrick Hoge, "Stephen Quake is 'out to hunt death down and punch him in the face,'" *San Francisco Business Times,* June 8, 2017, https://www.bizjournals.com/sanfrancisco/news/2017/06/08/biotech-2017-stephen-quake-chan-zuckerberg-biohub.html.
7 D. Pushkarev et al., "Single-molecule sequencing of an individual human genome," *Nature Biotechnology* 27, (2009): 847–850, https://www.ncbi.nlm.nih.gov/pmc/articles/PMC4117198/.
8 Kevin Davies, "Quake Sequences Personal Genome Using Helicos Single-Molecule Sequencing," *Bio-IT World,* August 10, 2019, http://www.bio-itworld.com/news/08/10/09/stephen-quake-personal-genome-single-molecule-sequencing.html.
9 E. A. Ashley et al., "Clinical evaluation incorporating a personal genome," *Lancet* 375, (2010): 1525–1535, https://www.ncbi.nlm.nih.gov/pmc/articles/PMC2937184/.
10 Kevin Davies, "The Medical Utility of Genome Sequencing," *Bio-IT World,* June 8, 2011, http://www.bio-itworld.com/2011/issues/may-jun/medical-utility-genome-sequencing.html.
11 Steve Lombardi, phone interview, July 23, 2019.
12 H. C. Fan et al., "Noninvasive diagnosis of fetal aneuploidy by shotgun sequencing DNA from maternal blood," *Proceedings of the National Academy of Sciences* 105, (2008): 16266–16271, https://www.ncbi.nlm.nih.gov/pmc/articles/PMC2562413/.
13 N. Jiang, J. He, J. A. Weinstein et al., "Lineage Structure of the Human Antibody Repertoire in Response to Influenza Vaccination," *Science Translational Medicine* 5, (2013): 171ra19, https://www.ncbi.nlm.nih.gov/pmc/articles/PMC3699344/.
14 Kevin Davies, "The bedrock of BGI: Huanming Yang," *Bio-IT World,* September 27, 2011, http://www.bio-itworld.com/issues/2011/sept-oct/bedrock-bgi-huanming-yang.html.
15 Allison Proffitt, "Sequencing the human secret," *Bio-IT World,* September 28, 2010, http://www.bio-itworld.com/2010/issues/sep-oct/bgi-hk.html.
16 R. Li et al., "The sequence and *de novo* assembly of the giant panda genome," *Nature* 463, (2010): 311–317, https://www.nature.com/articles/nature08696.
17 Julianna LeMieux, "MGI Delivers the $100 Genome at AGBT," *Genetic Engineering & Biotechnology News,* February 26, 2020, https://www.genengnews.com/topics/omics/mgi-delivers-the-100-genome-at-agbt-conference/.
18 Michael Specter, "The Gene Factory," *New Yorker,* December 29, 2013, https://www.newyorker.com/magazine/2014/01/06/the-gene-factory.
19 Karen Zhang, "Before gene-editing controversy, Chinese scientist He Jiankui was rising star who received 41.5 million yuan in government grants," *South China Morning Post,* December 3, 2018, https://www.scmp.com/news/hong-kong/health-environment/article/2176131/gene-editing-controversy-chinese-scientist-he?onboard=true.
20 Aaron Krol, "Direct Genomics' new clinical sequencer revives a forgotten DNA technology," *Bio-IT World,* October 29, 2015, http://www.bio-itworld.com/2015/10/29/direct-genomics-new-clinical-sequencer-revives-forgotten-dna-technology.html.
21 Ibid.
22 Ibid.
23 Steve Lombardi, phone interview, July 23, 2019.
24 Hannah Devlin, "Britain's House of Lords approves conception of three-person babies," *Guardian,* February 24, 2015, https://www.theguardian.com/politics/2015/feb/24/uk-house-of-lords-approves-conception-of-three-person-babies.
25 Jessica Hamzelou, "Exclusive: World's first baby born with new '3 parent' technique," *New Scientist,* September 27, 2016, https://www.newscientist.com/article/2107219-exclusive-worlds-first-baby-born-with-new-3-parent-technique/.
26 Ariana Eunjung Cha, "This fertility doctor is pushing the boundaries of human reproduction, with little regulation," *Washington Post,* May 14, 2018, https://www.washingtonpost.com/national

/health-science/this-fertility-doctor-is-pushing-the-boundaries-of-human-reproduction-with-little-regulation/2018/05/11/ea9105dc-1831-11e8-8b08-027a6ccb38eb_story.html.

27 Megan Molteni, "A Controversial Fertility Treatment Gets Its First Big Test," *WIRED*, January 28, 2019, https://www.wired.com/story/a-controversial-fertility-treatment-gets-its-first-big-test/.

28 Emily Mullin, "Despite Calls for a Moratorium, More 'Three-Parent' Babies Expected Soon," *One Zero*, September 16, 2019, https://onezero.medium.com/despite-calls-for-a-moratorium-more-three-parent-babies-expected-soon-8a2464165423.

29 Ian Sample, "World's first baby born from new procedure using DNA of three people," *Guardian*, September 27, 2016, https://www.theguardian.com/science/2016/sep/27/worlds-first-baby-born-using-dna-from-three-parents.

30 Emily Mullin, "U.S. researcher says he's ready to start four pregnancies with 'three-parent' embryos," *STAT*, April 18, 2019, https://www.statnews.com/2019/04/18/new-york-researcher-ready-to-start-pregnancies-with-three-parent-embryos/.

31 Kang Ning, "The Village That AIDS Tore Apart," *Sixth Tone*, May 31, 2016, http://www.sixthtone.com/news/897/village-aids-tore-apart.

32 Roger Gosden, "Robert Edwards (1925–2013)," *Nature* 497, (2013): 318, https://www.nature.com/articles/497318a.

33 R. G. Edwards et al., "Early stages of Fertilization *in vitro* of Human Oocytes Matured *in vitro*," *Nature* 221, (1969): 632–635, https://www.nature.com/articles/221632a0.

34 R. G. Edwards and D. J. Sharpe, "Social Values and Research in Human Embryology," *Nature* 231, (1971): 87–91, https://www.nature.com/articles/231087a0?proof=trueInJun.

35 K. Dow, "'The men who made the breakthrough': How the British press represented Patrick Steptoe and Robert Edwards in 1978," *Reproductive Biomedicine & Society Online* 4, (2017): 59–67, https://www.sciencedirect.com/science/article/pii/S2405661817300199.

36 Nicole Karlis, "More than 8 million IVF infants have been born worldwide: report," *Salon*, July 5, 2018, https://www.salon.com/2018/07/05/more-than-8-million-ivf-infants-born-worldwide-report/.

37 Gina Kolata, "Robert G. Edwards dies at 87; Changed rules of conception with first 'test tube baby,'" *New York Times*, April 10, 2013, https://www.nytimes.com/2013/04/11/us/robert-g-edwards-nobel-winner-for-in-vitro-fertilization-dies-at-87.html.

38 J. Benjamin Hurlbut, "Imperatives of Governance," *Perspectives in Biology and Medicine* 63, (2020): 177–194, https://muse.jhu.edu/article/748059/pdf.

Chapter 16: Breaking the Glass

1 Pam Belluck, "Gene-edited babies: What a Chinese scientist told an American mentor," *New York Times,* April 14, 2019, https://www.nytimes.com/2019/04/14/health/gene-editing-babies.html.

2 He Jiankui, "Cold Spring Harbor Gene Editing Conference," *Science Net* (blog), August 24, 2016, http://blog.sciencenet.cn/home.php?mod=space&uid=514529&do=blog&id=998292.

3 M. DeWitt et al., "Selection-free Genome Editing of the Sickle Mutation in Human Adult Hematopoietic Stem/Progenitor Cells," *Science Translational Medicine* 8, (2016): 360ra134, https://www.ncbi.nlm.nih.gov/pmc/articles/PMC5500303/.

4 Mark DeWitt, phone interview, July 2, 2019.

5 He Jiankui, "The safety of gene-editing of human embryos to be resolved," *Science Net* (blog), February 19, 2017, http://blog.sciencenet.cn/home.php?mod=space&uid=514529&do=blog&id=1034671.

6 George Church, "Future, Human, Nature: Reading, Writing, Revolution," IGI, January 26, 2017, last viewed June 30, 2020, https://vimeo.com/209623759.

7 J. Benjamin Hurlbut, "Imperatives of Governance: Human Genome Editing and the Problem of Progress," *Perspectives in Biology and Medicine* 63, (2020): 177–194, https://muse.jhu.edu/article/748059.

8 Ibid.

9 Preetika Ran, Amy Dockser Marcus, and Wenxin Fan, "China, unhampered by rules, races ahead in gene-editing trials," *Wall Street Journal,* January 21, 2018, https://www.wsj.com/articles/china-unhampered-by-rules-races-ahead-in-gene-editing-trials-1516562360.

10 P. Liang et al., "CRISPR/Cas9-mediated gene editing in human tripronuclear zygotes," *Protein & Cell* 6, (2015): 363–372, https://www.ncbi.nlm.nih.gov/pmc/articles/PMC4417674/.

11 X. Kang et al., "Introducing precise genetic modification into human 3PN embryos by CRISPR/Cas-mediated genome editing," *Journal of Assisted Reproduction and Genetics* 33, (2016): 581–586, https://www.ncbi.nlm.nih.gov/pmc/articles/PMC4870449/.

12 L. Tang et al., "CRISPR/Cas9-mediated gene editing in human zygotes using Cas9 protein," *Molecular Genetics and Genomics* 292, (2017): 525–533, https://link.springer.com/article/10.1007%2Fs00438-017-1299-z.

13 G. Li et al., "Highly efficient and precise base editing in discarded human tripronuclear embryos," *Protein & Cell* 8, (2017): 776–779, https://link.springer.com/article/10.1007/s13238-017-0458-7.

14 P. Liang et al., "Correction of B-thalassemia mutant by base editor in human embryos," *Protein & Cell* 8, (2017): 811–822, https://link.springer.com/article/10.1007/s13238-017-0475-6.

15 C. Zhou et al., "Highly efficient base editing in human triplonuclear zygotes," *Protein & Cell* 8, (2017): 772–775, https://www.ncbi.nlm.nih.gov/pmc/articles/PMC5636752/.

16 L. Tang et al., "Highly efficient ssODN-mediated homology-directed repair of DBSs generated by CRISPR/Cas9 in human 3PN zygotes," *Molecular Reproduction and Development* 85, (2018): 461–463, https://onlinelibrary.wiley.com/doi/abs/10.1002/mrd.22983.

17 Y. Zeng et al., "Correction of the Marfan syndrome pathogenic FBN1 mutation by base editing in human cells and heterozygous embryos," *Molecular Therapy* 26, 2631–2637, https://www.cell.com/molecular-therapy-family/molecular-therapy/fulltext/S1525-0016(18)30378-2.

18 K. S. Bosley et al., "CRISPR-Germline Editing—The Community Speaks," *Nature Biotechnology* 33, (2015): 478–486, https://www.nature.com/articles/nbt.3227.

19 Kathy Niakan, "Human embryo genome editing license," The Francis Crick Institute, https://www.crick.ac.uk/research/labs/kathy-niakan/human-embryo-genome-editing-licence.

20 N. M. E. Fogarty et al., "Genome editing reveals a role for OCT4 in human embryogenesis," *Nature* 550, (2017): 67–73, https://www.nature.com/articles/nature24033.

21 Steve Connor, "First human embryos edited in U.S.," *MIT Technology Review,* July 26, 2017, https://www.technologyreview.com/s/608350/first-human-embryos-edited-in-us/.

22 H. Ma et al., "Correction of a pathogenic gene mutation in human embryos," *Nature* 548, (2017): 413–419, https://www.nature.com/articles/nature23305.

23 Megan Molteni, "US scientists edit a human embryo—but superbabies won't come easy," *WIRED,* August 2, 2017, https://www.wired.com/story/first-us-crispr-edited-embryos-suggest-superbabies-wont-come-easy/.

24 Ewan Callaway, "Did CRISPR really fix a genetic mutation in these human embryos?," *Nature* 8, August 2018, https://www.nature.com/articles/d41586-018-05915-2.

25 Pam Belluck, "In breakthrough, scientists edit a dangerous mutation from genes in human embryos," *New York Times*, August 2, 2017, https://www.nytimes.com/2017/08/02/science/gene-editing-human-embryos.html.

26 T. Schmid, "A Portlander was the first scientist to successfully edit human embryos. You can hear how he did it," *Willamette Week*, March 5, 2018, https://www.wweek.com/promotions/2018/03/05/a-portlander-was-the-first-scientist-to-successfully-edit-human-embryos-you-can-hear-how-he-did-it/.

27 J. Benjamin Hurlbut, "Imperatives of Governance: Human Genome Editing and the Problem of Progress," *Perspectives in Biology and Medicine* 63, (2020): 177–194, https://muse.jhu.edu/article/748059.

28 Steven Lombardi, phone interview, July 23, 2019.

29 He Jiankui, "Informed consent form," https://www.sciencemag.org/sites/default/files/crispr_informed-consent.pdf.

30 Jon Cohen, "The untold story of the 'circle of trust' behind the world's first gene-edited babies," *Science*, August 1, 2019, https://www.sciencemag.org/news/2019/08/untold-story-circle-trust-behind-world-s-first-gene-edited-babies.

31 Shu-Ching Jean Chen, "Genomic Dreams Coming True In China," *Forbes,* August 28, 2013, https://www.forbes.com/sites/forbesasia/2013/08/28/genomic-dreams-coming-true-in-china/#85db5162760a.

32 JK He, "Jiankui He talking about human genome editing," YouTube video, 14:45, July 29, 2017, https://www.youtube.com/watch?v=llxNRGMxyCc&t=723s.

33 Sharon Begley, "'CRISPR babies' lab asked U.S. scientist for help to disable cholesterol gene in human embryos," *STAT,* December 4, 2018, https://www.statnews.com/2018/12/04/crispr-babies-cholesterol-gene-editing/.

34 Teng Jing Xuan, "Found: CCTV's glowing 2017 coverage of gene-editing pariah He Jiankui," *CX Live*, November 30, 2018, https://www.caixinglobal.com/2018-11-30/found-cctvs-glowing-2017-coverage-of-gene-editing-pariah-he-jiankui-101353981.html.

35 A. Lash, "'JK Told Me He Was Planning This': A CRISPR Baby Q&A with Matt Porteus," *Xconomy*, December 4, 2018, https://xconomy.com/national/2018/12/04/jk-told-me-he-was-planning-this-a-crispr-baby-qa-with-matt-porteus/.

36 Jon Cohen, "The untold story of the 'circle of trust' behind the world's first gene-edited babies," *Science*, August 1, 2019, https://www.sciencemag.org/news/2019/08/untold-story-circle-trust-behind-world-s-first-gene-edited-babies.

37 Elena Shao, "Former Stanford postdoc criticized for creating the world's first gene-edited babies," *The Stanford Daily*, December 5, 2018, https://www.stanforddaily.com/2018/12/05/former-stanford-postdoc-criticized-for-creating-the-worlds-first-gene-edited-babies/.

38 Pam Belluck, "Gene-edited babies: What a Chinese scientist told an American mentor," *New York Times,* April 14, 2019, https://www.nytimes.com/2019/04/14/health/gene-editing-babies.html?module=inline.

39 Candice Choi and Marilynn Marchione, "AP exclusive: US Nobelist was told of gene-edited babies," AP, January 28, 2019, https://apnews.com/3f3bdc73e7c84fe685f2813510329d62.

40 A. Joseph, R. Robbins, and S. Begley, "An Outsider Claimed to Make Genome-Editing History—And the World Snapped to Attention," *STAT,* November 26, 2018, https://www.statnews.com/2018/11/26/he-jiankui-gene-edited-babies-china/.

41 Chengzu Long, interview TRI-CON, San Francisco, March 2, 2019.

42 Hunt Willard, email, June 29, 2019.

Chapter 17: A Maculate Conception

1 Broad Institute, "Genome editing and the germline: A conversation," YouTube video, 1:01:03, last viewed May 2, 2020, https://www.youtube.com/watch?v=POIeIILDo7k&t=.

2 Lap-Chee Tsui, email, July 6, 2019.

3 Jon Cohen, "After last week's shock, scientists scramble to prevent more gene-edited babies," *Science*, December 4, 2018, https://www.sciencemag.org/news/2018/12/after-last-weeks-shock-scientists-scramble-prevent-more-gene-edited-babies.
4 Sharon Begley and Andrew Joseph, "The CRISPR shocker: How genome-editing scientist He Jiankui rose from obscurity to stun the world," *STAT,* December 17, 2018, https://www.statnews.com/2018/12/17/crispr-shocker-genome-editing-scientist-he-jiankui/.
5 Jon Cohen, "After last week's shock, scientists scramble to prevent more gene-edited babies," *Science*, December 4, 2018, https://www.sciencemag.org/news/2018/12/after-last-weeks-shock-scientists-scramble-prevent-more-gene-edited-babies.
6 Patrick Tam, email, July 29, 2019.
7 Qiu Renzong, 2nd International Human Genome Editing Summit, Hong Kong, November 28, 2018.
8 Jing-Bao Nie, "He Jiankui's genetic misadventure: Why him? Why China?," *Hastings Center,* December 5, 2018, https://www.thehastingscenter.ojrg/jiankuis-genetic-misadventure-china/.
9 "Chinese ministry to investigate gene-edited babies claim," *People's Daily*, November 28, 2018, http://en.people.cn/n3/2018/1128/c90000-9523145.html.
10 Open letter, "122 scientists issued a joint statement: strongly condemned 'the first immune AIDS gene editor,'" November 26, 2018, https://www.yicai.com/news/100067069.html.
11 Bloomberg QuickTake News, "CRISPR Co-Inventor Slams Claims of Gene-Edited Babies," YouTube video, 2:02, last viewed June 2, 2020, https://www.youtube.com/watch?v=bk-1UEkxzVo.
12 Emmanuelle Charpentier, email, December 7, 2018.
13 Al Jazeera English, "#CRISPRbabies: What's the future of gene editing? | The Stream," YouTube video, 25:40, last viewed June 2, 2020, https://www.youtube.com/watch?v=E989S8P0pKc.
14 Antonio Regalado, "CRISPR inventor Feng Zhang calls for moratorium on gene-edited babies," *MIT Technology Review*, November 26, 2018, https://www.technologyreview.com/2018/11/26/66361/crispr-inventor-feng-zhang-calls-for-moratorium-on-baby-making/.
15 Kathy Niakan, "Expert reaction to Jiankui He's defence of his work," Science Media Centre, November 28, 2018, http://www.sciencemediacentre.org/expert-reaction-to-jiankui-hes-defence-of-his-work/.
16 Francis Collins, "Statement on claim of first gene-edited babies by Chinese researcher," *The NIH Director* (blog), November 28, 2018, https://www.nih.gov/about-nih/who-we-are/nih-director/statements/statement-claim-first-gene-edited-babies-chinese-researcher.
17 C. Wang et al., "Gene-edited babies: Chinese Academy of Medical Sciences' response and action," *Lancet* 393, (2019): 25–26, https://www.thelancet.com/journals/lancet/article/PIIS0140-6736(18)33080-0/fulltext.
18 Paul Knoepfler, "He Jiankui didn't really gene edit those girls; he mutated them," *The Niche*, December 4, 2018, https://ipscell.com/2018/12/he-jiankiu-didnt-really-gene-edit-those-girls-he-mutated-them/.
19 J. Savulescu, "Monstrous gene editing experiment," *Practical Ethics* (blog), November 26, 2018, http://blog.practicalethics.ox.ac.uk/2018/11/press-statement-monstrous-gene-editing-experiment/.
20 Derek Lowe, "After Such Knowledge," *In the Pipeline* (blog), November 28, 2018, https://blogs.sciencemag.org/pipeline/archives/2018/11/28/after-such-knowledge.
21 Kevin Davies, "CRISPR's China Crisis," *Genetic Engineering & Biotechnology News*, January 11, 2019, https://www.genengnews.com/insights/crisprs-china-crisis/.
22 Organizing Committee of the Second International Summit on Human Genome Editing, "On Human Genome Editing II," National Academies of Sciences, Engineering, Medicine, November 29, 2018, http://www8.nationalacademies.org/onpinews/newsitem.aspx?RecordID=11282018b.

23 Luke W. Vrotsos, "Chinese Researcher Who Said He Gene-Edited Babies Breaks Week of Silence, Vows to Defend Work," *Harvard Crimson*, December 7, 2018, https://www.thecrimson.com/article/2018/12/7/harvard-profs-react-to-human-gene-edit/.

24 E. Chen and P. Mozur, "Chinese scientist who claimed to make genetically edited babies is kept under guard," *New York Times*, December 28, 2018, https://www.nytimes.com/2018/12/28/world/asia/he-jiankui-china-scientist-gene-editing.html.

25 Steve Lombardi, phone interview, July 23, 2019.

26 Kevin Davies, "He Jiankui Fired in Wake of CRISPR Babies Investigation," *Genetic Engineering & Biotechnology News*, January 21, 2019, https://www.genengnews.com/news/he-jiankui-fired-in-wake-of-crispr-babies-investigation/.

27 Zach Coleman and Michelle Chan, "Chinese 'baby editing' scientist retreats from flagship company," *Nikkei Asian Review*, July 16, 2019, https://asia.nikkei.com/Business/Biotechnology/Chinese-baby-editing-scientist-retreats-from-flagship-company.

28 Dan Cloer, "Genetically modified babies: An insider's view," *Vision*, August 13, 2019, https://www.vision.org/interview-william-hurlbut-genetically-modified-babies-insiders-view-8975.

29 Jennifer Doudna, "He Jiankui," *Time* 100, April 2019, http://time.com/collection/100-most-influential-people-2019/5567707/he-jiankui/.

Chapter 18: Crossing the Germline

1 Ed Yong, "The CRISPR baby scandal gets worse by the day," *Atlantic*, December 3, 2018, https://www.theatlantic.com/science/archive/2018/12/15-worrying-things-about-crispr-babies-scandal/577234/.

2 The He Lab, "Why we chose HIV and CCR5 first," YouTube video, 2:59, last viewed June 2, 2020, https://www.youtube.com/watch?v=aezxaOn0efE.

3 P. Tebas, "Gene Editing of *CCR5* in Autologous CD4 T Cells of Persons Infected with HIV," *New England Journal of Medicine* 370, (2014): 901–910, https://www.nejm.org/doi/full/10.1056/NEJMoa1300662.

4 Rick Mullin, "On crossing an ethical line in human genome editing," *Chemical & Engineering News*, December 12, 2018, https://cen.acs.org/biological-chemistry/genomics/crossing-ethical-line-human-genome/96/web/2018/12.

5 Sean Ryder, "#CRISPRbabies: Notes on a scandal," *CRISPR Journal* 1, (2018): 355–357, https://doi.org/10.1089/crispr.2018.29039.spr.

6 Ibid.

7 R. Lovell-Badge, "CRISPR babies: a view from the centre of the storm," *Development*, (February 2019): 146, http://dev.biologists.org/content/146/3/dev175778.

8 H. Greely, "He Jiankui, embryo editing, CCR5, the London patient, and jumping to conclusions," *STAT*, April 15, 2019, https://www.statnews.com/2019/04/15/jiankui-embryo-editing-ccr5/.

9 Antonio Regalado, "China's CRISPR twins might have had their brains inadvertently enhanced," *MIT Technology Review*, February 21, 2019, https://www.technologyreview.com/s/612997/the-crispr-twins-had-their-brains-altered/.

10 M. T. Joy et al., "CCR5 Is a Therapeutic Target for Recovery after Stroke and Traumatic Brain Injury," *Cell* 176, (2019): 1143–1157, https://www.cell.com/cell/pdf/S0092-8674(19)30107-2.pdf.

11 X. Wei and R. Nielsen, "Retraction Note: CCR5-Δ32 is deleterious in the homozygous state in humans," *Nature Medicine* 25, (2019): 1796, https://www.nature.com/articles/s41591-019-0637-6.

12 Rebecca Robbins, "Major error undermines study suggesting change introduced in the CRISPR babies experiment shortens lives," *STAT*, September 27, 2019, https://www.statnews.com/2019/09/27/major-error-undermines-study-suggesting-change-introduced-in-the-crispr-babies-experiment-shortens-lives/.

13 George Church, interview, Boston, August 3, 2019.

14 Dominic Wilkinson in Editorial, "CRISPR-Cas9: a world first?," *Lancet,* December 8, 2018, https://www.thelancet.com/journals/lancet/article/PIIS0140-6736(18)33111-8/fulltext.

15 Arthur Caplan, "He Jiankui's Moral Mess," *PLOS Biologue* (blog), December 3, 2018, https://blogs.plos.org/biologue/2018/12/03/he-jiankuis-moral-mess/.

16 J. B. Hurlbut, "Imperatives of Governance: Human Genome Editing and the Problem of Progress," *Perspectives in Biology and Medicine* 63, (2020): 177–194, https://muse.jhu.edu/article/748059/pdf.

17 Hank Greely, "How can we decide if a biomedical advance is ethical?," *Leapsmag*, February 1, 2019, https://leapsmag.com/how-can-we-decide-if-a-biomedical-advance-is-ethical/.

18 Pam Belluck, "Gene-edited babies: What a Chinese scientist told an American mentor," *New York Times*, April 14, 2019, https://www.nytimes.com/2019/04/14/health/gene-editing-babies.html.

19 Pam Belluck, "Stanford Clears Professor of Helping With Gene-Edited Babies Experiment," *New York Times*, April 16, 2019, https://www.nytimes.com/2019/04/16/health/stanford-gene-editing-babies.html.

20 Alex Lash, "'JK told me he was planning this': A CRISPR baby Q&A with Matt Porteus," *Xconomy*, December 4, 2018, https://xconomy.com/national/2018/12/04/jk-told-me-he-was-planning-this-a-crispr-baby-qa-with-matt-porteus/?single_page=true.

21 Mark DeWitt, phone interview, July 2, 2019.

22 Candice Choi and Marilynn Marchione, "US Nobelist was told of gene-edited babies, Emails show," *US News*, January 28, 2019, https://www.usnews.com/news/world/articles/2019-01-28/chinese-scientist-told-us-nobelist-about-gene-edited-babies.

23 Hannah Osborne, "China's He Jiankui told Nobel winner Craig Mello about gene-edited babies months before birth," *Newsweek,* January 30, 2019, https://www.newsweek.com/craig-mello-he-jiankui-gene-editing-experiment-babies-nobel-prize-1311524.

24 N. Wade, "In new way to edit DNA, hope for treating disease," *New York Times*, December 28, 2009, https://www.nytimes.com/2009/12/29/health/research/29zinc.html.

25 Michael Specter, "The Gene Hackers," *New Yorker*, November 9, 2015, https://www.newyorker.com/magazine/2015/11/16/the-gene-hackers.

26 Ibid.

27 D. Baltimore et al., "A prudent path forward for genomic engineering and germline gene modification," *Science* 348, (2015): 36–38, https://www.ncbi.nlm.nih.gov/pmc/articles/PMC4394183/.

28 Jennifer Kahn, "The Crispr Quandary," *New York Times Magazine*, November 9, 2015, https://www.nytimes.com/2015/11/15/magazine/the-crispr-quandary.html.

29 R. Pollack, "Eugenics lurk in the shadow of CRISPR," *Science* 348, (2015): 871, https://science.sciencemag.org/content/348/6237/871.1.full.

30 H. Miller, "Germline gene therapy: We're ready," *Science* 348, (2015): 1325, https://science.sciencemag.org/content/348/6241/1325.1.

31 David Baltimore, "Why we need a summit on human gene editing," *Issues in Science and Technology,* Spring 2016, https://issues.org/why-we-need-a-summit-on-human-gene-editing/.

32 B. Baker, "The ethics of changing the human genome," *Bioscience* 66, (2016): 267–273, https://academic.oup.com/bioscience/article/66/4/267/2464097.

33 C. Brokowski, "Do CRISPR germline editing statements cut it?," *CRISPR Journal* 1, (2018): 115125, https://www.liebertpub.com/doi/10.1089/crispr.2017.0024.

34 John Holdren, "A note on genome editing," The White House, May 26, 2015, https://obamawhitehouse.archives.gov/blog/2015/05/26/note-genome-editing.

35 NASEM, *Human Genome Editing: Science, Ethics and Governance* (Washington, DC: The National Academies Press, 2017), https://www.nap.edu/catalog/24623/human-genome-editing-science-ethics-and-governance.

36 J. B. Hurlbut, "Imperatives of Governance: Human Genome Editing and the Problem of Progress," *Perspectives in Biology and Medicine* 63, (2020): 177–194, https://muse.jhu.edu/article/748059/pdf.

37 C. Brokowski, "Do CRISPR germline editing statements cut it?," *CRISPR Journal* 1, (2018): 115125, https://www.liebertpub.com/doi/10.1089/crispr.2017.0024.

38 "Genome editing and human reproduction: social and ethical issues," Nuffield Council on Bioethics, July 17, 2018, http://nuffieldbioethics.org/wp-content/uploads/Genome-editing-and-human-reproduction-FINAL-website.pdf.

39 Editorial, "Genome editing: proceed with caution," *Lancet* 392, (2018): 253, https://www.thelancet.com/journals/lancet/article/PIIS0140-6736(18)31653-2/fulltext.

40 Sarah Chan, 2nd International Human Genome Editing Summit, Hong Kong, November 28, 2018.

41 Michael Specter, "He Jiankui and the implications of experimenting with genetically edited babies," *New Yorker*, 2018, https://www.newyorker.com/news/daily-comment/he-jiankui-and-the-implications-of-experimenting-with-genetically-edited-babies.

42 Julianna LeMieux, "He Jiankui's Germline Editing Ethics Article Retracted by *The CRISPR Journal*," *Genetic Engineering & Biotechnology News*, February 20, 2019, https://www.genengnews.com/insights/he-jiankuis-germline-editing-ethics-article-retracted-by-the-crispr-journal/.

43 National Academies Genome Editing commission, http://nationalacademies.org/gene-editing/international-commission/index.htm.

44 Andrew Joseph, "Following 'CRISPR babies' scandal, senators call for international gene editing guidelines," *STAT*, July 16, 2019, https://www.statnews.com/2019/07/15/crispr-scandal-senators-guidelines/.

45 Rob Stein, "New U.S. Experiments Aim to Create Gene-Edited Human Embryos," *NPR*, February 1, 2019, https://www.npr.org/sections/health-shots/2019/02/01/689623550/new-u-s-experiments-aim-to-create-gene-edited-human-embryos.

46 Jon Cohen, "'I feel an obligation to be balanced.' Noted biologist comes to defense of gene editing babies," *Science*, November 28, 2018, https://www.sciencemag.org/news/2018/11/i-feel-obligation-be-balanced-noted-biologist-comes-defense-gene-editing-babies.

47 Samira Kiana, phone interview, January 30, 2019.

Chapter 19: Going Rogue

1 Antonio Regalado, "China's CRISPR babies: Read exclusive excerpts from the unseen original research," *MIT Technology Review*, December 2, 2019, https://www.technologyreview.com/f/614779/chinas-crispr-babies-read-exclusive-excerpts-from-the-unseen-original-research/.

2 NAS Colloquium, "Fyodor Urnov: The next generation of edited humans," YouTube video, 25:53, last viewed March 2, 2020, https://youtu.be/XzSWVzRSfnYt.

3 Ibid.

4 Editorial, "Brave new dialogue," *Nature Genetics* 51, (2019): 365, https://www.nature.com/articles/s41588-019-0374-2.

5 "Chinese court sentences 'gene-editing' scientist to three years in prison," *Reuters*, December 29, 2019, https://www.reuters.com/article/us-china-health-babies/chinese-court-sentences-gene-editing-scientist-to-three-years-in-prison-idUSKBN1YY06R.

6 Hannah Osborne, "Chinese scientist He Jiankui jailed for creating world's first gene edited babies 'in the pursuit of personal fame and gain,'" *Newsweek*, December 30, 2019, https://www.newsweek.com/he-jiankui-jailed-gene-editing-1479614.

7 Ken Moritsugu, "China convicts 3 researchers involved in gene-edited babies," AP, December 30, 2019, https://apnews.com/7bf5ad48696d24628e49254df504e3ee.

8 Dennis Normille, "Chinese scientist who produced genetically altered babies sentenced to 3 years in jail," *Science*, December 30, 2019, https://www.sciencemag.org/news/2019/12/chinese-scientist-who-produced-genetically-altered-babies-sentenced-3-years-jail.

9 Josiah Zayner, "CRISPR babies scientist He Jiankui should not be villainized—or headed to prison," *STAT*, January 2, 2020, https://www.statnews.com/2020/01/02/crispr-babies-scientist-he-jiankui-should-not-be-villainized/.

10 Eric Topol, "Editing Babies? We Need to Learn a Lot More First," *New York Times*, November 27, 2018, https://www.nytimes.com/2018/11/27/opinion/genetically-edited-babies-china.html.

11 Jane Qiu, "Chinese government funding may have been used for 'CRISPR babies' project, documents suggest," *STAT*, February 25, 2019, https://www.statnews.com/2019/02/25/crispr-babies-study-china-government-funding/.

12 X. Zhai et al., "Chinese Bioethicists Respond to the Case of He Jiankui," *Hastings Center*, February 7, 2019, https://www.thehastingscenter.org/chinese-bioethicists-respond-case-jiankui/.

13 Steve Lombardi, phone interview, July 23, 2019.

14 George Church, interview, Boston, August 3, 2019.

15 Michael Specter, "The gene factory," *New Yorker*, December 30, 2013, https://www.newyorker.com/magazine/2014/01/06/the-gene-factory.

16 Steven Lee Myers, "China's Moon Landing: Lunar Rover Begins Its Exploration," *New York Times*, January 3, 2019, https://www.nytimes.com/2019/01/03/world/asia/china-change-4-moon.html.

17 Preetika Rana, Amy Dockser Marcus, and Wenxin Fan, "China, unhampered by rules, races ahead in gene-editing trials," *Wall Street Journal*, January 21, 2018, https://www.wsj.com/articles/china-unhampered-by-rules-races-ahead-in-gene-editing-trials-1516562360.

18 Rob Stein, "First U.S. patients treated with CRISPR as human gene-editing experiments get underway," *NPR*, April 16, 2019, https://www.npr.org/sections/health-shots/2019/04/16/712402435/first-u-s-patients-treated-with-crispr-as-gene-editing-human-trials-get-underway.

19 Yangyang Cheng, "China will always be bad at bioethics," *Foreign Policy*, April 13, 2018, https://foreignpolicy.com/2018/04/13/china-will-always-be-bad-at-bioethics/.

20 Z. Liu et al., "Cloning of Macaque Monkeys by Somatic Cell Nuclear Transfer," *Cell* 172, (2018): 881–887, https://www.cell.com/fulltext/S0092-8674(18)30057-6.

21 Didi Kirsten Tatlow, "A Scientific Ethical Divide Between China and West," *New York Times*, June 29, 2015, https://www.nytimes.com/2015/06/30/science/a-scientific-ethical-divide-between-china-and-west.html.

22 Yangyang Cheng, "China will always be bad at bioethics," *Foreign Policy*, April 13, 2018, https://foreignpolicy.com/2018/04/13/china-will-always-be-bad-at-bioethics/.

23 Ibid.

24 Lin Yang, "Exploring the Future Of Life Economy With BGI Co-Founder Wang Jian," *Forbes*, June 16, 2016, https://www.forbes.com/sites/linyang/2016/06/16/exploring-the-future-of-life-economy-with-bgi-co-founder-wang-jian/#6ed2f3cd75e0.

25 Mei Fong, "Before the Claims of CRISPR Babies, There Was China's One-Child Policy," *New York Times*, November 28, 2019, https://www.nytimes.com/2018/11/28/opinion/china-crispr-babies.html.

26 Yangyang Cheng, "China will always be bad at bioethics," *Foreign Policy*, April 13, 2018, https://foreignpolicy.com/2018/04/13/china-will-always-be-bad-at-bioethics/.

27 J. B. Hurlbut, "Imperatives of governance," *Perspectives in Biology and Medicine* 63, (2020): 177–194, https://muse.jhu.edu/article/748059.

28 Sharon Begley, "He took a crash course in bioethics. Then he created CRISPR babies," *STAT*, November 27, 2018, https://www.statnews.com/2018/11/27/crispr-babies-creator-soaked-up-bioethics/.

29 Broad Institute, "Genome editing and the germline: A conversation," YouTube video, 1:01:03, last viewed June 3, 2020, https://youtu.be/POIeIILDo7k.

30 Sandy Macrae, "Workshop on Genome Editing," National Academy of Sciences, Washington, DC, August 2019.

31 David Cyranoski, "Russian biologist plans more CRISPR-edited babies," *Nature*, June 10, 2019, https://www.nature.com/articles/d41586-019-01770-x.

32 "Expert reaction to *New Scientist* exclusive reporting that five deaf Russian couples (and scientist Denis Rebrikov) want to try CRISPR to have a child who can hear," Science Media Centre, July 4, 2019, https://www.sciencemediacentre.org/expert-reaction-to-new-scientist-exclusive-reporting-that-five-deaf-russian-couples-and-scientist-denis-rebrikov-want-to-try-crispr-to-have-a-child-who-can-hear/.

33 Jon Cohen, "Russian geneticist answers challenges to his plan to make gene-edited babies," *Science*, June 13, 2019, https://www.sciencemag.org/news/2019/06/russian-geneticist-answers-challenges-his-plan-make-gene-edited-babies.

34 Michael Le Page, "Exclusive: Five couples lined up for CRISPR babies to avoid deafness," *New Scientist*, July 4, 2019, https://www.newscientist.com/article/2208777-exclusive-five-couples-lined-up-for-crispr-babies-to-avoid-deafness/.

35 Stepan Kravchenko, "Future of Genetically Modified Babies May Lie in Putin's Hands," *Bloomberg*, September 29, 2019, https://www.bloomberg.com/news/articles/2019-09-29/future-of-genetically-modified-babies-may-lie-in-putin-s-hands.

36 Jon Cohen, "Embattled Russian scientist sharpens plans to create gene-edited babies," *Science*, October 21, 2019, https://www.sciencemag.org/news/2019/10/embattled-russian-scientist-sharpens-plans-create-gene-edited-babies.

37 Jon Cohen, "Deaf couple may edit embryo's DNA to correct hearing mutation," *Science*, October 21, 2019, https://www.sciencemag.org/news/2019/10/deaf-couple-may-edit-embryo-s-dna-correct-hearing-mutation.

38 David Cyranoski, "Russian 'CRISPR-baby' scientist has started editing genes in human eggs with goal of altering deaf gene," *Nature*, October 18, 2019, https://www.nature.com/articles/d41586-019-03018-0.

39 Russia Insight, "SCARY: Putin Warns Of GM Super Human Soldiers That Are Worse Than Nukes," YouTube video, 2:13, October 24, 2017, https://www.youtube.com/watch?v=9v3TNGmbArs.

40 E. Lander, F. Baylis, F. Zhang et al., "Adopt a moratorium on heritable genome editing," *Nature* 367, (2019): 165–168, https://www.nature.com/articles/d41586-019-00726-5.

41 Eric S. Lander, "The Heroes of CRISPR," *Cell* 164, (2016): 18–28, https://www.cell.com/cell/pdf/S0092-8674(15)01705-5.pdf.

42 C. D. Wolinetz and F. S. Collins, "NIH supports call for moratorium on clinical use of germline gene editing," *Nature* 567:175 (2019), https://www.ncbi.nlm.nih.gov/pmc/articles/PMC6688589/.

43 ASGCT, "Scientific leaders call for global moratorium on germline gene editing," April 24, 2019, https://www.asgct.org/research/news/april-2019/scientific-leaders-call-for-global-moratorium-on-g.

44 Joel Achenbach, "NIH and top scientists call for moratorium on gene-edited babies," *Washington Post*, March 13, 2019, https://www.washingtonpost.com/science/2019/03/13/nih-top-scientists-call-moratorium-gene-edited-babies/.

45 Sharon Begley, "Leading scientists, backed by NIH, call for a global moratorium on creating 'CRISPR babies,'" *STAT*, March 13, 2019, https://www.statnews.com/2019/03/13/crispr-babies-germline-editing-moratorium/.

46 George Church, interview, Boston, August 3, 2019.
47 Jennifer Doudna, "CRISPR's unwanted anniversary," *Science* 366, (2019): 777, https://science.sciencemag.org/content/366/6467/777.
48 Giro Studio, "Editorial Humility: A Moratorium on Heritable Genome Editing?," YouTube video, 2:17:43, May 6, 2019, https://www.youtube.com/watch?v=s7p4D31aLTI&list=PLdT7Y4C6bsoRyqh-mpBZBCIPsOKey5YsJ&index=11&t=994s.
49 Françoise Baylis, Keystone Symposium, Banff, Canada, February 9, 2020.
50 Mei Fong, "Before the Claims of CRISPR Babies, There Was China's One-Child Policy," *New York Times*, November 28, 2019, https://www.nytimes.com/2018/11/28/opinion/china-crispr-babies.html.
51 H. Wang and H. Yang, "Gene-edited babies: What went wrong and what could go wrong," *PLOS Biology* 17: e3000224, https://doi.org/10.1371/journal.pbio.3000224.
52 M. Zhang et al., "Human cleaving embryos enable robust homozygotic nucleotide substitutions by base editors," *Genome Biology* 20, (2019): 101, https://genomebiology.biomedcentral.com/articles/10.1186/s13059-019-1703-6#Decs.
53 Stephen Chen, "Gene-editing breakthrough in China comes with urgent call for global rules," *South China Morning Post*, June 1, 2019, https://www.scmp.com/news/china/science/article/3012615/gene-editing-breakthrough-china-comes-urgent-call-global-rules.
54 Heidi Ledford, "CRISPR gene editing in human embryos wreaks chromosomal mayhem," *Nature*, June 25, 2020, https://www.nature.com/articles/d41586-020-01906-4.

Chapter 20: To Extinction and Beyond

1 Ben Mezrich, *Woolly: The True Story of the Quest to Revive One of History's Most Iconic Extinct Creatures* (New York: Atria Books, 2017).
2 Ozzy Osbourne, "The Osbourne Identity," *Sunday Times*, October 24, 2010, https://www.thetimes.co.uk/article/the-osbourne-identity-d9kjh3cxmql.
3 G. M. Church, "Sponsored sequencing: our vision for the future of genomics," *Genetic Engineering & Biotechnology News*, June 11, 2019, https://www.genengnews.com/commentary/sponsored-sequencing-our-vision-for-the-future-of-genomics/.
4 Stephen Colbert, "George Church," *Time*, April 20, 2017, https://time.com/collection/2017-time-100/4742749/george-church/.
5 Constance Grady, "It's Margaret Atwood's dystopian future, and we're just living in it," *Vox*, June 8, 2016, https://www.vox.com/2016/6/8/11885596/margaret-atwood-dystopian-future-handmaids-tale-maddaddam-pigoons.
6 Michael Specter, "How the DNA revolution is changing us," *National Geographic*, August 2016, https://www.nationalgeographic.com/magazine/2016/08/dna-crispr-gene-editing-science-ethics/.
7 D. Niu et al., "Inactivation of porcine endogenous retrovirus in pigs using CRISPR-Cas9," *Science* 357, (2017): 1303–1307, https://science.sciencemag.org/content/357/6357/1303.
8 Alice George, "The sad, sad story of Laika, the space dog, and her one-way trip into orbit," *Smithsonianmag.com*, April 11, 2018, https://www.smithsonianmag.com/smithsonian-institution/sad-story-laika-space-dog-and-her-one-way-trip-orbit-1-180968728/.
9 Eriona Hysolli, "An American-Russian collaboration to repopulate Siberia with woolly mammoths . . . or something similar," *Medium*, December 31, 2018, https://medium.com/@eriona.hysolli/an-american-russian-collaboration-to-repopulate-siberia-with-woolly-mammoths-or-something-similar-9cbac4e985cb.
10 George Church, interview HMS, Boston, August 3, 2019.
11 Ross Andersen, "The Arctic Mosquito Swarms Large Enough to Kill a Baby Caribou," *Atlantic*, September 16, 2015, https://www.theatlantic.com/science/archive/2015/09/arctic-mosquitoes-and-the-chaos-of-climate-change/405322/.

12 Beth Shapiro, *How to Clone a Mammoth* (Princeton, N.J.: Princeton University Press, 2015).

13 The Royal Institution, "How to Clone a Mammoth: The Science of De-Extinction—with Beth Shapiro," YouTube video, 54:10, last viewed June 15, 2020, https://www.youtube.com/watch?v=xO043PSBnKU.

14 Ben Jacob Novak, "De-Extinction," *Genes* 9, (2018): 548, https://www.mdpi.com/2073-4425/9/11/548/htm.

15 Steven Salzberg, "The Loneliest Word, And The Extinction Crisis," *Forbes*, July 8, 2019, https://www.forbes.com/sites/stevensalzberg/2019/07/08/is-this-the-loneliest-word/#1d02e2bd2367.

16 Ed Yong, "The last of its kind," *Atlantic*, July 2019, https://www.theatlantic.com/magazine/archive/2019/07/extinction-endling-care/590617/.

17 Stewart Brand, "The dawn of de-extinction. Are you ready?," TED, February 2013, https://www.ted.com/talks/stewart_brand_the_dawn_of_de_extinction_are_you_ready?language=en.

18 Elizabeth Kolbert, *The Sixth Extinction: An Unnatural History* (New York: Henry Holt & Co., 2014).

19 B. J. Novak et al., "Advancing a New Toolkit for Conservation: From Science to Policy," *CRISPR Journal* 1, (2018): 11–15, https://www.liebertpub.com/doi/10.1089/crispr.2017.0019.

20 Gabriel Popkin, "To save iconic American chestnut, researchers plan introduction of genetically engineered tree into the wild," *Science*, August 29, 2018, https://www.sciencemag.org/news/2018/08/save-iconic-american-chestnut-researchers-plan-introduction-genetically-engineered-tree.

21 G. G. R. Murray et al., "Natural selection shaped the rise and fall of passenger pigeon genomic diversity," *Science* 358, (2017): 951–954, https://science.sciencemag.org/content/358/6365/951.

22 Amy Dockser Marcus, "Meet the Scientists Bringing Extinct Species Back from the Dead," *Wall Street Journal*, October 11, 2018, https://www.wsj.com/articles/meet-the-scientists-bringing-extinct-species-back-from-the-dead-1539093600.

23 Ibid.

24 Charles Q. Choi, "First extinct-animal clone created," *National Geographic*, February 10, 2009, https://www.nationalgeographic.com/science/2009/02/news-bucardo-pyrenean-ibex-deextinction-cloning/.

25 Timothy Winegard, "People v mosquitoes: what to do about our biggest killer," *Guardian*, September 20, 2019, https://www.theguardian.com/environment/2019/sep/20/man-v-mosquito-biggest-killer-malaria-crispr.

26 E. Tognotti, "Program to eradicate Malaria in Sardinia, 1946–1950," *Emerging Infectious Diseases* 15, (2009): 1460–1466, https://www.ncbi.nlm.nih.gov/pmc/articles/PMC2819864/.

27 K. Davies and K. M. Esvelt, "Gene Drives, White-Footed Mice, and Black Sheep: An Interview with Kevin Esvelt," *CRISPR Journal* 1, (2018): 319–324, https://www.liebertpub.com/doi/abs/10.1089/crispr.2018.29031.kda.

28 K. M. Esvelt et al., "A system for the continuous directed evolution of biomolecules," *Nature* 472, (2011): 499–503, https://www.ncbi.nlm.nih.gov/pmc/articles/PMC3084352/.

29 A. Burt, "Site-specific selfish genes as tools for the control and genetic engineering of natural populations," *Proceedings of the Royal Society Biological Sciences* 270, (2003): 921–928, https://www.ncbi.nlm.nih.gov/pmc/articles/PMC1691325/.

30 N. Windbichler et al., "A synthetic homing endonuclease-based gene drive system in the human malaria mosquito," *Nature* 473, (2011): 212–215, https://www.ncbi.nlm.nih.gov/pmc/articles/PMC3093433/.

31 K. M. Esvelt et al., "Emerging technology: Concerning RNA-guided gene drives for the alteration of wild populations," *eLife* 3, (2014): e03401, https://elifesciences.org/articles/03401.

32 V. Gantz et al., "Highly efficient Cas9-mediated gene drive for population modification of the malaria vector mosquito *Anopheles stephensi*," *Proceedings of the National Academy of Sciences USA* 112, (2015): E6736–6743, https://www.pnas.org/content/112/49/E6736.

33 A. Hammond et al., "A CRISPR-Cas9 gene drive system targeting female reproduction in the malaria mosquito vector *Anopheles gambiae*," *Nature Biotechnology* 34, (2016): 78–83, https://www.nature.com/articles/nbt.3439.

34 Kevin M. Esvelt, "Gene drive should be a nonprofit technology," *STAT*, November 27, 2018, https://www.statnews.com/2018/11/27/gene-drive-should-be-nonprofit-technology/.

35 Sharon Begley, "In a lab pushing the boundaries of biology, an embedded ethicist keeps scientists in check," *STAT*, February 23, 2017, https://www.statnews.com/2017/02/23/bioethics-harvard-george-church/.

36 K. Kyrou et al., "A CRISPR–Cas9 gene drive targeting *doublesex* causes complete population suppression in caged *Anopheles gambiae* mosquitoes," *Nature Biotechnology* 36, (2018): 1062–1066, https://www.nature.com/articles/nbt.4245.

37 Anna Pujol-Mazzini, "'We don't want to be guinea pigs': How one African community is fighting genetically modified mosquitoes," *Telegraph*, October 8, 2019, https://www.telegraph.co.uk/global-health/science-and-disease/dont-want-guinea-pigs-one-african-community-fighting-genetically/.

38 Martin Fletcher, "Mutant mosquitoes: Can gene editing kill off malaria?," *Telegraph*, August 11, 2018, https://www.telegraph.co.uk/news/0/mutant-mosquitoes-can-gene-editing-kill-malaria/.

39 Feng Zhang, CRISPRcon, Boston, June 4, 2018.

40 J. E. DiCarlo et al., "Safeguarding CRISPR-Cas9 gene drives in yeast," *Nature Biotechnology* 33, (2015): 1250–1255, https://www.ncbi.nlm.nih.gov/pmc/articles/PMC4675690/.

41 Michael Eisenstein, "A toolbox for keeping CRISPR in check," *Genetic Engineering & Biotechnology News*, July 11, 2019, https://www.genengnews.com/insights/a-toolbox-for-keeping-crispr-in-check/.

42 B. Maji et al., "A High-Throughput Platform to Identify Small-Molecule Inhibitors of CRISPR-Cas9," *Cell* 177, (2019): 1067–1079, https://www.cell.com/cell/pdf/S0092-8674(19)30395-2.pdf.

43 Julianna LeMieux, "CRISPR-accelerated gene drives pump the brakes," *Genetic Engineering & Biotechnology News*, July 1, 2019, https://www.genengnews.com/topics/genome-editing/crispr-accelerated-gene-drives-pump-the-brakes/.

44 "Synthetic Biology," Convention on Biological Diversity, November 28, 2018, https://www.cbd.int/doc/c/2c62/5569/004e9c7a6b2a00641c3af0eb/cop-14-l-31-en.pdf.

45 Nicholas Wade, "Giving Malaria a Deadline," *New York Times*, September 24, 2018, https://www.nytimes.com/2018/09/24/science/gene-drive-mosquitoes.html.

46 Martin Fletcher, "Mutant mosquitoes: Can gene editing kill off malaria?," *Telegraph*, August 11, 2018, https://www.telegraph.co.uk/news/0/mutant-mosquitoes-can-gene-editing-kill-malaria/.

47 K. Davies and K. M. Esvelt, "Gene Drives, White-Footed Mice, and Black Sheep: An Interview with Kevin Esvelt," *CRISPR Journal* 1, (2018): 319–324, https://www.liebertpub.com/doi/abs/10.1089/crispr.2018.29031.kda.

48 J. Buchtal et al., "Mice against ticks: an experimental community-guided effort to prevent tick-borne disease by altering the shared environment," *Proceedings of the Royal Society Biological Sciences* 374, (2019): 20180105, https://www.ncbi.nlm.nih.gov/pmc/articles/PMC6452264/.

49 Allison Snow, "Gene editing to stop Lyme disease: caution is warranted," *STAT*, August 22, 2019, https://www.statnews.com/2019/08/22/gene-editing-to-stop-lyme-disease-caution-is-warranted/.

50 Charles Darwin, Letter to Asa Gray, May 22, 1860, https://www.darwinproject.ac.uk/letter/DCP-LETT-2814.xml.

51 Rebecca Kreston, "The special brand of horror that is the New World Screwworm," *Discover Body Horrors* (blog), July 22, 2013, http://blogs.discovermagazine.com/bodyhorrors/2013/07/22/screwworm-myiasis/#.XSqhFuhKg2w.

52 C. Noble et al., "Daisy-chain gene drives for the alteration of local populations," *Proceedings of the National Academy of Sciences* 116, (2019): 8275–8282, https://www.ncbi.nlm.nih.gov/pmc/articles/PMC6486765/.
53 Antonio Regalado, "Top U.S. Intelligence Official Calls Gene Editing a WMD Threat," *MIT Technology Review,* February 9, 2016, https://www.technologyreview.com/2016/02/09/71575/top-us-intelligence-official-calls-gene-editing-a-wmd-threat/.
54 Ewen MacAskill, "Bill Gates warns tens of millions could be killed by bio-terrorism," *Guardian,* February 18, 2017, https://www.theguardian.com/technology/2017/feb/18/bill-gates-warns-tens-of-millions-could-be-killed-by-bio-terrorism.
55 Daniel M. Gerstein, "Can the bioweapons convention survive Crispr?," *Bulletin of the Atomic Scientists,* July 25, 2016, https://thebulletin.org/2016/07/can-the-bioweapons-convention-survive-crispr/.

Chapter 21: Farm Aid

1 Mark Lynas, *Seeds of Science* (New York: Bloomsbury Sigma, 2018).
2 Kelly Servick, "Once again, U.S. expert panel says genetically engineered crops are safe to eat," *Science,* May 17, 2016, https://www.sciencemag.org/news/2016/05/once-again-us-expert-panel-says-genetically-engineered-crops-are-safe-eat#.
3 Brad Plumer, "More than 100 Nobel laureates are calling on Greenpeace to end its anti-GMO campaign," *Vox,* June 30, 2016, https://www.vox.com/2016/6/30/12066826/greenpeace-gmos-nobel-laureates.
4 Issues Ink Media, "Mark Lynas on his conversion to supporting GMOs—Oxford Lecture on Farming," YouTube video, 51:52, last viewed June 22, 2020, https://www.youtube.comwatch?v=vf86QYf4Suo.
5 Will Storr, "Mark Lynas: Truth, Treachery and GM food," *Observer,* March 9, 2013, https://www.theguardian.com/environment/2013/mar/09/mark-lynas-truth-treachery-gm.
6 Erin Brodwin, "We'll be eating the first Crispr'd foods within 5 years, according to a geneticist who helped invent the blockbuster gene-editing tool," *Business Insider,* April 20, 2019, https://www.businessinsider.com/first-crispr-food-5-years-berkeley-scientist-inventor-2019-4.
7 Sarah Webb, "Plants in the CRISPR," *BioTechniques,* March 16, 2018, https://www.future-science.com/doi/10.2144/000114583.
8 Jon Cohen, "To feed its 1.4 billion, China bets big on genome editing of crops," *Science,* July 29, 2019, https://www.sciencemag.org/news/2019/07/feed-its-14-billion-china-bets-big-genome-editing-crops.
9 Matt Ridley, "Editing Our Genes, One Letter at a Time," *Wall Street Journal,* January 11, 2013, http://www.mattridley.co.uk/blog/precision-editing-of-dna/.
10 Charles Mann, *The Wizard and the Prophet* (New York: Knopf, 2018).
11 Paula Park, "Mary-Dell Chilton," *The Scientist,* April 29, 2002, https://www.the-scientist.com/news-profile/mary-dell-chilton-53389.
12 Steven Salzberg, "Surprise! Your Beer and Tea Are Actually Transgenic GMOs," *Forbes,* January 20, 2020, https://www.forbes.com/sites/stevensalzberg/2020/01/20/surprise-here-are-12-organic-foods-that-are-transgenic-gmos/#6a9e43ab427f.
13 Susan Hockfield, *The Age of Living Machines* (Norton: New York, 2019).
14 R. Oliva et al., "Broad-spectrum resistance to bacterial blight in rice using genome editing," *Nature Biotechnology* 37, (2019): 1344–1350, https://www.nature.com/articles/s41587-019-0267-z.
15 Phil Edwards, "A Renaissance painting reveals how breeding changed watermelons," *Vox,* August 3, 2016, https://www.vox.com/2015/7/28/9050469/watermelon-breeding-paintings.
16 Stephen S. Hall, "Crispr Can Speed Up Nature—and Change How We Grow Food," *WIRED,* July 17, 2018, https://www.wired.com/story/crispr-tomato-mutant-future-of-food/.

17 S. Soyk et al., "Variation in the flowering gene SELF PRUNING 5G promotes day-neutrality and early yield in tomato," *Nature Genetics* 49, (2017): 162–168, https://www.nature.com/articles/ng.3733.
18 Z. H. Lemmon et al., "Rapid improvement of domestication traits in an orphan crop by genome editing," *Nature Plants* 4, (2018): 776–770, https://www.nature.com/articles/s41477-018-0259-x.
19 C-T. Kwon et al., "Rapid customization of Solanaceae fruit crops for urban agriculture," *Nature Biotechnology* 38, (2019): 182–188, https://www.nature.com/articles/s41587-019-0361-2.
20 Rodolphe Barrangou, "CRISPR craziness: A response to the EU Court ruling," *CRISPR Journal* 1, (2018): 4, https://www.liebertpub.com/doi/full/10.1089/crispr.2018.29025.edi.
21 Jon Cohen, "To feed its 1.4 billion, China bets big on genome editing of crops," *Science*, July 29, 2019, https://www.sciencemag.org/news/2019/07/feed-its-14-billion-china-bets-big-genome-editing-crops.
22 Oxford Nanopore Technologies, "Mick Watson | The MinION: Applications in Animal Health and Food Security," YouTube video, 25:30, last viewed February 11, 2019, https://youtu.be/UK8KHlkHhhc.
23 Megan Molteni, "The First Gene Edited Food Is Now Being Served," *WIRED*, March 20, 2019, https://www.wired.com/story/the-first-gene-edited-food-is-now-being-served/.
24 Y. Wang et al., "Simultaneous editing of three homoeoalleles in hexaploid bread wheat confers heritable resistance to powdery mildew," *Nature Biotechnology* 32, (2014): 947-951, https://www.nature.com/articles/nbt.2969.
25 R. Zhang et al., "Generation of herbicide tolerance traits and a new selectable marker in wheat using base editing," *Nature Plants* 5, (2019): 480–485, https://www.nature.com/articles/s41477-019-0405-0.
26 Roger Williams, "Green is Florida's new orange," *Florida Weekly*, August 23, 2018, https://charlottecounty.floridaweekly.com/articles/green-is-floridas-new-orange/.
27 Ibid.
28 Cici Zhang, "Citrus greening is killing the world's orange trees. Scientists are racing to help," *Chemical & Engineering News*, June 9, 2019, https://cen.acs.org/biological-chemistry/biochemistry/Citrus-greening-killing-worlds-orange/97/i23.
29 L. Sun et al., "Citrus Genetic Engineering for Disease Resistance: Past, Present and Future," *International Journal of Molecular Sciences* 20, (2019): 5256, https://www.ncbi.nlm.nih.gov/pmc/articles/PMC6862092/.
30 Cici Zhang, "Citrus greening is killing the world's orange trees. Scientists are racing to help," *Chemical & Engineering News*, June 9, 2019, https://cen.acs.org/biological-chemistry/biochemistry/Citrus-greening-killing-worlds-orange/97/i23.
31 Amy Harmon, "A race to save the orange by altering its DNA," *New York Times*, July 27, 2013, https://www.nytimes.com/2013/07/28/science/a-race-to-save-the-orange-by-altering-its-dna.html.
32 Paul Voosen, "Can genetic engineering save the Florida orange?," *National Geographic*, September 13, 2014, https://news.nationalgeographic.com/news/2014/09/140914-florida-orange-citrus-greening-gmo-environment-science/.
33 Dan Koeppel, "Yes, we will have no bananas," *New York Times*, June 18, 2008, https://www.nytimes.com/2008/06/18/opinion/18koeppel.html.
34 Myles Karp, "The banana is one step closer to disappearing," *National Geographic*, August 12, 2019, https://www.nationalgeographic.com/environment/2019/08/banana-fungus-latin-america-threatening-future/.
35 Emiko Terazono and Clive Cookson, "Gene editing: how agritech is fighting to shape the food we eat," *Financial Times*, February 10, 2019, https://www.ft.com/content/74fb67b8-2933-11e9-a5ab-ff8ef2b976c7.
36 Sam Bloch, "At CRISPRcon, an organic luminary embraces gene editing. Will the industry follow?," *The Counter*, June 6, 2018, https://thecounter.org/klaas-martens-organic-gene-editing-crispr-gmo/.

37 Ruramiso Mashumba, CRISPRcon 2018, Boston, June 4, 2018.
38 M. A. Gomez et al., "Simultaneous CRISPR/Cas9-mediated editing of cassava eIF4E isoforms nCBP-1 and nCBP-2 reduces cassava brown streak disease symptom severity and incidence," *Plant Biotechnology Journal* (2018): 1–14, doi: 10.1111/pbi.12987.
39 Emily Willingham, "Seralini Paper Influences Kenya Ban of GMO Imports," *Forbes*, December 9, 2012, https://www.forbes.com/sites/emilywillingham/2012/12/09/seralini-paper-influences-kenya-ban-of-gmo-imports/#1d951b2268a0.
40 Marcel Kuntz, "The Seralini Affair. The Dead-End of an Activist Science," *Fondation pour l'Innovation Politique*, September 2019, https://www.supportprecisionagriculture.org/165_LaffaireSERALINI_GB_2019-09-25_w.pdf.
41 Anon, "Smelling a rat," *Economist*, December 7, 2013, https://www.economist.com/science-and-technology/2013/12/07/smelling-a-rat.
42 "Controversial Seralini study linking GM to cancer in rats is republished," *Guardian*, June 24, 2014, https://www.theguardian.com/environment/2014/jun/24/controversial-seralini-study-gm-cancer-rats-republished.
43 Verenardo Meeme, "Kenya picks 1,000 farmers to grow GMO cotton," Cornell Alliance for Science, March 9, 2020, https://allianceforscience.cornell.edu/blog/2020/03/kenya-picks-1000-farmers-to-grow-gmo-cotton/.
44 Nnimmo Bassey, CRISPRcon, Boston, June 4, 2018.
45 K. Davies and N. Gutterson, "Planting Progress: An Interview with Neal Gutterson," *CRISPR Journal* 1, (2018): 270–273, https://www.liebertpub.com/doi/full/10.1089/crispr.2018.29023.int.
46 Y. Zhang et al., "A CRISPR way for accelerating improvement of food crops," *Nature Food* 1, (2020): 200–205, https://www.nature.com/articles/s43016-020-0051-8.
47 Nick Stockton, "The FDA Wants to Regulate Edited Animal Genes as Drugs," *WIRED*, January 24, 2017, https://www.wired.com/2017/01/fda-wants-regulate-edited-animal-genes-drugs/.
48 D. Carroll et al., "Regulate genome-edited products, not genome editing itself," *Nature Biotechnology* 34, (2016): 477–479, https://www.nature.com/articles/nbt.3566.
49 D. F. Carlson et al., "Production of hornless dairy cattle from genome-edited cell lines," *Nature Biotechnology* 34, (2016): 479–481, https://www.nature.com/articles/nbt.3560.
50 Carolyn Y. Johnson, "Gene-edited farm animals are coming. Will we eat them?," *Washington Post*, December 17, 2018, https://www.washingtonpost.com/news/national/wp/2018/12/17/feature/gene-edited-farm-animals-are-coming-will-we-eat-them/.
51 A. L. Norris et al., "Template plasmid integration in germline genome-edited cattle," *Nature Biotechnology* 38, (2020): 163–164, https://www.nature.com/articles/s41587-019-0394-6.
52 Antonio Regalado, "Gene-edited cattle have a major screwup in their DNA," *MIT Technology Review*, August 29, 2019, https://www.technologyreview.com/s/614235/recombinetics-gene-edited-hornless-cattle-major-dna-screwup/.
53 C. Tait-Burkhard et al., "Livestock 2.0—genome editing for fitter, healthier, and more productive farmed animals," *Genome Biology* 19, (2018): 204, https://genomebiology.biomedcentral.com/articles/10.1186/s13059-018-1583-1.
54 Q. Zheng et al., "Reconstitution of *UCP1* Using CRISPR/Cas9 in the White Adipose Tissue of Pigs Decreases Fat Deposition and Improves Thermogenic Capacity," *Proceedings of the National Academy of Sciences USA* 114, (2017): E9474–9482, https://www.ncbi.nlm.nih.gov/pmc/articles/PMC5692550/.

Chapter 22: CRISPR Prime

1 Victor A. McKusick, "The defect in Marfan syndrome," *Nature* 352, (1991): 279–281, https://www.nature.com/articles/352279a0.pdf.

2 Steve Jones, *The Language of Genes* (New York: Anchor Books, 1994).
3 K. O'Brien, "He is blazing his own molecular trials," *Boston Globe*, February 13, 2006, http://archive.boston.com/news/science/articles/2006/02/13/he_is_blazing_his_own_molecular_trails/.
4 Megan Molteni, "Inside a chemist's quest to hack evolution and cure genetic disease," *WIRED*, June 12, 2018, https://www.wired.com/story/inside-a-chemists-quest-to-hack-evolution-and-cure-genetic-disease/.
5 K. M. Esvelt et al., "A System for the Continuous Directed Evolution of Biomolecules," *Nature* 472, (2011): 499–503, https://www.ncbi.nlm.nih.gov/pmc/articles/PMC3084352/.
6 Asher Mullard, "An audience with David Liu," *Nature Reviews Drug Discovery* 18, (2019): 330–331, https://www.nature.com/articles/d41573-019-00067-y.
7 CEN Online, "David Liu—Advice to the future of chemistry," YouTube video, 30:12, last viewed March 3, 2020, https://www.youtube.com/watch?v=cnf1C8Qf4hw.
8 Ibid.
9 Ibid.
10 Ryan Cross, "Inventor, chemist, and CRISPR craftsman: Inside David Liu's evolution workshop," *Chemical & Engineering News*, April 16, 2018, https://cen.acs.org/biological-chemistry/biotechnology/Inventor-chemist-CRISPR-craftsman-Inside/96/i16.
11 K. Davies, N. Gaudelli, and A. C. Komor, "The Beginning of Base Editing: An Interview with Alexis C. Komor and Nicole M. Gaudelli," *CRISPR Journal* 2, (2019): 81–90, https://www.liebertpub.com/doi/full/10.1089/crispr.2019.29050.kda.
12 A. C. Komor et al., "Programmable editing of a target base in genomic DNA without double-stranded DNA cleavage," *Nature* 533, (2016): 420–424, https://www.nature.com/articles/nature17946.
13 N. M. Gaudelli et al., "Programmable base editing of A•T to G•C in genomic DNA without DNA cleavage," *Nature* 551, (2017): 464–471, https://www.nature.com/articles/nature24644.
14 H. A. Rees and D. R. Liu, "Base editing: precision chemistry on the genome and transcriptome of living cells," *Nature Reviews Genetics* 19, (2018): 770–788, https://www.nature.com/articles/s41576-018-0059-1.
15 Megan Molteni, "Inside a chemist's quest to hack evolution and cure genetic disease," *WIRED*, June 12, 2018, https://www.wired.com/story/inside-a-chemists-quest-to-hack-evolution-and-cure-genetic-disease/.
16 Kevin Davies, "All about that base editing," *Genetic Engineering & Biotechnology News*, May 1, 2019, https://www.genengnews.com/insights/all-about-that-base-editing/.
17 David Liu, phone interview, March 8, 2019.
18 M. F. Richter et al., "Phage-assisted evolution of an adenine base editor with improved Cas domain compatibility and activity," *Nature Biotechnology*, March 20, 2020, https://www.nature.com/articles/s41587-020-0453-z.
19 Kevin Davies, "Base Editing Promise in Treating a Mouse Model of Progeria," *Genetic Engineering & Biotechnology News*, February 14, 2020, https://www.genengnews.com/news/base-editing-promise-in-treating-a-mouse-model-of-progeria/.
20 Sharon Begley, "In its first tough test, CRISPR base editing slashes cholesterol levels in monkeys," *STAT*, June 27, 2020, https://www.statnews.com/2020/06/27/crispr-base-editing-slashes-cholesterol-in-monkeys/.
21 Kary Mullis, Nobel lecture, December 8, 1993, https://www.nobelprize.org/prizes/chemistry/1993/mullis/lecture/.
22 Andrew Anzalone, *Nature* press conference, October 17, 2019.
23 David Liu, *Nature* press conference, October 17, 2019.

24 A. Anzalone et al., "Search-and-replace genome editing without double-strand breaks or donor DNA," *Nature* 576, (2019): 149–157, https://www.nature.com/articles/s41586-019-1711-4.
25 Megan Molteni, "A New Crispr Technique Could Fix Almost All Genetic Diseases," *WIRED*, October 21, 2019, https://www.wired.com/story/a-new-crispr-technique-could-fix-many-more-genetic-diseases/.
26 Sharon Begley, "New CRISPR tool has the potential to correct almost all disease-causing DNA glitches, scientists report," *STAT*, October 21, 2019, https://www.statnews.com/2019/10/21/new-crispr-tool-has-potential-to-correct-most-disease-causing-dna-glitches/.
27 Julianna LeMieux, "Genome Editing Heads to Primetime," *Genetic Engineering & Biotechnology News*, October 21, 2019, https://www.genengnews.com/insights/genome-editing-heads-to-primetime/.
28 David Liu, ESGCT conference, Barcelona, Spain, October 24, 2019.
29 Antonio Regalado, "The newest gene editor radically improves on CRISPR," *MIT Techology Review*, October 21, 2019, https://www.technologyreview.com/s/614599/the-newest-gene-editor-radically-improves-on-crispr/.
30 Megan Molteni, "This Company Wants to Rewrite the Future of Genetic Disease," *WIRED,* July 7, 2020, https://www.wired.com/storythis-company-wants-to-rewrite-the-future-of-genetic-disease/.
31 B. Y. Mok, M. H. de Moraes, et al., "A bacterial cytidine deaminase toxin enables CRISPR-free mitochondrial base editing," *Nature,* 2020, https://doi.org/10.1038/s41586-020-2477-4.
32 A. V. Anzalone, L. W. Koblan, and D. R. Liu, "Genome editing with CRISPR–Cas nucleases, base editors, transposases and prime editors," *Nature Biotechnology*, 2020, https://www.nature.com/articles/s41587-020-0561-9.
33 Katie Jennings, "This Startup Might Finally Cure Sickle Cell Disease—After A Century Of Racist Neglect," *Forbes,* July 10, 2020, https://www.forbes.com/sites/katiejennings/2020/07/10/this-startup-might-finally-cure-sickle-cell-disease-after-a-century-of-racist-neglect/#43b401104d3e.

Chapter 23: Volitional Evolution

1 Kevin Davies, "GenePeeks' Sperm Bank Acquisition Heralds Genome Screening of 'Virtual Progeny,'" *Bio-IT World*, January 4, 2013, http://www.bio-itworld.com/2013/1/4/sperm-bank-acquisition-heralds-genome-screening-virtual-progeny.html.
2 Fyodor Urnov, Keystone symposium, Banff, Canada, February 9, 2020.
3 Eric Lander, International Summit on Human Gene Editing, Washington, DC, December 1, 2015.
4 A. H. Handyside et al., "Pregnancies From Biopsied Human Preimplantation Embryos Sexed by Y-specific DNA Amplification," *Nature* 344, (1990): 768–770, https://www.nature.com/articles/344768a0.
5 M. Viotti et al., "Estimating Demand for Germline Genome Editing: An In Vitro Fertilization Clinic Perspective," *CRISPR Journal* 2, (2019): 304–315, https://doi.org/10.1089/crispr.2019.0044.
6 Ibid.
7 Jeffrey Steinberg, "Choose Your Baby's Eye Color," The Fertility Institutes, https://www.fertility-docs.com/*programs*-and-services/pgd-screening/choose-your-babys-eye-color.php (accessed January 27, 2020).
8 Rob Stein, "Scientists attempt controversial experiment to edit DNA in human sperm using CRISPR," *NPR*, August 22, 2019, https://www.npr.org/sections/health-shots/2019/08/22/746321083/scientists-attempt-controversial-experiment-to-edit-dna-in-human-sperm-using-cri.

9 Philip R. Reilly, *Abraham Lincoln's DNA* (Cold Spring Harbor, NY: Cold Spring Harbor Lab Press, 2000).

10 M. Pagel, "Designer Humans," *The Edge*, 2016, https://www.edge.org/response-detail/26605.

11 Leon Kass, "Preventing a Brave New World," *The New Republic*, June 21, 2001, https://web.stanford.edu/~mvr2j/sfsu09/extra/Kass3.pdf.

12 Derek So, "The Use and Misuse of Brave New World in the CRISPR Debate," *CRISPR Journal* 2, (2019): 316–322, https://www.liebertpub.com/doi/10.1089/crispr.2019.0046.

13 J. J. Cox et al., "An SCN9A channelopathy causes congenital inability to experience pain," *Nature* 444, (2006): 894–898, https://www.ncbi.nlm.nih.gov/pubmed/17167479.

14 Matthew Shaer, "The family that feels almost no pain," *Smithsonian Magazine*, May 2019, https://www.smithsonianmag.com/science-nature/family-feels-almost-no-pain-180971915/.

15 Michael Segalov, "Meet the super humans," *Observer*, January 26, 2020, https://www.theguardian.com/society/2020/jan/26/meet-the-super-humans-four-people-describe-their-extraordinary-powers.

16 Helen Branswell, "Experts search for answers in limited information about mystery pneumonia outbreak in China," *STAT*, January 4, 2020, https://www.statnews.com/2020/01/04/mystery-pneumonia-outbreak-china/.

17 J. F. Arboleda-Velasquez et al., "Resistance to autosomal dominant Alzheimer's disease in an APOE3 Christchurch homozygote: a case report," *Nature Medicine* 25, (2019): 1680–1683, https://www.nature.com/articles/s41591-019-0611-3.

18 Tony Kettle, "Christchurch discovery holds promise for Alzheimer's disease," *Stuff*, December 16, 2019, https://www.stuff.co.nz/science/118119856/christchurch-discovery-holds-promise-for-alzheimers-disease.

19 Stephen S. Hall, "Genetics: A gene of rare effect," *Nature* 9, (April 2013), https://www.nature.com/news/genetics-a-gene-of-rare-effect-1.12773.

20 knoepflerp, "The Science and Ethics of Genetically Engineered Human DNA," YouTube video, 1:43:45, June 19, 2015, https://www.youtube.com/watch?v=FLne4CnMXzo.

21 M. Sulak et al., "TP53 copy number expansion is associated with the evolution of increased body size and an enhanced DNA damage response in elephants," *eLife* 5, (2016): e11994, https://elifesciences.org/articles/11994.

22 Emily Mullin, "The Defense Department Plans to Build Radiation-proof CRISPR Soldiers," *OneZero*, September 27, 2019, https://onezero.medium.com/the-government-aims-to-use-crispr-to-make-soldiers-radiation-proof-3e18b00c9553.

23 Anjana Ahuja, "Crossing ethical red lines in gene editing," *Financial Times*, December 27, 2019, https://www.ft.com/content/6218346c-258d-11ea-9f81-051dbffa088d.

24 Eric Lander, International Summit on Human Gene Editing, Washington, DC, December 1, 2015.

25 Alissa Poh, "My Cell Phone Rings in A Minor," *Science Notes*, 2008, http://sciencenotes.ucsc.edu/0801/pages/pitch/pitch.html.

26 E. Theusch et al., "Genome-wide Study of Families with Absolute Pitch Reveals Linkage to 8q24.21 and Locus Heterogeneity," *American Journal of Human Genetics* 85, (2009): 112–119, https://www.ncbi.nlm.nih.gov/pmc/articles/PMC2706961/.

27 Stephen Hsu, "G: Unnatural selection," *Radiolab*, July 25, 2019, https://www.wnycstudios.org/story/g-unnatural-selection.

28 Julianna LeMieux, "The risky business of embryo selection," *Genetic Engineering & Biotechnology News*, April 1, 2019, https://www.genengnews.com/magazine/april-2019-vol-39-no-4/the-risky-business-of-embryo-selection/.

29 A. V. Khera et al., "Genome-wide polygenic scores for common diseases identify individuals with risk equivalent to monogenic mutations," *Nature Genetics* 50, (2018): 1219–1224, https://www.ncbi.nlm.nih.gov/pmc/articles/PMC6128408/.

30 L. Lello et al., "Genomic Prediction of 16 Complex Disease Risks Including Heart Attack, Diabetes, Breast and Prostate Cancer," *Scientific Reports* 9, (2019), 15286, https://www.nature.com/articles/s41598-019-51258-x.

31 G. Huguet, "Estimating the effect-size of gene dosage on cognitive ability across the coding genome," *bioRxiv,* April 5, 2020, https://www.biorxiv.org/content/10.1101/2020.04.03.024554v1.

32 Julianna LeMieux, "Polygenic risk scores and Genomic Prediction: Q&A with Stephen Hsu," *Genetic Engineering & Biotechnology News*, April 1, 2019, https://www.genengnews.com/topics/omics/polygenic-risk-scores-and-genomic-prediction-qa-with-steven-hsu/.

33 N. R. Treff et al., "Preimplantation Genetic Testing for Polygenic Disease Relative Risk Reduction: Evaluation of Genomic Index Performance in 11,883 Adult Sibling Pairs," *Genes* 11, 648 (2020), https://doi.org/10.3390/genes11060648.

34 Julianna LeMieux, "Polygenic risk scores and Genomic Prediction: Q&A with Stephen Hsu," *Genetic Engineering & Biotechnology News*, April 1, 2019, https://www.genengnews.com/topics/omics/polygenic-risk-scores-and-genomic-prediction-qa-with-steven-hsu/.

35 Erik Parens, Paul S. Appelbaum, and Wendy Chung, "Embryo editing for higher IQ is a fantasy. Embryo profiling for it is almost here," *STAT,* February 12, 2019, https://www.statnews.com/2019/02/12/embryo-profiling-iq-almost-here/.

36 Julianna LeMieux, "The risky business of embryo selection," *Genetic Engineering & Biotechnology News*, April 1, 2019, https://www.genengnews.com/magazine/april-2019-vol-39-no-4/the-risky-business-of-embryo-selection/.

37 E. Karavani et al., "Screening Human Embryos for Polygenic Traits has Limited Utility," *Cell* 179, (2019): P1424–1435, https://www.cell.com/cell/pdf/S0092-8674(19)31210-3.pdf.

38 Sek Kathiresan, "Settling the Score with Genetic Diseases," GEN Keynote webinar, April 16, 2020, https://www.genengnews.com/resources/webinars/settling-the-score-with-genetic-diseases/.

39 Julianna LeMieux, "The risky business of embryo selection," *Genetic Engineering & Biotechnology News*, April 1, 2019, https://www.genengnews.com/magazine/april-2019-vol-39-no-4/the-risky-business-of-embryo-selection/.

40 UNESCO, "Universal Declaration on the Human Genome and Human Rights," November 11, 1997, https://en.unesco.org/themes/ethics-science-and-technology/human-genome-and-human-rights.

41 Henry T. Greely, "Human Germline Genome Editing: An Assessment," *CRISPR Journal* 2, (2019): 253–265, https://www.liebertpub.com/doi/abs/10.1089/crispr.2019.0038.

42 Misha Angrist, *Here Is a Human Being* (New York: Harper, 2010).

43 K. Davies and G. Church, "Radical Technology Meets Radical Application: An interview with George Church," *CRISPR Journal* 2, (2019): 346–351, https://www.liebertpub.com/doi/full/10.1089/crispr.2019.29074.gch.

44 Doha Debates, "Gene Editing & the Future of Genetics. FULL DEBATE. Doha Debates," YouTube video, 1:38:50, March 31, 2020, https://www.youtube.com/watch?v=6-imA51Qk0M.

45 Editorial, "Genome editing: proceed with caution," *Lancet* 392, (2019): 253, https://www.thelancet.com/journals/lancet/article/PIIS0140-6736(18)31653-2/fulltext.

46 Kenan Malik, "Fear of dystopian change should not blind us to the potential of gene editing," *Guardian*, July 22, 2018, https://www.theguardian.com/commentisfree/2018/jul/21/designer-babies-gene-editing-curing-disease.

47 Henry T. Greely, "Human Germline Genome Editing: An Assessment," *CRISPR Journal* 2, (2019): 253–265, https://www.liebertpub.com/doi/abs/10.1089/crispr.2019.0038.

48 Steven Pinker, "The moral imperative for bioethics," *Boston Globe*, August 1, 2015, https://www.bostonglobe.com/opinion/2015/07/31/the-moral-imperative-for-bioethics/JmEkoyzlTAu9oQV76JrK9N/story.html.

49 Michael J. Sandel, "The case against perfection," *Atlantic*, April 2004, https://www.theatlantic.com/magazine/archive/2004/04/the-case-against-perfection/302927/.

50 Lev Facher, "NIH director says there's work to do on regulating genome editing globally," *STAT*, November 29, 2018, https://www.statnews.com/2018/11/29/nih-director-says-theres-work-to-do-on-regulating-genome-editing-globally/.

51 Michael J. Sandel, "The case against perfection," *Atlantic*, April 2004, https://www.theatlantic.com/magazine/archive/2004/04/the-case-against-perfection/302927/.

52 Lord Moynihan, House of Lords, Hansard, January 30, 2020, https://hansard.parliament.uk/lords/2020-01-30/debates/637E3108-D287-445B-8460-4B739A24CCF8/GeneEditing.

53 K. Davies and G. Church, "Radical Technology Meets Radical Application: An interview with George Church," *CRISPR Journal* 2, (2019): 346–351, https://www.liebertpub.com/doi/full/10.1089/crispr.2019.29074.gch.

54 Gulzaar Barn, "Don't genetically enhance people—improve society instead," *Economist*, April 30, 2019, https://www.economist.com/open-future/2019/04/30/dont-genetically-enhance-people-improve-society-instead.

55 C. Lippert et al., "Identification of individuals by trait prediction using whole-genome sequencing data," *Proceedings of the National Academy of Sciences* 114, (2017): 10166–10171, https://www.pnas.org/content/114/38/10166.

56 Y. Erlich, "Major flaws in 'Identification of individuals by trait prediction using whole-genome sequencing data,'" *bioRxiv*, September 7, 2017, https://doi.org/10.1101/18533.

57 Editorial, "Genome editing: proceed with caution," *Lancet* 392, (2019): 253, https://www.thelancet.com/journals/lancet/article/PIIS0140-6736(18)31653-2/fulltext.

58 George Will, "The real Down syndrome problem: Accepting genocide," *Washington Post*, March 14, 2018, https://www.washingtonpost.com/opinions/whats-the-real-down-syndrome-problem-the-genocide/2018/03/14/3c4f8ab8-26ee-11e8-b79d-f3d931db7f68_story.html.

59 Rebecca Cokley, "Please don't edit me out," *Washington Post*, August 10, 2017, https://www.washingtonpost.com/opinions/if-we-start-editing-genes-people-like-me-might-not-exist/2017/08/10/e9adf206-7d27-11e7-a669-b400c5c7e1cc_story.html.

60 Ethan J. Weiss, "Billy Idol," *Project Muse*, vol. 63, Winter 2020, https://muse.jhu.edu/article/748051.

61 John Harris, *The Value of Life* (Abingdon-on-Thames, UK: Routledge, 1985).

62 Helen C. O'Neill, "Clinical Germline Genome Editing," *Perspectives in Biology and Medicine* 63, (2020): 101–110, https://muse.jhu.edu/article/748054.

Chapter 24: Bases Loaded

1 Tim Hunt, Keystone Symposium, Banff, Canada, February 9, 2020.

2 R. C. Wilson and D. Carroll, "The Daunting Economics of Therapeutic Genome Editing," *CRISPR Journal* 2, (2019): 280–284, https://www.liebertpub.com/doi/full/10.1089/crispr.2019.0052.

3 Kevin Davies, "NIH Director Backs Moratorium for Heritable Genome Editing," *Genetic Engineering & Biotechnology News*, November 8, 2019, https://www.genengnews.com/topics/genome-editing/nih-director-backs-moratorium-for-heritable-genome-editing/.

4 Sharon Begley, "As calls mount to ban embryo editing with CRISPR, families hit by inherited diseases say, not so fast," *STAT*, April 17, 2019, https://www.statnews.com/2019/04/17/crispr-embryo-editing-ban-opposed-by-families-carrying-inherited-diseases/.

5 Gina Kolata, "Scientists Designed a Drug for Just One Patient. Her Name Is Mila," *New York Times*, October 9, 2019, https://www.nytimes.com/2019/10/09/health/mila-makovec-drug.html.

6 Erika Check Hayden, "If DNA is like software, can we just fix the code?," *MIT Technology Review*, February 26, 2020, https://www.technologyreview.com/2020/02/26/905713/dna-is-like-software-fix-the-code-personalized-medicine/.

7 Fyodor Urnov, Keystone Symposium, Banff, Canada, February 9, 2020.

8 Ed Yong, "How the Pandemic Will End," *Atlantic*, March 25, 2020, https://www.theatlantic.com/health/archive/2020/03/how-will-coronavirus-end/608719/.

9 Megan Molteni and Gregory Barber, "How a Crispr Lab Became a Pop-Up Covid Testing Center," *WIRED*, April 2, 2020, https://www.wired.com/story/crispr-lab-turned-pop-up-covid-testing-center/.

10 Matthew Herper, "CRISPR pioneer Doudna opens lab to run Covid-19 tests," *STAT*, March 30, 2020, https://www.statnews.com/2020/03/30/crispr-pioneer-doudna-opens-lab-to-run-covid-19-tests/.

11 Megan Molteni and Gregory Barber, "How a Crispr Lab Became a Pop-Up Covid Testing Center," *WIRED*, April 2, 2020, https://www.wired.com/story/crispr-lab-turned-pop-up-covid-testing-center/

12 "First rounders: Feng Zhang," *Nature Biotechnology*, podcast audio, October 1, 2018, https://www.nature.com/articles/nbt0918-784.

13 J. Achenbach and L. McGinley, "FDA gives emergency authorization for CRISPR-based diagnostic tool for coronavirus," *Washington Post*, May 7, 2020, https://www.washingtonpost.com/health/fda-gives-emergency-authorization-for-crispr-based-diagnostic-tool-for-coronavirus/2020/05/07/f98029bc-9082-11ea-a9c0-73b93422d691_story.html.

14 Carl Zimmer, "With Crispr, a Possible Quick Test for the Coronavirus," *New York Times*, May 5, 2020, https://www.nytimes.com/2020/05/05/health/crispr-coronavirus-covid-test.html.

15 Darrell Etherington, "Pinterest CEO and a team of leading scientists launch a self-reporting COVID-19 tracking app," *TechCrunch*, April 2, 2020, https://techcrunch.com/2020/04/02/pinterest-ceo-and-a-team-of-leading-scientists-launch-a-self-reporting-covid-19-tracking-app/.

16 Rob Copeland, "The Secret Group of Scientists and Billionaires Pushing a Manhattan Project for Covid-19," *Wall Street Journal*, April 27, 2020, https://www.straitstimes.com/world/united-states/scientists-and-billionaires-drive-manhattan-project-seeking-to-combat-covid-19.

17 Alex Philippidis, "COVID-19 Drug & Vaccine Candidate Tracker," *Genetic Engineering & Biotechnology News*, May 18, 2020, https://www.genengnews.com/covid-19-candidates/covid-19-drug-and-vaccine-tracker/.

18 T. R. Abbott et al., "Development of CRISPR as a prophylactic strategy to combat novel coronavirus and influenza," *bioRxiv*, March 14, 2020, https://www.biorxiv.org/content/10.1101/2020.03.13.991307v1.

19 Jennifer Doudna, "Biochemist Explains How CRISPR Can Be Used to Fight COVID-19," *Amanpour*, March 30, 2020, http://www.pbs.org/wnet/amanpour-and-company/video/biochemist-explains-how-crispr-can-be-used-to-fight-covid-19/.

20 Lorrie Moore, "Bioperversity," *New Yorker*, May 12, 2003, https://www.newyorker.com/magazine/2003/05/19/bioperversity.

21 Christof Koch, *The Quest for Consciousness* (Englewood, CO: Roberts & Co., 2004).

INDEX

T